Solutions manual to accompany

Organic Chemistry

Second Edition

Jonathan Clayden, Nick Greeves, and Stuart Warren

Jonathan Clayden
University of Manchester

Stuart Warren
University of Cambridge

OXFORD
UNIVERSITY PRESS

OXFORD
UNIVERSITY PRESS

Great Clarendon Street, Oxford, OX2 6DP,
United Kingdom

Oxford University Press is a department of the University of Oxford.
It furthers the University's objective of excellence in research, scholarship,
and education by publishing worldwide. Oxford is a registered trade mark of
Oxford University Press in the UK and in certain other countries

First edition published 2001

Impression: 6

British Library Cataloguing in Publication Data
Data available

978-0-19-966334-7

Printed in Great Britain by
Bell and Bain Ltd, Glasgow

Suggested solutions for Chapter 2

2

> **PROBLEM 1**
> Draw good diagrams of saturated hydrocarbons with seven carbon atoms having (a) linear, (b) branched, and (c) cyclic structures. Draw molecules based on each framework having both ketone and carboxylic acid functional groups in the same molecule.

Purpose of the problem

To get you drawing simple structures realistically and to steer you away from rules and names towards more creative and helpful ways of representing molecules.

Suggested solution

There is only one linear hydrocarbon but there are many branched and cyclic options. We offer some possibilities, but you may have thought of others.

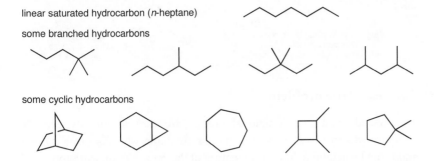

linear saturated hydrocarbon (*n*-heptane)

some branched hydrocarbons

some cyclic hydrocarbons

We give you a few examples of keto-carboxylic acids based on these structures. A ketone has to have a carbonyl group not at the end of a chain; a carboxylic acid functional group by contrast *has* to be at the end of a chain. You will notice that no carboxylic acid based on the first three cyclic structures is possible without adding another carbon atom.

linear molecules containing ketone and carboxylic acid

some branched keto-acids

some cyclic keto-acids

PROBLEM 2

Draw for yourself the structures of amoxicillin and Tamiflu shown on page 10 of the textbook. Identify on your diagrams the functional groups present in each molecule and the ring sizes. Study the carbon framework: is there a single carbon chain or more than one? Are they linear, branched, or cyclic?

SmithKline Beecham's amoxycillin
β-lactam antibiotic
for treatment of bacterial infections

Tamiflu (oseltamivir):
invented by
Gilead Sciences;
marketed by Roche

Purpose of the problem

To persuade you that functional groups are easy to identify even in complicated structures: an ester is an ester no matter what company it keeps and it can be helpful to look at the nature of the carbon framework too.

Suggested solution

The functional groups shouldn't have given you any problem except perhaps for the sulfide (or thioether) and the phenol (or alcohol). You should have seen that both molecules have an amide as well as an amine.

amine NH$_2$ sulfide   ether   ester
H$_3$C   CH$_3$
HO   HN   NH$_2$ amine
phenol or alcohol   amide   amide   CO$_2$H   carboxylic acid   H$_3$C   O   amide

The ring sizes are easy and we hope you noticed that one bond between the four- and the five-membered ring in the penicillin is shared by both rings.

six-membered   NH$_2$ five-membered
HO   H$_3$C   CH$_3$
HO   HN   six-membered
four-membered   CO$_2$H   H$_3$C   O   NH$_2$

The carbon chains are quite varied in length and style and are broken up by N, O, and S atoms.

cyclic C$_6$ NH$_2$ cyclic C$_3$ linear C$_5$ linear C$_2$
H$_3$C   CH$_3$
HO   H$_3$C   HN   cyclic C$_6$
linear C$_2$ linear C$_2$ H$_3$C   O   NH$_2$
CO$_2$H   branched C$_5$

PROBLEM 3

Identify the functional groups in these two molecules

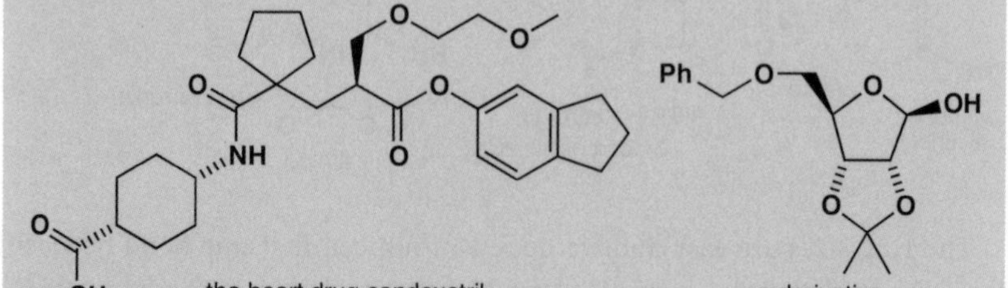

the heart drug candoxatril

a derivative
of the sugar ribose

Purpose of the problem

Identifying functional groups is not just a sterile exercise in classification: spotting the difference between an ester, an ether, an acetal and a hemiacetal is the first stage in understanding their chemistry.

Suggested solution

The functional groups are marked on the structures below. Particularly important is to identify an acetal and a hemiacetal, in which both 'ether-like' oxygens are bonded to a single carbon, as a single functional group.

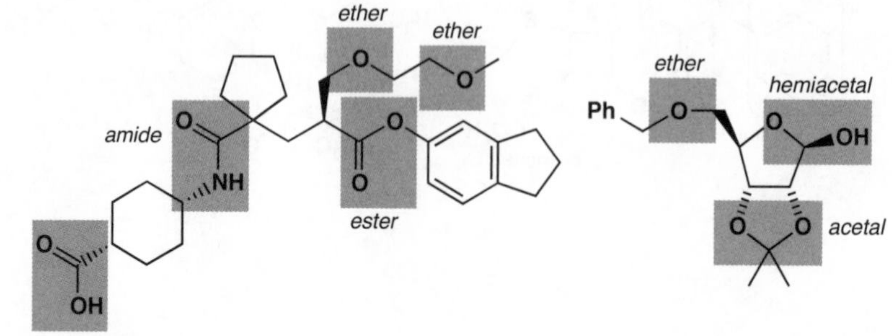

PROBLEM 4

What is wrong with these structures? Suggest better ways to represent these molecules

Purpose of the problem

To shock you with two dreadful structures and to try to convince you that well drawn realistic structures are more attractive to the eye as well as easier to understand and quicker to draw.

Suggested solution

The bond angles are grotesque with square planar saturated carbon atoms, bent alkynes with 120° bonds, linear alkenes with bonds at 90° or 180°, bonds coming off a benzene ring at the wrong angles and so on. If properly drawn, the left hand structure will be clearer without the hydrogen atoms. Here are better structures for each compound but you can think of many other possibilities.

PROBLEM 5

Draw structures for the compounds named systematically here. In each case suggest alternative names that might convey the structure more clearly if you were speaking to someone rather than writing.

(a) 1,4-di-(1,1-dimethylethyl)benzene

(b) 1-(prop-2-enyloxy)prop-2-ene

(c) cyclohexa-1,3,5-triene

Purpose of the problem

To help you appreciate the limitations of systematic names, the usefulness of part structures and, in the case of (c), to amuse.

Suggested solution

(a) A more helpful name would be *para*-di-*t*-butyl benzene. It is sold as 1,4-di-*tert*-butyl benzene, an equally helpful name. There are two separate numerical relationships.

(b) This name conveys neither the simple symmetrical structure nor the fact that it contains two allyl groups. Most chemists would call it 'diallyl ether' though it is sold as 'allyl ether'.

(c) This is of course simply benzene!

> **PROBLEM 6**
> Translate these very poor structural descriptions into something more realistic. Try to get the angles about right and, whatever you do, don't include any square planar carbon atoms or any other bond angles of 90°.
>
> (a) $C_6H_5CH(OH)(CH_2)_4COC_2H_5$
>
> (b) $O(CH_2CH_2)_2O$
>
> (c) $(CH_3O)_2CH=CHCH(OCH_3)_2$

Purpose of the problem

An exercise in interpretation and composition. This sort of 'structure' is sometimes used in printed text. It gives no clue to the shape of the molecule.

Suggested solution

You probably need a few 'trial and error' drawings first but simply drawing out the carbon chain gives you a good start. The first is straightforward—the (OH) group is a substituent joined to the chain and not part of it. The second compound must be cyclic—it is the ether solvent commonly known as dioxane. The third gives no hint as to the shape of the alkene and we have chosen *trans*. It also has two ways of representing a methyl group. Either is fine, but it is better not to mix the two in one structure.

$C_6H_5CH(OH)(CH_2)_4COC_2H_5$ $O(CH_2CH_2)_2O$ $(CH_3O)_2CH=CHCH(OMe)_2$

OH

PROBLEM 7

Identify the oxidation level of all the carbon atoms of the compounds in problem 6.

Purpose of the problem

This important exercise is one you will get used to very quickly and, before long, do without thinking. If you do it will save you from many trivial errors. Remember that the oxidation *state* of all the carbon atoms is +4 or C(IV). The oxidation *level* of a carbon atom tells you to which oxygen-based functional group it can be converted without oxidation or reduction.

Suggested solution

Just count the number of bonds between the carbon atom and heteroatoms (atoms which are not H or C). If none, the atom is at the hydrocarbon level (□), if one, the alcohol level (O), if two the aldehyde or ketone level, if three the carboxylic acid level (●) and, if four, the carbon dioxide level.

■ Why alkenes have the alcohol oxidation level is explained on page 33 of the textbook.

PROBLEM 8

Draw full structures for these compounds, displaying the hydrocarbon framework clearly and showing all the bonds in the functional groups. Name the functional groups.

(a) $AcO(CH_2)_3NO_2$

(b) MeO_2CCH_2OCOEt

(c) $CH_2=CHCONH(CH_2)_2CN$

Purpose of the problem

This problem extends the purpose of problem 6 as more thought is needed and you need to check your knowledge of the 'organic elements' such as Ac.

Suggested solution

For once the solution can be simply stated as no variation is possible. In the first structure 'AcO' represents an acetate ester and that the nitro group can have only four bonds (not five) to N. The second has two ester groups on the central carbon, but one is joined to it by a C–O and the other by a C–C bond. The last is straightforward.

PROBLEM 9

Draw structures for the folllowing molecules, and then show them again using at least one 'organic element' symbol in each.

(a) ethyl acetate

(b) chloromethyl methyl ether

(c) pentanenitrile

(d) *N*-acetyl *p*-aminophenol

(e) 2,4,6,-tri-(1,1-dimethylethyl)phenylamine

Purpose of the problem

Compound names mean nothing unless you can visualize their structures. More practice using 'organic elements'.

Suggested solution

The structures are shown below—things to look out for are the difference between acetyl and acetate, the fact that the carbon atom of the nitrile group is included in the name, and the way that a *tert*-butyl group can be named as '1,1-dimethylethyl'.

■ There is a list of the abbreviations known as 'organic elements' on page 42 of the textbook.

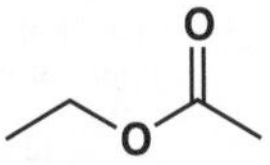

EtOAc

ethyl acetate

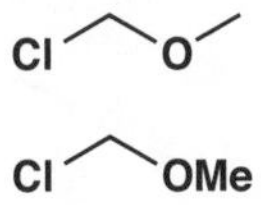

chloromethyl methyl ether

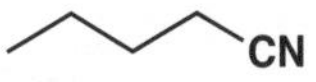

BuCN

pentanenitrile

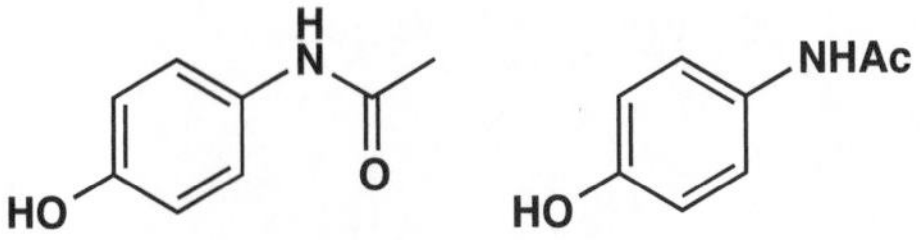

N-acetyl *p*-aminophenol

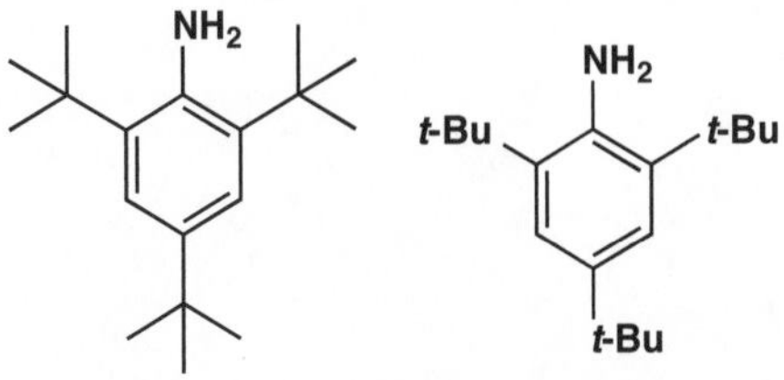

2,4,6,-tri-(1,1-dimethylethyl)phenylamine

PROBLEM 10
Suggest at least six different structures that would fit the formula C₄H₇NO. Make good realistic diagrams of each one and say which functional groups are present.

Purpose of the problem

The identification and naming of functional groups is more important than the naming of compounds, because the names of functional groups tell you about their chemistry. This was your chance to experiment with different groups and different carbon skeletons and to experience the large number of compounds you could make from a formula with few atoms.

Suggested solution

We give twelve possible structures – there are of course many more. You need not have used the names in brackets as they are ones more experienced chemists might use.

alkyne, primary amine
primary alcohol

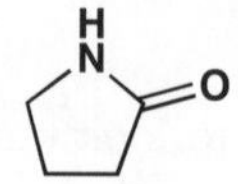

(cyclic) amide
(lactam)

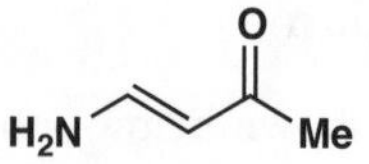

ketone, alkene,
primary amine (enamine)

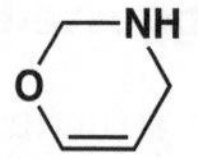

ether, alkene
secondary amine

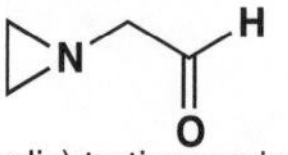

(cyclic) tertiary amine
aldehyde

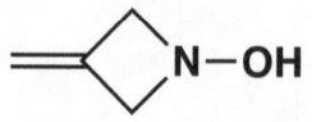

alkene, amine, alcohol
(cyclic hydroxylamine)

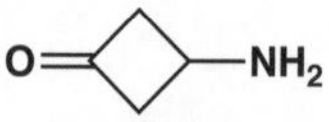

(cyclic) ketone
primary amine

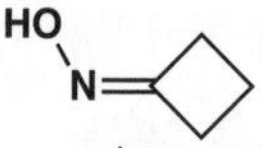

oxime
imine and alcohol

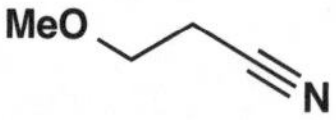

ether, nitrile

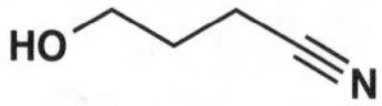

primary alcohol,
nitrile

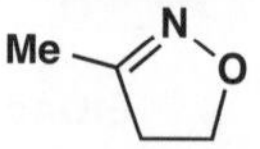

imine, ether
(isoxazoline)

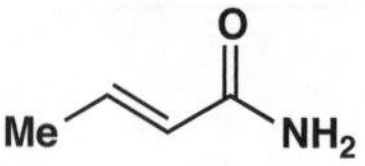

alkene, primary amide

3

Suggested solutions for Chapter 3

PROBLEM 1

Assuming that the molecular ion is the base peak (100% abundance) what peaks would appear in the mass spectrum of each of these molecules:

(a) C_2H_5BrO

(b) C_{60}

(c) C_6H_4BrCl

In cases (a) and (c) suggest a possible structure of the molecule. What is (b)?

Purpose of the problem

To give you some practice with mass spectra and, in particular, at interpreting isotopic peaks. The molecular ion is the most important ion in the spectrum and often the only one that interests us.

Suggested solution

Bromine has two isotopes, ^{79}Br and ^{81}Br in about a 1:1 ratio. Chlorine has two isotopes ^{35}Cl and ^{37}Cl in about a 3:1 ratio. There is about 1.1% ^{13}C in normal compounds.

(a) C_2H_5BrO will have two main molecular ions at 124 and 126. There will be very small (2.2%) peaks at 125 and 126 from the 1.1% of ^{13}C at each carbon atom.

(b) C_{60} has a molecular ion at 720 with a strong peak at 721 of 60 x 1.1 = 66%, more than half as strong as the ^{12}C peak at 720. This compound is buckminsterfullerene.

■ Buckminsterfullerene is on page 25 of the textbook.

(c) This compound is more complicated. It will have a 1:1 ratio of ^{79}Br and ^{81}Br and a 3:1 ratio of ^{35}Cl and ^{37}Cl in the molecular ion. There are four peaks from these isotopes (ratios in brackets) $C_6H_4{}^{79}Br{}^{35}Cl$ (3), $C_6H_4{}^{81}Br{}^{35}Cl$ (3), $C_6H_4{}^{79}Br{}^{37}Cl$ (1), and $C_6H_4{}^{81}Br{}^{37}Cl$ (1), the masses of these peaks being 190, 192, 192, and 194. So the complete molecular ion will have three main peaks at 190, 192, and 194 in a ratio of 3:4:1 with peaks at 191, 193, and 194 at 6.6% of the peak before it.

Compounds (a) and (c) might be isomers of compounds such as these:

PROBLEM 2

Ethyl benzoate $PhCO_2Et$ has these peaks in its ^{13}C NMR spectrum: 17.3, 61.1, 100–150 (four peaks) and 166.8 ppm. Which peak belongs to which carbon atom? You are advised to make a good drawing of the molecule before you answer.

Purpose of the problem

■ These regions are described on page 56 of the textbook.

To familiarize you with the four regions of the spectrum.

Suggested solution

It isn't possible to say which aromatic carbon is which and it doesn't matter. The rest are straightforward.

PROBLEM 3

Methoxatin was mentioned on page 44 of the textbook where we said 'it proved exceptionally difficult to solve the structure by NMR.' Why is it so difficult? Could anything be gained from the ^{13}C or ^1H NMR? What information could be gained from the mass spectrum and the infra red?

Purpose of the problem

To convince you that this structure really needs an X-ray solution but also to get you to think about what information is available by the other methods. Certainly mass spectroscopy, NMR, and IR would have been tried first.

Suggested solution

There are only two hydrogens on carbon atoms and they are both on aromatic rings. There are only two types of carbon atom: carbonyl groups and unsaturated ring atoms. This information is mildly interesting but is essentially negative—it tells us what is not there but gives us no information on the basic skeleton, where the carboxylic acids are, nor does it reveal the 1,2-diketone in the middle ring.

The mass spectrum would at least give the molecular formula $C_{14}H_6N_2O_8$ and the infra-red would reveal an N–H group, carboxylic acids, and perhaps the 1,2-diketone. The X-ray was utterly convincing and the molecule has now been synthesized, confirming the structure.

■ The synthesis of methoxatin is described in J. A. Gainor and S. M. Weinreb, *J. Org. Chem.*, 1982, **47**, 2833.

PROBLEM 4

The solvent formerly used in some correcting fluids is a single compound $C_2H_3Cl_3$, having ^{13}C NMR peaks at 45.1 and 95.0 ppm. What is its structure? How would you confirm it spectroscopically? A commercial paint thinner gives two spots on chromatography and has ^{13}C NMR peaks at 7.0, 27.5, 35.2, 45.3, 95.6, and 206.3 ppm. Suggest what compounds might be used in this thinner.

Purpose of the problem

To start you on the road to structure identification with one very simple problem and some deductive reasoning. It is necessary to think about the size of the chemical shifts to solve this problem.

Suggested solution

With the very small molecule $C_2H_3Cl_3$ it is best to start by drawing all the possible structures. In fact there are only two.

The first would have a peak for the methyl group in the 0–50 region and one for the CCl_3 group at a very large chemical shift because of the three chlorine atoms. The second isomer would have two peaks in the 50–100 region, not that far apart. The second structure looks better but it would be easily confirmed by proton NMR as the first structure would have one peak only but the second would have two peaks for different CHs. The solvent is indeed the second structure 1,1,2-trichloroethane.

Two of the peaks (45.3 and 95.6) in the paint thinner are much the same as those for this compound (chemical shifts change slightly in a mixture as the two compounds dissolve each other). The other compound has a carbonyl group at 206.3 and three saturated carbon atoms, two close to the carbonyl group (larger shifts) and one further away. Butanone fits the bill perfectly. You were not expected to decide which CH_2 group belongs to which molecule—that can be found out by running a spectrum of pure butanone.

butanone

PROBLEM 5

The 'normal' O–H stretch in the infrared (i.e. without hydrogen bonding) comes at about 3600 cm^{-1}. What is the reduced mass (μ) for O–H? What happens to the reduced mass when you double the mass of each atom in turn, i.e. what is μ for O–D and what is μ for S–H? In fact, both O–D and S–H stretches come at about 2,500 cm^{-1}. Why?

Purpose of the problem

To get you thinking about the positions of IR bands in terms of the two main influences: reduced mass and bond strength.

Suggested solution

Using the equation on page 64 of the textbook we find that the reduced mass of OH is 16/17 or about 0.94. When you double the mass of H, the reduced mass of OD becomes 32/18 or about 1.78—nearly double that of OH. But when you double the mass of O, the reduced mass of SH is 32/33 or about 0.97 – hardly changed from OH! The change in the reduced mass from OH to OD is enough to account for the change in stretching frequency—a change of about $\sqrt{2}$. But the change in reduced mass from OH to SH cannot account for the change in frequency. The explanation is that the S–H bond is weaker than the O–H bond by a factor of about 2. So both both O–D and S–H absorb at about the same frequency.

There is an important principle to be deduced from this problem. Very roughly, all the reduced masses of all bonds involving the heavier elements (C, N, O, S etc.) differ by relatively small amounts and the differences in stretching frequency are mainly due to changes in bond strength, though it can be significant in comparing, say, C–O with C–Cl. With bonds involving hydrogen the reduced mass becomes by far the most important factor.

PROBLEM 6

Three compounds, each having the formula C_3H_5NO, have the IR data summarized here. What are their structures? Without ^{13}C NMR data it might be easier to draw some or all possible structures before trying to decide which is which. In what ways would ^{13}C NMR data help?

(a) One sharp band above 3000 cm^{-1} and one strong band at about 1700 cm^{-1}

(b) Two sharp bands above 3000 cm^{-1} and two bands between 1600 and 1700 cm^{-1}

(c) One strong broad band above 3000 cm^{-1} and a band at about 2200 cm^{-1}

Purpose of the problem

To show that IR alone does have some use but that NMR data are usually essential as well. In answers to exam questions of this type it is important to show how you interpret the data as well as to give a structure. If you get the structure right, this doesn't matter, but if you get it wrong, you may still get credit for your interpretation.

Suggested solution

(a) One sharp band above 3000 cm^{-1} must be an N–H and one strong band at about 1700 cm^{-1} must be a carbonyl group. That leaves C_2H_4, so we might have one of the structures shown below, though other less likely structures are possible too. ^{13}C NMR data would help as it would definitely show two types of saturated carbon (along with the carbonyl group) for the first compound, but only one for the second.

(b) Two sharp bands above 3000 cm^{-1} must be an NH_2 group and two bands between 1600 and 1700 cm^{-1} suggest a carbonyl group and an alkene. This leaves us with three hydrogen atoms so we must have something like the molecules below. ^{13}C NMR data would help as it would show an alkene carbon shifted downfield by being joined to electronegative nitrogen in the second case.

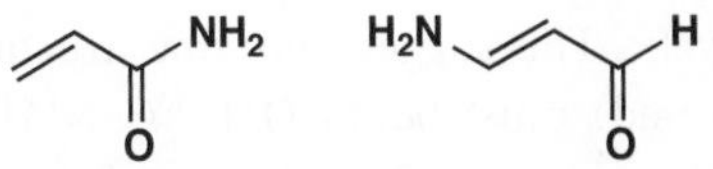

■ You will meet other ways of distinguishing these compounds in chapters 13 and 18.

(c) One strong broad band above 3000 cm^{-1} must be an OH group and a band at about 2200 cm^{-1} must be a triple bond, presumably CN as otherwise we have nowhere to put the nitrogen atom. This means structures of this sort.

> **PROBLEM 7**
>
> Four compounds having the formula $C_4H_6O_2$ have the IR and NMR data given below. How many DBEs (double bond equivalents—see p. 75 in the textbook) are there in $C_4H_6O_2$? What are the structures of the four compounds? You might again find it useful to draw a few structures to start with.
>
> (a) IR: 1745 cm^{-1}; ^{13}C NMR 214, 82, 58, and 41 ppm
>
> (b) IR: 3300 cm^{-1} (broad); ^{13}C NMR 62 and 79 ppm.
>
> (c) IR: 1770 cm^{-1}; ^{13}C NMR 178, 86, 40, and 27 ppm.
>
> (d) IR: 1720 and 1650 cm^{-1} (strong); ^{13}C NMR 165, 133, 131, and 54 ppm.

Purpose of the problem

First steps in identifying a compound from two sets of data. Because the molecules are so small (only four carbon atoms) drawing out a few trial structures is a good way to start.

Suggested solution

Here are some possible structures for $C_4H_6O_2$. It is clear that there are two double bond equivalents and that double bonds and rings are likely to feature. Functional groups are likely to include alcohol, aldehyde, ketone and carboxylic acid.

(a) IR: 1745 cm^{-1} must be a carbonyl group; ^{13}C NMR 214 must be an aldehyde or ketone, 82 and 58 look like two carbons next to oxygen and 41 is a carbon not next to oxygen but not far away. As the second oxygen doesn't show up in the IR, it must be an ether. As there is only one double bond, the compound must be cyclic. This suggests just one structure.

(b) IR: 3300 cm^{-1} (broad) must be an OH; ^{13}C NMR 62 and 79 show a symmetrical molecule and no C=O so it must have a triple bond and a saturated carbon next to oxygen. This again gives only one structure.

■ The alkyne does not show up in the IR as it is symmetrical: see p. 71 of the textbook.

(c) IR: 1770 cm^{-1} must be some sort of carbonyl group; ^{13}C NMR 178 suggests an acid derivative, 86 is a saturated carbon next to oxygen, 40 and 27 are saturated carbons not next to oxygen. There is only one double bond so it must be a ring. Looks like a close relative of (a).

(d) IR 1720 and 1650 cm^{-1} (strong) must be C=C and C=O; ^{13}C NMR 165 is an acid derivative, 131 and 133 must be an alkene, and 54 is a saturated carbon next to oxygen. That defines all the carbon atoms. It is not significant that we cannot say which alkene carbon is which.

(a) 82 58 214 41   (b) 62 HO 79 OH   (c) 86 27 178 40   (d) 165 Me 133/131 O 54

PROBLEM 8

You have dissolved *tert*-butanol in MeCN with an acid catalyst, left the solution overnight, and found crystals in the morning with the following characteristics. What are the crystals?

IR: 3435 and 1686 cm^{-1}; ^{13}C NMR: 169, 50, 29, and 25 ppm; ^{1}H NMR: 8.0, 1.8, and 1.4 ppm; Mass spectrum (%): 115 (7), 100 (10), 64 (5), 60 (21), 59 (17), 58 (100), and 56 (7). Don't try to assign all the peaks in the mass spectrum.

Purpose of the problem

This is a common situation: you carry out a reaction and find a product that is not starting material, but what is it? You'll need to use all the information and some logic. What you must *not* do is to decide in advance what the product is from your (limited) knowledge of chemistry and make the data fit.

Suggested solution

The molecular ion in the mass spectrum is 115 and is presumably C$_6$H$_{13}$NO—the sum of the two reagents *t*-BuOH and MeCN. It appears that they have added together but the IR shows that neither OH nor CN has survived. So what do we know?

• The IR tells us we have an N–H and a C=O group, accounting for both heteroatoms.

• The ^{13}C NMR shows a carbonyl group (169) and three types of saturated carbon.

- There must be a lot of symmetry, suggesting that the *t*-Bu group has survived.

This leaves four fragments: NH, C=O, Me, and *t*-Bu, confirmed also by the ^{1}H NMR spectrum, which tells us that we have three types of H atoms. We can join these fragments up in two ways:

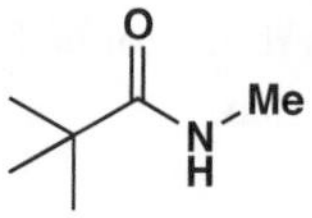

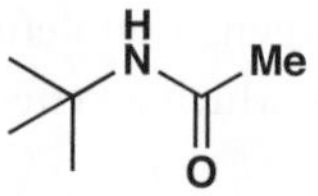

We might prefer the second as it retains the skeleton of MeCN, but a better reason is the base (100%) peak in the mass spectrum at 58. This is $Me_2C=NH_2^+$ which could easily come from the second structure but only by extensive reorganization of the first structure.

The second structure is in fact correct but we need further analysis of the proton NMR (chapter 13) to be sure.

■ If you are solving this problem after having already studied the more detailed description of ^{1}H NMR spectroscopy in chapter 13, it will help you to know that all three signals in the ^{1}H spectrum are singlets: no two types of H atom can be adjacent to each other.

PROBLEM 9

How many signals would you expect in the ^{13}C NMR spectrum of these compounds?

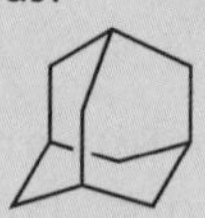

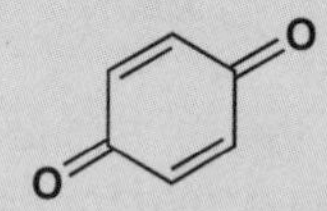

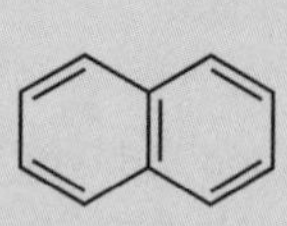

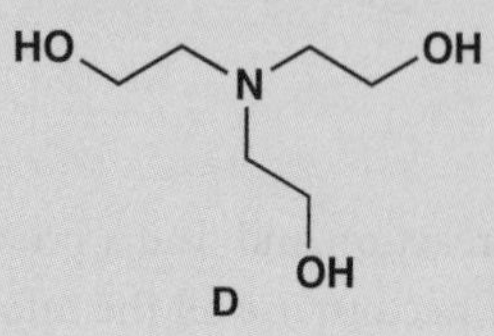

Purpose of the problem

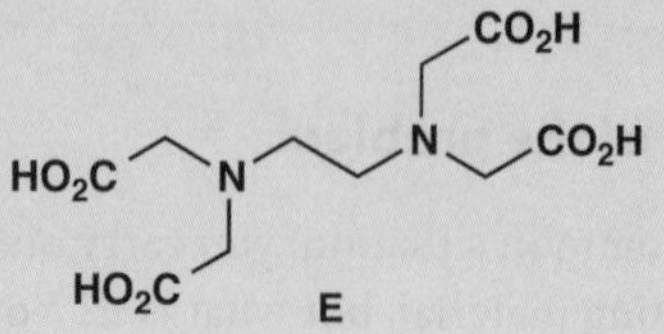

To get you thinking about symmetry.

Suggested solution

Compound **A** has tetrahedral symmetry and there are only two types of carbon: every CH_2 is the same, as is every CH, so it has two signals. This is the famous compound adamantane—a crystalline solid in spite of its being a hydrocarbon with only ten carbon atoms. If you do not see the symmetry, make a model—it is a beautiful structure.

Compound **B** is symmetrical too: the two C=O groups are the same and so are all the other carbon atoms in the ring. It is an orange crystalline solid called quinone. Two signals.

Compound **C** is naphthalene and has high symmetry: the two benzene rings are the same and there are only three types of carbon atom. Three signals.

Compound **D** is 'triethanolamine' used a lot by biochemists. It has threefold symmetry and only two types of carbon atom. Two signals.

Compound **E** is 'EDTA' (ethylenediaminetetraacetic acid) an important chelating agent for metals. This time there are three types of carbon atom. Three signals.

PROBLEM 10

When benzene is treated with *tert*-butyl chloride and aluminium trichloride, a crystalline product **A** is formed that contains only C and H. Mass spectrometry tells us the molecular mass is 190. The ^{1}H NMR spectrum looks like this:

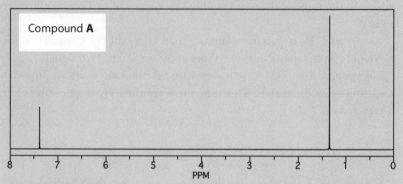

Compound **A**

If crystals of **A** are treated again with more *tert*-butyl chloride and aluminium chloride, a new oily compound **B** may be isolated, this time with a molecular mass of 246. Its ^{1}H NMR spectrum is similar to that of **A**, but not quite the same:

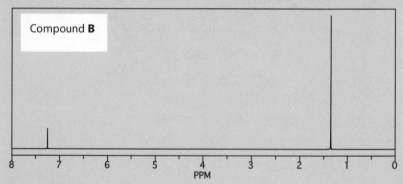

Compound **B**

What are the two compounds? How many signals do you expect in the ^{13}C NMR spectrum of each compound?

Purpose of the problem

Identifying compounds from spectroscopic data, whether you know the reaction or not, is a key skill you must develop.

Suggested solution

The ^{1}H NMR spectrum is so simple that both compounds must have a lot of symmetry. Each of the two signals is in a different region of the spectrum (see p. 60 of the textbook), so both compounds have one type of H attached to sp^2-hybridized C atoms (presumably from the benzene starting material) and one type of H attached to sp^3-hybridized C atoms (presumably from the *tert*-butyl chloride starting material).

Often a good place to start with this sort of problem is to use the molecular mass to work out approximately how many of each of the starting molecules have been incorporated into the product: benzene has a mass of 78 and the *tert*-butyl group a mass of 57 (the chloride must be lost as there is no chlorine in the product), so it looks as though **A** is made up from one benzene molecule plus two *tert*-butyl groups and **B** from one benzene molecule and three *tert*-butyl groups (with loss of two or three hydrogen atoms where the *tert*-butyls are bonded to the benzene ring).

So, the only question left is how the substituents are arranged. Two *tert*-butyl groups could be arranged *ortho*, *meta* or *para* to each other, but only the *para* arrangement is possible for **A** because only when the two groups are *para* are all the protons of the aromatic ring identical (check for yourself).

With three *tert*-butyl groups, there are three possible arrangements (again, draw them for yourself to check), but as before only one of these allows all the protons on the benzene ring to be identical, each sandwiched between two *tert*-butyl groups. We have our structures for **A** and **B**. Both of them will show four signals in the ^{13}C NMR spectrum, two of them between 100 and 150 ppm (C atoms in aromatic rings) and two of them between 0 and 50 ppm (saturated C atoms).

all four hydrogens on the ring are identical

all three hydrogens on the ring are identical

A   **B**

4

Suggested solutions for Chapter 4

PROBLEM 1

Textbooks sometimes describe the structure of sodium chloride like this 'an electron is transferred from the valence shell of a sodium atom to the valence shell of a chlorine atom.' Why would this not be a sensible way to make sodium chloride?

Purpose of the problem

To make you think about genuine ways to make compounds rather than theoretical ways.

Suggested solution

Of course sodium chloride consists of arrays of sodium cations without their 2s electron and chloride anions that have eight electrons in the 2s and 2p orbitals, but that is not how sodium chloride is made. Sodium atoms are present in sodium metal but where would you get the chlorine atoms? Mixing sodium and chlorine (Cl_2) would undoubtedly give sodium chloride but these are two aggressive reagents that would probably explode. Indeed, you would be more likely to make sodium and chlorine by the electrolysis of sodium chloride than the other way round. In any case, why make sodium chloride? Salt mines and the oceans are full of it.

PROBLEM 2

The H–C–H bond angle in methane is 109.5°. The H–O–H bond angle of water is close to this number but the H–S–H bond angle of H_2S is near 90°. What does this tell us about the bonding in water and H_2S? Draw a diagram of the molecular orbitals in H_2S.

Purpose of the problem

An exploration of hybridization.

Suggested solution

If the bond angle in water is close to the tetrahedral angle of perfectly symmetrical methane, water must be more or less tetrahedral (with respect to the arrangement of its electrons) too. We can think of the 2s and 2p electrons in water as hybridized into four pairs of electrons, two in H–O bonds and two as lone pairs on the oxygen atom. But H_2S has a near right angle for its H–S–H bond. This suggests that the bonds are formed with p orbitals on the sulfur atom and that H_2S is not hybridized. Orbital diagram of H_2S: you might have drawn something like this:

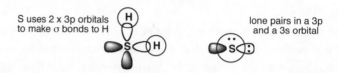

PROBLEM 3

Though the helium molecule He_2 does not exist (p. 91 of the textbook explains why), the cation He_2^+ does exist. Why?

Purpose of the problem

To encourage you to think about the filling of molecular orbitals and to accept surprising conclusions.

Suggested solution

He_2 does not exist because the number of anti-bonding electrons is the same as the number of bonding electrons. The bond order is zero. But if we remove an electron from the diagram on p. 91 of the textbook we have He_2^+, with two bonding electrons and only one anti-bonding electron. The bond order is one half. He_2^+ does exist.

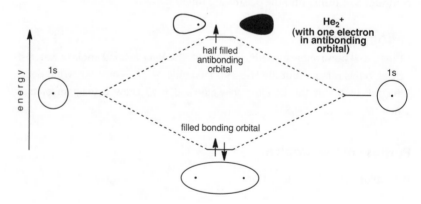

PROBLEM 4
Construct an MO diagram for LiH and suggest what type of bond it might have.

Purpose of the problem

To demonstrate that a simple MO treatment can be applied to to ionic as well as covalent structures.

Suggested solution

H has of course only one electron in a 1s orbital. Li has three – a full 1s shell and one electron in the 2s orbital. Li is very electropositive so its 2s orbital is high in energy—much higher than that of the 1s orbital of H. An electron is more stable in the 1s orbital of H than in the 2s orbital of Li, and the molecule is ionic. Both ions have the same electronic configuration: $1s^2$.

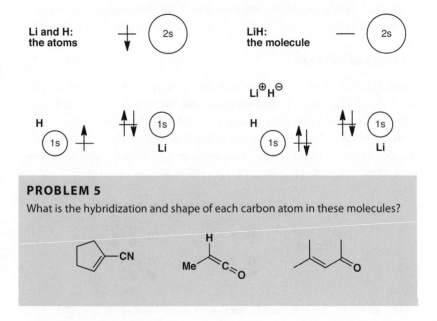

PROBLEM 5
What is the hybridization and shape of each carbon atom in these molecules?

Purpose of the problem

To give you practice at selecting the correct hybridization of carbon atoms.

Suggested solution

Simply count the number of σ-bonds at each carbon atom (not forgetting the hydrogens that may not be shown). Two σ-bonds means sp and linear, three means sp^2 and trigonal, and four means sp^3 and tetrahedral. In each case the bonds stay as far from each other as they can.

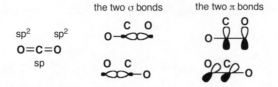

- □ sp³ tetrahedral
- ○ sp² trigonal
- ● sp linear

PROBLEM 6

Draw detailed structures for these molecules and predict their shapes. We have deliberately made non-committal drawings to avoid giving away the answer to the question. Don't use these sorts of drawings in your answer.

$$CO_2, CH_2=NCH_3, CHF_3, CH_2=C=CH_2, (CH_2)_2O$$

Purpose of the problem

To give you practice at selecting the right hybridization state for carbon atoms and translating this information into three-dimensional structures for the molecules.

Suggested solution

Carbon dioxide is linear as it has only two C–C σ-bonds and no lone pairs on C. The C atom must be sp hybridized and the only trick is to get the two π-bonds orthogonal to each other. They must be like that because the p orbitals on C used to make the two π-bonds are themselves orthogonal (p_x and p_y). Most people draw the O atoms as sp² hybridized rather than sp or even unhybridized but this doesn't matter as you really can't tell.

the two σ bonds   the two π bonds

sp²   sp²

O=C=O

sp

The imine has a C=N double bond so it must have sp² hybridized C and N. This means that the lone pair on nitrogen is in an sp² orbital and not in a p orbital. The molecule is planar (except for the methyl group which is, of course, tetrahedral) and is bent at the nitrogen atom.

Trifluoromethane is, of course, tetrahedral with an sp^3 hybridized carbon atom. The arrangement of the lone pairs round the fluorine (not shown) can also be assumed to be tetrahedral.

The next molecule $CH_2=C=CH_2$ is allene and it has the same shape as CO_2, and for the same reasons. We can now be sure that the end carbons are sp^2 hybridized as they are planar, with the hydrogen substituents at 120°. to each other and to the rest of the molecule. As with CO_2, the two π bonds are orthogonal, meaning that the planes of the two terminal carbon atoms are also orthogonal, meaning that the molecule as a whole is not planar.

Finally, $(CH_2)_2O$ must be a three-membered ring and therefore the C–C–O skeleton must be planar (three points are always in a plane!). The two carbon atoms are sp^3 hybridized (four σ bonds) and are tetrahedral (though very distorted as the ring angle is 60°) with the H atoms above and below the ring. The oxygen atom is presumably also sp^3 hybridized, but it's hard to tell experimentally.

PROBLEM 7

Draw the shapes, showing estimated bond angles, of the following molecules:

(a) hydrogen peroxide, H_2O_2

(b) methyl isocyanate CH_3NCO

(c) hydrazine, NH_2NH_2

(d) diimide, N_2H_2

(e) the azide anion, N_3^-

Purpose of the problem

To think about shape and bond angles at elements other than carbon.

Suggested solution

Hydrogen peroxide, H_2O_2, has only single bonds: each oxygen atom has two lone pairs and the electron pairs, both bonding and non-bonding, are arranged tetrahedrally. The bond angles at oxygen will be approximately the tetrahedral angle of 109°.

In methyl isocyanate, CH_3NCO, the interesting atoms are the N and the C. The N atom must have a double bond to C, for which it must use a p orbital, leaving an s and two p orbitals for the remainder of the electrons. The N atom is sp^2 and trigonal, with one lone pair, so the bond angle at N is about 120°. The C atom is double bonded to both N and O, so the C atom is like the one in CO_2—linear, and sp hybridized.

In hydrazine, NH_2NH_2, there are only single bonds: both nitrogens are like amine nitrogens, pyramidal and sp^3 hybridized.

In diimide the only reasonable structure has a double bond between the two nitrogen atoms, HN=NH, making the nitrogens trigonal (they must use a p orbital to make this double bond, leaving an s and two p orbitals for the remainder of the bonding, i.e. they are sp^2 hybridized). Each nitrogen is trigonal, with 120° bond angles. An interesting point about diimide is that, like an alkene, it can have a *cis* and a *trans* isomer.

Bonding in the azide anion N_3^- is identical with that in carbon dioxide: the two molecules are isoelectronic (count the electrons to make sure). The central nitrogen is sp hybridized and linear.

PROBLEM 8

Where would you expect to find the lone pairs in (a) water, (b) acetone ($Me_2C=O$), and (c) nitrogen (N_2)?

Purpose of the problem

More thinking about the arrangements of electrons at O and N. The location of lone pairs might not seem easy to determine, but it affects the way that molecules form coordination complexes, for example.

Suggested solution

The oxygen atom of water is surrounded by eight electrons in two bonding and two non-bonding orbitals: it is sp^3 hybridized and the electron pairs are arranged tetrahedrally, so the lone pairs point towards the remaining two vertices of a tetrahedron.

In acetone, the oxygen atom must use one of its p orbitals to form the π bond to carbon, so it is left with an s and two p orbitals to accommodate the six electrons making up the σ bond to C and the two lone pairs. These three electron pairs are presumably arranged trigonally, so the lone pairs will lie in the plane of the carbonyl group, about 120° apart.

The two nitrogen atoms of N_2 each need two of their p orbitals to form the two π bonds, so they are left with one s and one p orbital for the two remaining electron pairs: the σ bond and the lone pair. The nitrogen atoms are sp hybridized and the lone pairs are 180° from the other N.

5

Suggested solutions for Chapter 5

PROBLEM 1

Each of these molecules is electrophilic. Identify the electrophilic atom and draw a mechanism for a reaction with a generalized nucleophile Nu⁻, giving the structure of the product in each case.

Purpose of the problem

The recognition of electrophilic sites is half the battle in starting to understand mechanisms.

Suggested solution

We have two cations, two carbonyl compounds and two compounds with σ bonds only. One of the cations has three bonds to a positively charged carbon so that is the electrophilic site as it has an empty orbital. The nucleophile will attack here.

The other cation has a three-valent oxygen atom that cannot be the electrophilic site. The nucleophile must attack the proton instead. Some nucleophiles might attack the carbon atom joined to the cationic oxygen.

The two carbonyl compounds will be attacked at the carbonyl group by the nucleophile. In general, π-bonds are more easily broken than σ-bonds and the negative charge goes on to the electronegative oxygen atom. In fact these anions will not be the final products of the reactions. As we will explore in more detail in later chapters, the first will pick up a proton to give an alcohol but the second might decompose with the release of the stable carboxylate anion.

■ This type of *substitution reaction* is discussed in much more detail in chapter 10.

The remaining electrophiles have σ-bonds only, one of which must break. Chlorine is symmetrical so it doesn't matter which end you attack. You have more choice with MeSCl but the stability of the chloride ion wins the day: attack occurs at sulfur.

PROBLEM 2

Each of these molecules is nucleophilic. Identify the nucleophilic atom and draw a mechanism for a reaction with a generalized nucleophile E⁺, giving the structure of the product in each case.

Purpose of the problem

The recognition of nucleophilic sites is the other half of the battle in starting to understand mechanisms.

Suggested solution

This time there are three anions but two of them (the alkyne and the sulfur anions) have lone pair electrons. We should start our arrows from the negative charges and they are the points of attachment of the electrophile in the product.

The third anion is like the borohydride anion discussed on p. 119 of the textbook. The negative charge does not represent a pair of electrons on Al: all the electrons are in the Al–H bonds and we must start our arrow from one of those. The nucleophilic site is a hydrogen atom.

The remaining nucleophiles have lone pairs. The nitrogen-containing molecule is hydrazine: both nitrogens are the same, and the product is positively charged, so it will lose a proton to become more stable.

The phosphorus compound has four atoms with lone pairs, the P and three O atoms. The lone pairs on oxygen are in lower energy orbitals than the one on phosphorus (P is lower down the periodic table and less electronegative than O), so it is the lone pair on P that reacts. The product is positvely charged but this time it can't lose a proton.

PROBLEM 3

Complete these mechanisms by drawing the structure of the product(s).

Purpose of the problem

Practice in interpreting curly arrows and drawing the products. Once the arrows are drawn, there is no more scope for decision making, so just draw the products.

Suggested solution

Just break the bonds that are broken and make the bonds that are being formed. Don't forget to put in any charges and make sure you have neither created nor destroyed charge overall. You might straighten out the products a bit so that there are no funny angles.

PROBLEM 4

Put in the curly arrows on these starting materials to show how the product is formed. The compounds are drawn in a convenient arrangement to help you.

Purpose of the problem

To encourage you to be prepared to try and draw mechanisms for reactions you have never seen and to show you how easy it is.

Suggested solution

Just work out which bonds are lost and which are formed and draw arrows out of the one into the space for the other. Start your arrows on a source of electrons: an oxyanion in both these cases. End your arrows on an electronegative atom: oxygen in the first and bromine in the second example here.

Don't worry if your arrows are not exactly the same as ours – so long as they start and finish in the right place they're all right. The notes on the mechanisms are just to help you see what is going on: you would not normally include them. The second reaction looks more complicated than the first but it is actually easier: just move electrons through the molecule.

PROBLEM 5

Draw mechanisms for the reactions in the following sequence.

Purpose of the problem

Practice in drawing curly arrows for a simple sequence of reactions.

Suggested solution

First look for the bond being broken and the bond being formed. In the first reaction, the weak C–I bond is breaking and a C–O is forming. The new OH group must come from the hydroxide, so here we have our nucleophile: HO⁻. The electrophile is the alkyl iodide. Make your first arrow start on a source of electrons in the nucleophile—in this case that has to be the hydroxide's negative charge. Those electrons make the new C–O bond, so send them towards C. The old C–I bond must break at the same time, forming an iodide anion.

new bond

In the next step, we just lose H, or rather H⁺, from the hydroxyl group. Sodium hydride is the reagent, and hydride is a base. Bases are just nucleophiles which attack H, so we can start our arrow on the negative charge again, attack H and break the H–O bond, pushing the electrons onto oxygen to make the anion. You need to draw out the bond between O and H to represent this mechanism clearly. The other product is of course gaseous hydrogen.

new bond

In the final step, the oxyanion must act as the nucleophile and the electrophile is benzyl bromide. The arrows start on the negative charge, and show the new C–O bond forming and the old C–Br bond breaking.

new bond

PROBLEM 6

Each of these electrophiles could react with a nucleophile at one of (at least) two atoms. Identify these atoms and draw a mechanism and products for each reaction.

Purpose of the problem

Considering possible alternative reactions. One of the reactions might seem trivial, but it isn't.

Suggested solution

In each case one of the electrophilic sites is an acidic proton. There is also the electrophilic π bond (C=N⁺ or C=O). For the first case, we draw the two reactions separately.

In case you were seduced by the positively charged nitrogen atom (we hope you weren't), we should also remind you of a reaction that most definitely cannot happen: direct attack at N: the supposed product has an impossible five bonds to nitrogen.

In the second compound there are three possibilities. The acidic proton of the carboxylic acid and the electrophilic C=O bond are both possible reaction sites, but now so is the positively charged phosphorus. Phosphorus comes below nitrogen in the periodic table so, unlike N, it can have five bonds.

PROBLEM 7

These three reactions all give the products shown, but not by the mechanisms drawn! For each mechanism, explain what is wrong, and draw a better one.

Purpose of the problem

Getting a feel for how you can, and can't, use curly arrows to represent mechanisms.

Suggested solution

In the first reaction, the nucleophile is the amine and the electrophile is the methyl iodide, so the arrow is right in the sense that it starts on the nucleophile, where the electrons come from. However, we have stressed that a curly arrow should start on a *representation* of an electron pair, in other words on a lone pair, a bond or a negative charge. Here the arrow just starts on an atom: this is no good, and we must draw in the lone pair. The other problem is at the end of the arrow. It shows a new bond being formed to C, so unless a bond breaks then the C atom will have an impossible five bonds. We need another arrow to show the C–I bond breaking.

In the second reaction we form a cation by attack of a proton on an alkene. Which is the nucleophile? It can't be H$^+$: by definition a proton can't have a pair of electrons! The arrow must therefore start on the alkene and show the electrons moving towards the proton, not the other way round. The electrons come from the π bond, so the double bond is where we start the arrow. We only need one arrow, because as the new C–H bond forms, the C atom at other end of the old π bond is left with only 6 electrons, and becomes positively charged.

The last reaction forms a new S–Cl bond. The arrow we have drawn starts on a lone pair, but if electrons are moving from Cl to S, surely the S will become negatively charged? The mistake is that the arrow is the wrong way round: the sulfur atom is the nucleophile, and the more electronegative chlorine is the electrophile. Start the arrow on the sulfur lone pair, and break the old Cl–Cl bond as the electrons arrive at Cl.

PROBLEM 8

In your corrected mechanisms for problem 7, explain in each case which orbital is the HOMO of the nucleophile and which orbital is the LUMO of the electrophile.

Purpose of the problem

Reinforcing the link between curly arrows and molecular orbitals.

Suggested solution

The nucleophilic amine of the first reaction uses its lone pair to form the new bond: the HOMO is the sp^3 orbital containing this non-bonding electron pair. The empty orbital used by the electrophile must be an antibonding orbital, since as the electrons arrive there they cause the C–I bond to break: the LUMO is the C–I σ^* orbital.

When the alkene is protonated, it loses its π bond, and we have already pointed out that the electrons must come from this bond (that's why we start the arrow there). So the HOMO is the C=C π orbital. The LUMO is the only orbital the H⁺ ion has available: its empty 1s orbital.

In the third reaction, the LUMO of the electrophile is easy to spot: it must be the Cl–Cl σ* orbital, since that is the bond that breaks. The nucleophile uses the lone pair on sulfur to react, so the HOMO is the non-bonding orbital occupied by this lone pair.

PROBLEM 9

Draw a mechanism for the following reaction. (This is harder, but if you draw out the structures of the reactants first, and consider that one is an acid and one is a base, you will make a good start.)

$$PhCHBrCHBrCO_2H + NaHCO_3 \longrightarrow PhCH=CHBr + NaHCO_3$$

Purpose of the problem

Working out the mechanism for a more difficult transformation.

Suggested solution

Working out the structure of the starting material, even through it's written very unhelpfully, is straightforward if you take into account the fact that each carbon has four substituents. Bicarbonate is a base, so you expect the carboxylic acid to be deprotonated to form a carboxylate anion.

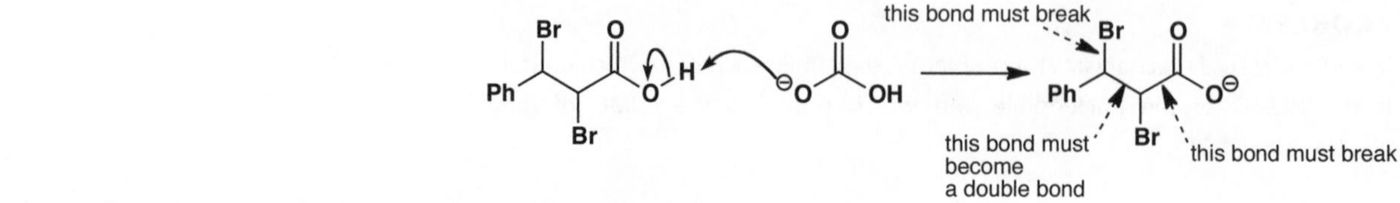

Now for the real reaction. Looking at the product tells us we have to lose CO_2, along with one of the bromine atoms, so the bonds that have to break are the ones shown in the structure above. The best thing to do in this case is to start 'pushing arrows'—one of the reasons they are so powerful is they often lead you through to the product. Start at the obvious place—the negative charge— and be guided by the bonds that have to break and form.

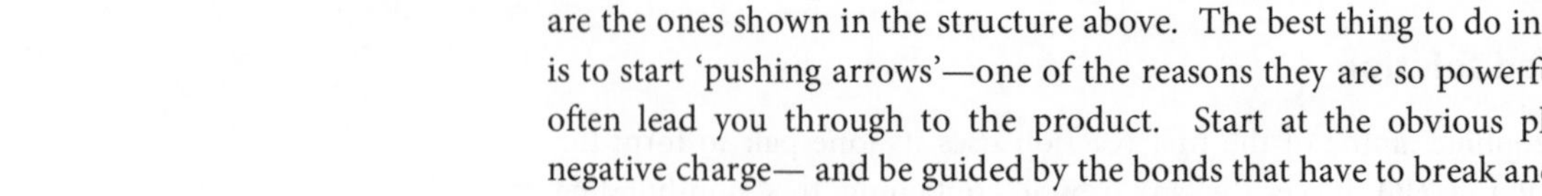

Everything works and the electrons end up happily on a bromide anion.

6

Suggested solutions for Chapter 6

PROBLEM 1

Draw mechanisms for these reactions:

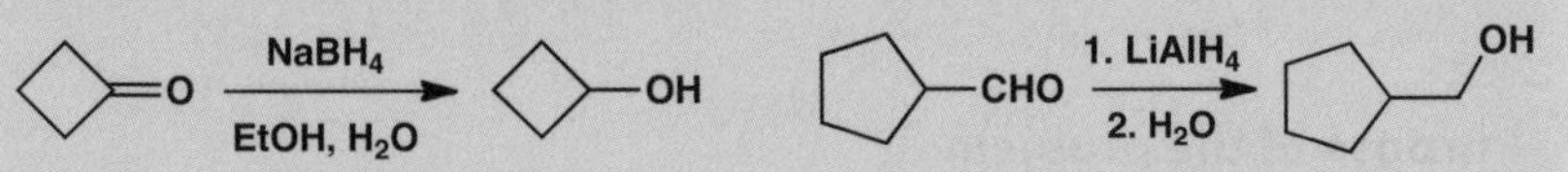

Purpose of the problem

Rehearsal of a simple but important mechanism that works for all aldehydes and ketones.

Suggested solution

Draw out the BH_4 and AlH_4 anions, with the carbonyl compound positioned so that one of the hydrogens can be transferred to the carbonyl group, and then transfer the hydrogen from B or Al to C. A proton transfer is needed to make the alcohol: from the solvent in the first case and during the work-up with water in the second.

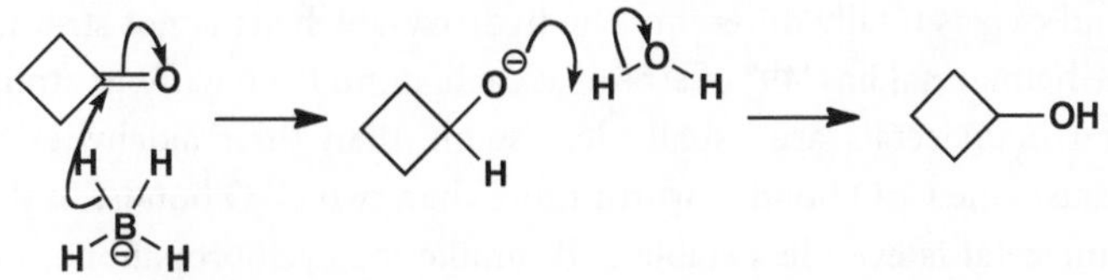

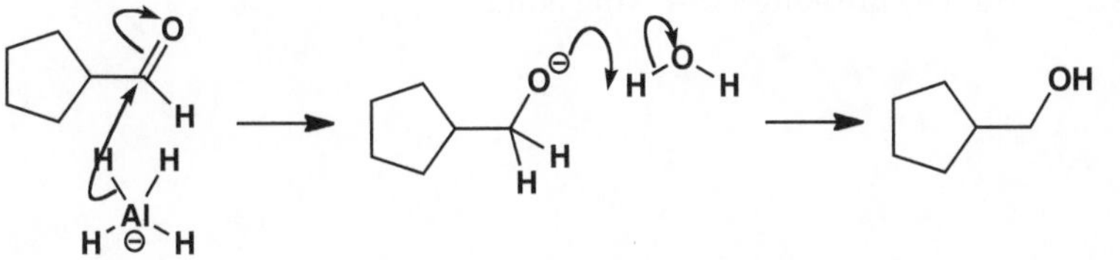

■ This reaction shows that you *can* reduce aldehydes with lithium aluminium hydride, even if you would usually prefer the more practical sodium borohydride.

PROBLEM 2

Cyclopropanone exists as the hydrate in water but 2-hydroxyethanal does not exist as the hemiacetal. Explain.

Purpose of the problem

To get you thinking about equilibria and hence the stability of compounds.

Suggested solution

Hydration is an equilibrium reaction so the mechanism is not strictly relevant to the question, though there is no shame in including mechanisms whenever you can. To answer the question we must consider the effect of the three-membered ring on the relative stability of starting material and product. All three-membered rings are very strained because the bond angles are 60° instead of 109° or 120°. Cyclopropanone is particularly strained because the sp^2 carbonyl carbon would like a bond angle of 120°—there is '60° of strain.' In the hydrate that carbon atom is sp^3 hybridized and so there is only about '49° of strain.' Not much gain, but the hydrate is more stable than the ketone.

The second case is totally different. The hydroxy-aldehyde is not strained at all but the hemiacetal has '49° of strain' at each atom. Even without strain, hydrates and hemiacetals are usually less stable than their aldehydes or ketones because one C=O bond is worth more than two C–O bonds. In this case the hemiacetal is even less stable and, unlike the cyclopropanone, can escape strain by breaking a C–O ring bond.

PROBLEM 3

One way to make cyanohydrins is illustrated here. Suggest a detailed mechanism for the process.

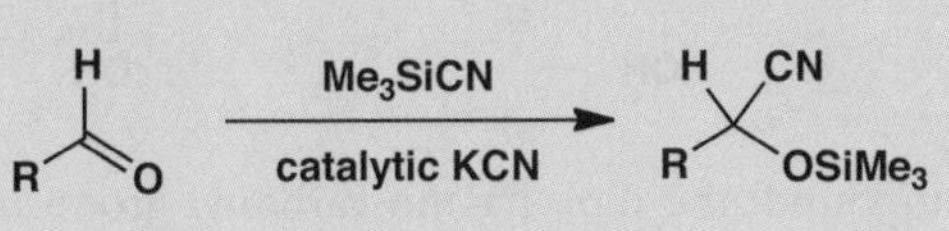

Purpose of the problem

To help you get used to mechanisms involving silicon and revise an important way to promote additions to the carbonyl group.

Suggested solution

The silyl cyanide is an electrophile while the cyanide ion in the catalyst is the nucleophile. Cyanide adds to the carbonyl group and the oxyanion product is captured by silicon, liberating another cyanide ion for the next cycle.

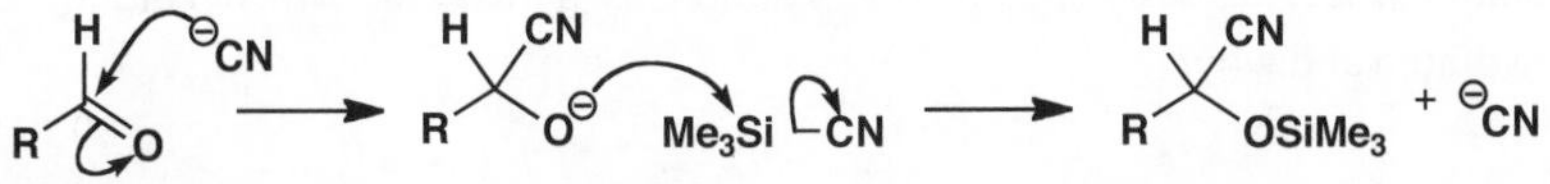

PROBLEM 4

There are three possible products from the reduction of this compound with sodium borohydride. What are their structures? How would you distinguish them spectroscopically, assuming you can isolate pure compounds?

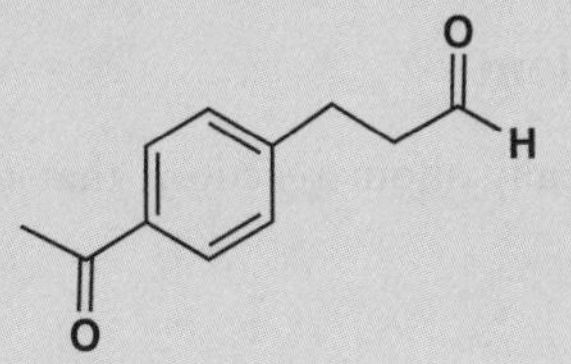

Purpose of the problem

To let you think practically about reactions that may give more than one product.

Suggested solution

The three compounds are easily drawn: one or other carbonyl group, or both, may be reduced.

■ Calculations from D. H. Williams and Ian Fleming (2007), *Spectroscopic methods in organic chemistry* (6th edn) McGraw Hill, London, 2007 suggest about 80 ppm for the C–OH carbon in the ketone and about 60 ppm for the aldehyde.

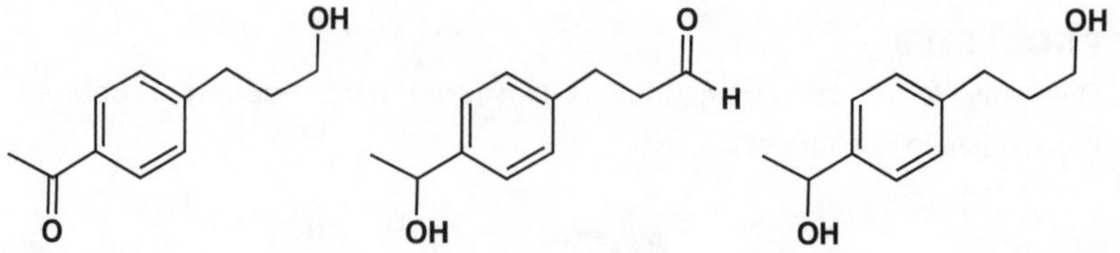

The third compound, the diol, has no carbonyl group in the ^{13}C NMR spectrum or the infrared and has a molecular ion two mass units higher than the other two products. Distinguishing those is more tricky, and needs techniques you will meet in detail in chapter 18. The hydroxyketone has a conjugated carbonyl group (C=O stretch at about 1680 cm^{-1} in the infrared spectrum) while the hydroxyaldehyde is not conjugated (C=O stretch at about 1730 cm^{-1} in the infrared). The chemical shift of the C–OH carbons will also be different because the benzene ring is joined to this carbon in the aldehyde but not in the ketone.

PROBLEM 5

The triketone shown here is called 'ninhydrin' and is used for the detection of amino acids. It exists in aqueous solution as a hydrate. Which ketone is hydrated and why?

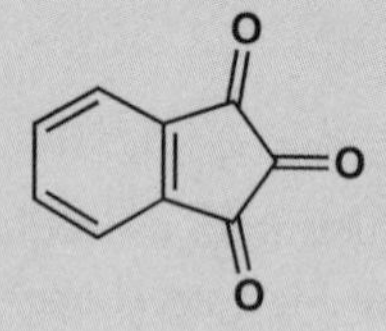

Purpose of the problem

To let you think practically about reactions that may give more than one product.

Suggested solution

The two ketones next to the benzene ring are stabilized by conjugation with it but also destabilized by the central ketone—two electron-withdrawing groups next to each other is a bad thing. The central carbonyl group is not stabilized by conjugation and is destabilized by *two* other ketones so it forms the hydrate. Did you remember that hydrate formation is thermodynamically controlled?

PROBLEM 6

This hydroxyketone shows no peaks in its infrared spectrum between 1600 and 1800 cm⁻¹, but it does show a broad absorption at 3000–3400 cm⁻¹. In the ^{13}C NMR spectrum there are no peaks above 150 ppm but there is a peak at 110 ppm. Suggest an explanation.

Purpose of the problem

Structure determination to solve a conundrum.

Suggested solution

The evidence shows that there is no carbonyl group in the molecule but that there is an OH group. The peak at 110 ppm looks at first sight like an alkene, but it could also be an unusual saturated carbon atom bonded to two oxygens. You might have argued that an alcohol and a ketone could combine to give a hemiacetal, and that is, of course, just what it is. The compound exists as a stable hemiacetal because it has a favourable five-membered ring.

■ P. 136 of the textbook explains why cyclic hemiacetals are stable.

PROBLEM 7

Each of these compounds is a hemiacetal and therefore formed from an alcohol and a carbonyl compound. In each case give the structures of the original materials.

Purpose of the problem

To give you practice in seeing the underlying structure of a hemiacetal.

Suggested solution

Each OH group represents a carbonyl group in disguise (marked with a grey circle). Just break the bond between this carbon and the other oxygen atom and you will see what the hemiacetal was made from. The first example shows how it is done.

The next is similar but the alcohol is a different molecule.

Do not be deceived by the third example. There is one hemiacetal (two oxygens joined to the same carbon atom) but the other OH is just a tertiary alcohol.

The last two examples are not quite the same. The first is indeed symmetrical but the second has one oxygen atom in a different position so that there is only one hemiacetal. Note that these hemiacetals may not be stable.

PROBLEM 8

Trichloroethanol my be prepared by the direct reduction of chloral hydrate in water with sodium borohydride. Suggest a mechanism for this reaction. Take note that sodium borohydride does not displace hydroxide from carbon atoms!

Purpose of the problem

To help you detect bad mechanisms and find concealed good ones.

Suggested solution

If sodium borohydride doesn't displace hydroxide from carbon atoms, then what does it do? We know it attacks carbonyl groups to give alcohols and to get trichloroethanol we should have to reduce chloral. Hemiacetals are in equilibrium with their carbonyl equivalents, so…

PROBLEM 9

It has not been possible to prepare the adducts from simple aldehydes and HCl. What would be the structure of such compounds, if they could be made, and what would be the mechanism of their formation? Why can't these compounds be made?

Purpose of the problem

More revision of equilibria to help you develop a judgement on stability.

Suggested solution

This time we need a mechanism so that we can work out what would be formed. Protonation of the carbonyl group and then nucleophilic addition of chloride ion would give the supposed products.

There's nothing wrong with the mechanism, it's just that the reaction is an equilibrium that will run backwards. Hemiacetals are unstable because they decompose back to carbonyl compounds. Chloride ion is very stable and decomposition will be faster than it is for hemiacetals.

PROBLEM 10

What would be the products of these reactions? In each case give a mechanism to justify your prediction.

Purpose of the problem

Giving you practice in the art of predicting products—more difficult than simply justifying a known answer.

Suggested solution

The Grignard reagent will add to the carbonyl group and the work-up will give a tertiary alcohol as the final product.

The second reaction should give you brief pause for thought as you need to recall that borohydride reduces ketones but not esters.

Suggested solutions for Chapter 7

7

PROBLEM 1

Are these molecules conjugated? Explain your answer in any reasonable way.

Purpose of the problem

Revision of the basic kinds of conjugation and how to show conjugation with curly arrows.

Suggested solution

The first compound is straightforward with one conjugated system (an enone) and a non-conjugated alkene. You could draw curly arrows to show the conjugation, like this, and/or give a diagram to show the distribution of the electrons.

The last three compounds obviously form a related group with the same skeleton and only the alkene moved round. There is of course ester conjugation in all three and this is the only conjugation in the last molecule. The first has extended conjugation between the nitrogen lone pair and the carbonyl group and the second has simple conjugation between the alkene and the ester.

The only conjugation in the last compound is the delocalization of the ester oxygen lone pair. This is of course there in all the other compounds too.

PROBLEM 2

How extensive is the conjugated system(s) in these compounds?

Purpose of the problem

To explore more extensive conjugated systems.

Suggested solution

Both compounds are completely conjugated: even including the nitrogen atom in the first and the carbonyl group in the second. You can draw the arrows going either way round the ring to give different ways of writing the same structure. The arrows on the second compound should end on the carbonyl oxygen.

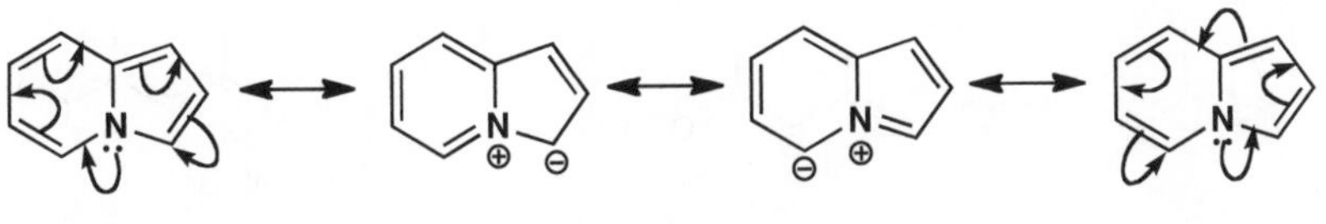

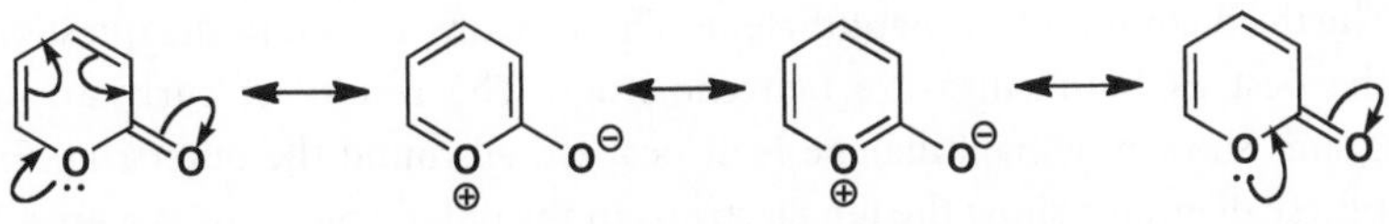

PROBLEM 3

Draw diagrams to represent the conjugation in these molecules. Draw two types of diagram:

(a) Show curly arrows linking at least two different ways of representing the molecule

(b) Indicate with dotted lines and partial charges (where necessary) the partial double bond (and charge) distribution.

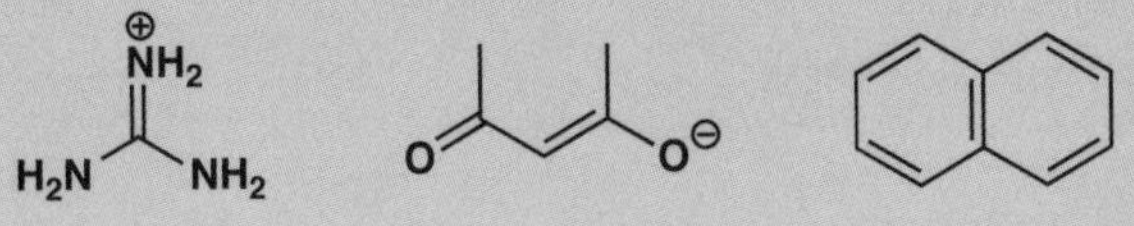

Purpose of the problem

A more exacting exploration of the precise details of conjugation.

Suggested solution

Treating each compound separately in the styles demanded by the question, the first (the guanidinium ion) is a very stable cation because of conjugation. The charge is delocalized onto all three nitrogen atoms as the first three structures show. Each nitrogen has an equal positive charge so our fourth diagram shows one third + on each.

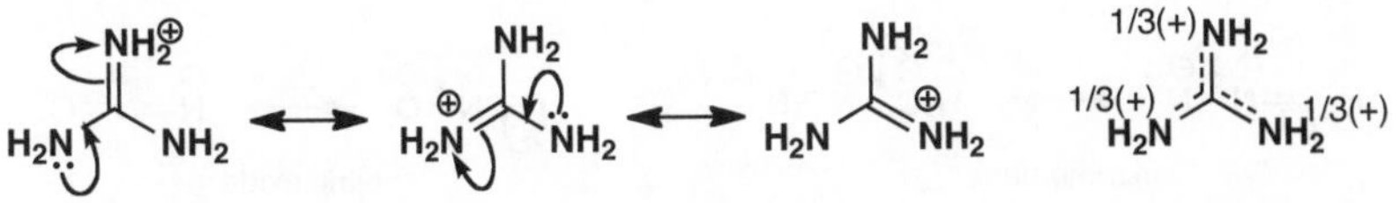

The second compound is what you will learn to call an enolate anion. The negative charge is delocalized throughout the molecule, mostly on the oxygens but some on carbon. It is difficult to represent this with partial charges but the charges on the oxygens will be nearly a half each.

The third compound is naphthalene. The structure drawn in the question is the best as both rings are benzene rings. The results of curly arrow diagrams show how naphthalene is delocalized all round the outer ring. In fact these diagrams show the ten electrons in the outer ring – this is a 4n + 2 number and all three diagrams show that naphthalene is aromatic.

PROBLEM 4

Draw curly arrows linking alternative structures to show the delocalization in
(a) diazomethane CH_2N_2
(b) nitrous oxide, N_2O
(c) dinitrogen tetroxide, N_2O_4

Purpose of the problem

Delocalization in some neutral molcules we nonetheless have to draw using charges.

Suggested solution

You saw on page 154 of the textbook that the nitro group, although it is neutral, can be represented as a pair of delocalized structures containing charges. The same is true for the explosive gas diazomethane. It has a linear structure, and we can draw two alternative structures, both with charges, even though it is a neutral compound. They're linked with the double headed arrow used for alternative representations for the same compound. We hope you remembered to avoid the trap of giving nitrogen five bonds!

diazomethane   nitric oxide

■ The idea of isoelectronic structures is introduced on p. 102 of the textbook. These compounds are also isoelectronic with carbon dioxide and azide (N_3^-).

Nitric oxide is very similar—in fact it is isoelectronic with diazomethane. You can think of is as a nitrogen molecule in which an oxygen atom has captured one of the lone pairs.

Dinitrogen tetroxide is a gas which decomposes to the more familiar brown air pollutant nitrogen dioxide (NO_2) at higher temperatures. The only way we can draw it seems most unsatisfactory: both nitrogens with positive charges! Even though these are not full positive charges, and this molecule does bring into focus the inadequacy of some valence bond representations, perhaps our discomfort with the structure is an indication of why this N–N bond is so weak…

PROBLEM 5

Which (parts) of these compounds are aromatic? Justify your answer with some electron counting. You may treat rings separately or together as you wish. You may notice that two of them are compounds we met in problem 2 of this chapter.

aklavinone: a tetracycline antibiotic

colchicine: a compound from the autumn crocus used to treat gout

Purpose of the problem

A simple exploration of the idea of aromaticity: can you count up to six? Remember: count only those π electrons in the ring and on no account put one electron on both atoms at either end of the bond: put the electrons where they are—in the bond.

Suggested solution

The numbers show how many π electrons there are in each bond or at each atom. The first compound has a lone pair on nitrogen in a p-orbital shared between both rings. Each ring has six electrons and the periphery of the whole molecule has ten electrons. Both rings and the entire molecule are aromatic. The second has four π electrons only so there is no aromaticity anywhere. The third has six π electrons in the ring including the lone pair on oxygen but not including the carbonyl group which is outside the ring. The compound is aromatic.

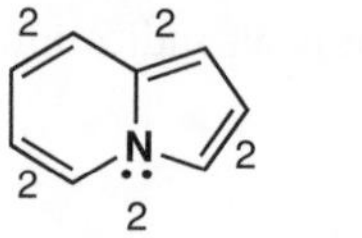

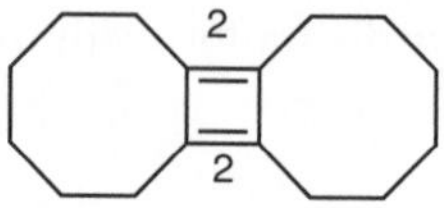

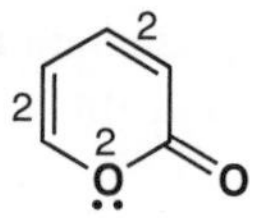

For the rest we have put the number of π electrons inside each ring and there are two aromatic rings in each compound. Again we don't count carbonyl group electrons as they are outside the ring. So one ring in aklavinone has only four electrons and is not aromatic while one of the seven-membered rings in colchicine is aromatic. Each compound has one saturated ring that cannot be aromatic.

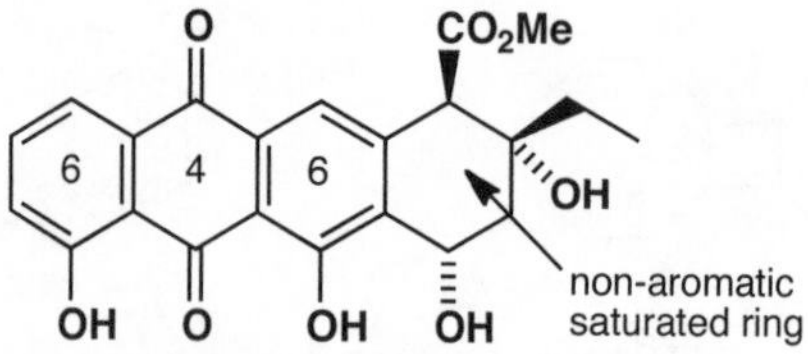

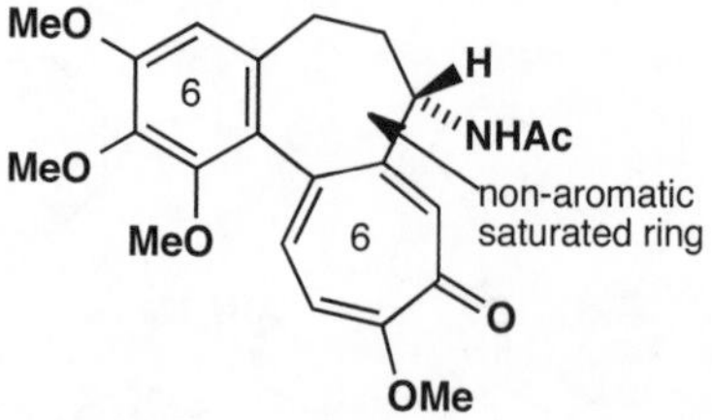

PROBLEM 6

The following compounds are considered to be aromatic. Account for this by identifying the appropriate number of delocalized electrons.

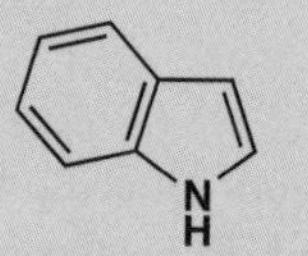

indole 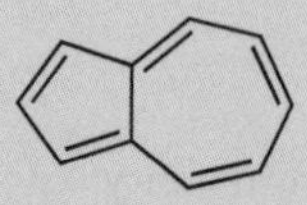   azulene 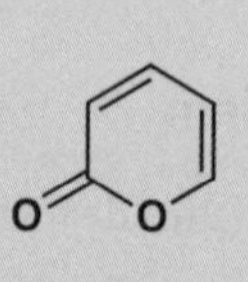   α-pyrone 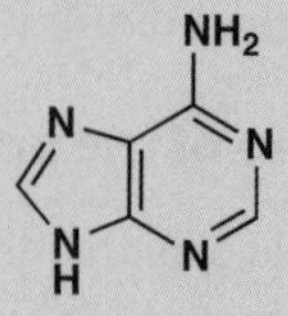   adenine

Purpose of the problem

Accounting for aromaticity in less familiar circumstances.

Suggested solution

Indole, as drawn here, has eight double bonds, which give eight delocalized electrons. To be aromatic, it needs 2n+2, so two more electrons must come from the lone pair of the nitrogen atom.

Azulene is isomeric with naphthalene, and it's quite easy to find the ten electrons – four in one ring and six in the other.

Pyridone has six electrons—two from the double bonds in the ring and two from nitrogen. That means that the carbonyl group, whose double bond is not part of the ring, does not contribute to the aromatic sextet. This is generally true for double bonds which stick out of the ring—see problem 2.

indole   azulene   4-pyridone

Adenine is one of the four bases which carry the genetic code in DNA. Its ten electrons arise as shown: eight from the double bonds and two from one of the nitrogen atoms in the five-membered ring. The other three nitrogens don't contribute their lone pairs, because they are not delocalized—like the lone pair in pyridine.

■ The details of bonding in pyridine will be explained in chapter 29, on page 724 of the textbook.

adenine

PROBLEM 7

Cyclooctatetraene (see p. 158 of the textbook) reacts readily with potassium metal to form a salt, K_2[cyclooctatetraene]. What shape do you expect the ring to have in this compound? A similar reaction of hexa(trimethylsilyl)benzene with lithium also gives a salt. What shape do you expect this ring to have?

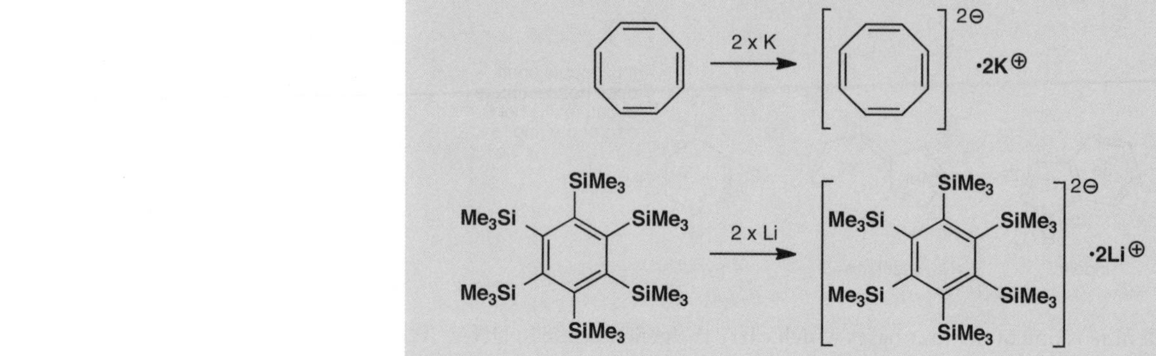

Purpose of the problem

The consequences of aromaticity and 'antiaromaticity'.

Suggested solution

Cyclooctatetraene, as explained on p. 158 of the textbook, is 'tub-shaped' and not planar, because its eight π electrons do not form a 2n+2 number. However, two atoms of potassium can reduce the cyclooctatetraene to a dianion by giving it two electrons, so now it has ten electrons, is aromatic, and is planar. Just one of the many possible delocalized structures of the product is shown here.

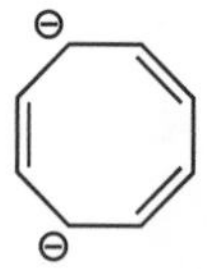

one way of drawing
the flat, aromatic dianion

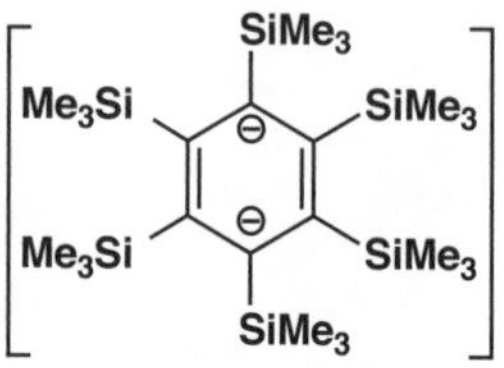

this dianion is no longer flat

When lithium reduces hexa(trimethylsilyl)benzene, the aromatic sextet is increased to a total of eight delocalized electrons, so the compound is no longer aromatic. The six membered ring in the salt is no longer flat.

PROBLEM 8

How would you expect the hydrocarbon below to react with bromine, Br_2?

Purpose of the problem

The consequences of aromaticity for reactivity.

Suggested solution

Aromatic rings typically react by substitution, so that they can retain the aromatic sextet. By contrast, alkenes react by electrophilic addition—the classic test for an alkene is that they decolourize bromine water. So, how will our hydrocarbon (known as indene) react? It contains an aromatic ring, but the five-membered ring is not aromatic—it contains a saturated carbon atom. So there is a choice of substitution on the six-membered ring or addition to the alkene in the five-membered ring. Alkenes are more reactive than benzene, so the alkene reacts first:

■ This contrasting behaviour was one of the pieces of evidence for aromaticity, and is on page 157 of the textbook.

product of addition to the alkene

PROBLEM 9

In aqueous solution, acetaldehyde (ethanal) is about 50% hydrated. Draw the structure of the hydrate of acetaldehyde. Under the same conditions, the hydrate of *N,N*-dimethylformamide is undetectable. Why the difference?

acetaldehyde   *N,N*-dimethylformamide

Purpose of the problem

The consequences of delocalization for reactivity.

Suggested solution

As you saw in chapter 6, aldehydes are readily hydrated. For amides, however, there is a price to pay: the delocalization that contributes to the stability of the amide would be lost on hydration, so dimethylformamide is not hydrated in aqueous solution.

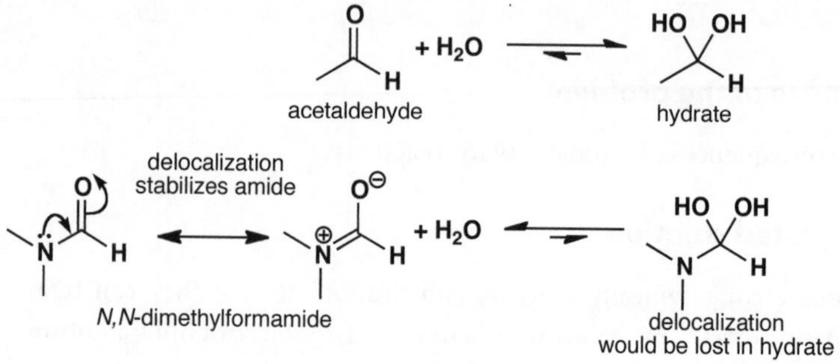

acetaldehyde

hydrate

delocalization stabilizes amide

N,N-dimethylformamide

delocalization would be lost in hydrate

Suggested solutions for Chapter 8

8

PROBLEM 1

How would you separate a mixture of these three compounds?

naphthalene   pyridine   *para*-toluic acid

Purpose of the problem

Revision of simple acidity and basicity in a practical situation.

Suggested solution

Pyridine is a weak base (pK_a of the pyridinium ion is about 5.5) and can be dissolved in aqueous acid. Naphthalene is neither an acid nor a base and is not soluble in water at any pH. *p*-Toluic acid is a weak acid (pK_a about 4.5) and can be dissolved in aqueous base. So dissolve the mixture in an organic solvent immiscible with water (say ether Et_2O or dichloromethane CH_2Cl_2) and extract with aqueous acid. This will dissolve the pyridine as its cation. Then extract the remaining organic layer with aqueous base such as $NaHCO_3$ which will remove the toluic acid as its water-soluble anion. You now have three solutions. Evaporate the organic solution to give crystalline naphthalene. Acidify the basic solution of *p*-toluic acid and the free acid will precipitate out and can be recrystallized. Add base to the pyridine solution, extract the pyridine with an organic solvent, and distil the pyridine. It doesn't matter if you extract the original solution with base first and acid second.

PROBLEM 2

In the separation of benzoic acid from toluene on p. 164 of the textbook we suggested using KOH solution. How concentrated a solution would be necessary to ensure that the pH was above the pK_a of benzoic acid (4.2)? How would you estimate how much KOH solution to use?

Purpose of the problem

To ensure you understand the relationship between pH and concentration.

Suggested solution

Even a very weak solution of KOH has a pH>4.2. If we want a pH of 5 (just above the pK_a of benzoic acid) we must ensure that we have $[H_3O^+] = 10^{-5}$ mol dm^{-3}. The ionic product of water is $[H_3O^+] \times [HO^-] = 10^{-14}$ and so we need 10^{-9} mol dm^{-3} of KOH. This is very dilute! The trouble would be that you need one hydroxide ion for each molecule of benzoic acid and so if you had, say, 1.22 g PhCO$_2$H (= 0.01 equiv.) you would need 1000 litres (dm^3) of KOH solution. It makes more sense to use a much more concentrated solution, say 0.1M. This would give an unnecessarily high pH (13) but you would need only 100 ml (0.1 dm^3) to extract your benzoic acid.

PROBLEM 3

What species would be present in a solution of this hydroxy-acid in (a) water at pH 7, (b) aqueous alkali at pH 12, and (c) in concentrated mineral acid?

Purpose of the problem

To get you thinking about what really is present in solution using rough pK_a as a guide in a practical situation.

Suggested solution

See page 173 of the textbook for the pK_a of phenol.

The CO$_2$H group will have a pK_a of about 4–5 and the phenolic OH a pK_a of about 10. So the carboxylic acid but not the phenol will be ionized at pH 7, they will both be ionized at pH 12, and there will be a mixture of free acid and protonated acid at very low pH. The proton will be on the carbonyl oxygen atom as this gives a delocalized cation.

PROBLEM 4

What would you expect to be the site of (a) protonation and (b) deprotonation if these compounds were treated with the appropriate acid or base? In each case suggest a suitable acid or base and give the structure of the products.

Purpose of the problem

Progressing to more taxing judgements on more interesting molecules.

Suggested solution

The simple amine piperidine will easily be protonated by even weak acids as the conjugate base has a pK_a of about 11. Any mineral acid such as HCl will do the job as would weaker acids such as RCO_2H. Deprotonation will remove the NH proton as nitrogen is more electronegative than carbon but a very strong base such as BuLi will be needed as the pK_a will be about 30–35. You could represent the product with an N–Li bond or as an anion.

The second example is more complicated but contains a normal tertiary amine so protonation will occur there with most acids as as the conjugate base has a pK_a of about 11. We use TsOH this time but that has no special significance. The tertiary amine cannot be deprotonated and in any case the alcohol is more acidic and a strong base will be needed, say NaH.

The third example is more complicated still. There is a normal OH group (pK_a of about 16) and a slightly acidic alkyne (pK_a of about 32). The basic group is not a simple amine but a delocalized amidine. Protonation occurs on the top (imine) nitrogen as the positive charge is then delocalized over both nitrogens. Protonation on the other nitrogen does not occur. The pK_a of the conjugate base is about 12.

The first proton to be removed by base will be from the alcohol and this will need a reasonably strong base such as NaH. Removal of the alkyne proton requires a much stronger base such as BuLi. You might represent the product as an alkyne anion or a covalently bonded alkyllithium.

PROBLEM 5

Suggest what species would be formed by each of these combinations of reagents. You are advised to use estimated pK_a values to help you and to beware of those cases where nothing happens.

Purpose of the problem

Learning to compare species of similar acidity or basicity.

Suggested solution

In each case one of the reagents might take a proton from the other. In example (a), would the phenolate anion remove a proton from acetic acid? The answer is *yes* because acetic acid is a much stronger acid than phenol The difference is five pH units so the equilibrium constant would be about 10^5 and the equilibrium would lie far across to the right.

(a)

pK_a about 5   pK_a about 10

Example (b) has a similar possible reaction but this time the pK_a difference is much smaller and the other way so the equilibrium constant is 100 and favours the starting materials.

(b)

pK_a about 7   pK_a about 5

Example (c) is rather different. We do have another carboxylic acid but this is a much stronger acid because of the three fluorine atoms and the equilibrium is far over to the left.

■ See page 178 of the textbook for the effect of fluorine on pK_a.

(c)

pK_a about 5   pK_a about –1

PROBLEM 6

What is the relationship between these two molecules? Discuss the structure of the anion that would be formed by the deprotonation of each compound.

Purpose of the problem

To help you recognize that conjugation may be closely related to tautomerism.

Suggested solution

They are tautomers: they differ only in the position of one hydrogen atom. It is on nitrogen in the first structure and on oxygen in the second. As it happens the first structure is more stable. They are both aromatic (check that you see why) but the first has a strong carbonyl group while the second has a weaker imine. Deprotonation may appear to give two different anions but they are actually the same because of delocalization. Note the different 'reaction' arrows: equilibrium sign for deprotonation and double headed arrow for delocalization.

PROBLEM 7

The carbon NMR spectrum of these compounds can be run in D_2O under the conditions shown. Why are these conditions necessary? What spectrum would you expect to observe?

Purpose of the problem

NMR revision and practice at judging the states of compounds at different pHs. Observation of hidden symmetry from conjugation.

Suggested solution

Both compounds are quite polar and not very soluble in the usual NMR solvents. In addition they have NH or OH protons that exchange in solution and broaden the spectrum. With acid or base catalysis the NH and OH protons are exchanged with deuterium and sharp signals appear. But in the strong acid or base used here, ions are formed. The first compound, a strongly basic guanidine (see p. 167 of the textbook) forms a cation in DCl. The cation is symmetrical, unlike the original guanidine, and a very simple spectrum results: just three types of carbon in the benzene ring and one very low field carbon (at large δ) for the carbon in the middle of the cation.

The second compound loses a proton from the OH group to give a delocalized symmetrical anion. There will be five signals in the NMR: the two methyl groups are the same (at small δ) as are the two CH_2 groups in the ring (at slightly larger δ). There is one unique carbon joined to the two methyl groups (at small δ) and another in the middle of the anion (at large δ). Finally both carbonyl groups are the same (at even larger δ).

PROBLEM 8

These phenols have approximate pK_a values of 4, 7, 9, 10, and 11. Suggest with explanations which pK_a value belongs to which phenol.

Purpose of the problem

Detailed examination of electronic effects to estimate pK_a values.

Suggested solution

Electron-withdrawing effects make phenols more acidic and electron-donating effects make them less acidic. Phenol itself (the fourth example) has a pK_a of 10. The only compound less acidic than phenol must be the third with three weakly electron-donating methyl groups. One chlorine atom has an inductive electron-withdrawing effect so the last compound has pK_a 9. The remaining two have the powerful electron-withdrawing nitro group. So the first compound, with two nitro groups, must have pK_a of 4 (making it as strong an acid as acetic acid) and the second, with one nitro group, must have pK_a of 7.

pK_a about 4   pK_a about 7   pK_a about 11

pK_a about 10   pK_a about 9

PROBLEM 9

The pK_a values of these two amino acids are as follows:

(a) cysteine: 1.8, 8.3, and 10.8

(b) arginine: 2.2, 9.0, and 13.2.

Assign these pK_a values to the functional groups in each amino acid and draw the most abundant structure that each molecule will have at pH 1, 7, 10, and 14.

cysteine   arginine

Purpose of the problem

Further revision in thinking about acidity and basicity of functional groups, and reinforcement of expected pK_a values for functional groups. Amino acids are particularly important.

Suggested solution

At high pH cysteine exists as a dianion as both the thiol and the carboxylic acid are anions. If we now add acid, at about pH 10 (actually 10.8) the amine will get protonated, then the thiol will be protonated at about pH 8 (actually 8.3) and finally the carboxylic acid will be protonated at low pH, rather lower than say $MeCO_2H$, as the electron-withdrawing ammonium group increases its acidity (actually at 1.8).

At high pH arginine exists as a monoanion—even the very basic guanidine group cannot be protonated at pH 14. If we now add acid, at about pH 13 the guanidine will get protonated, then the amino group will be protonated at about pH 10 (actually 9.0) and finally the carboxylic acid will be protonated at low pH. This carboxylic acid is rather more acidic than you might expect, but not surprisingly it is harder to protonate an anion in a molecule which already has an overall positive charge.

■ The basicity of arginine, and of the guanidine functional group it contains, was discussed on p. 175 of the textbook.

PROBLEM 10

Neither of these two methods for making pentan-1,4-diol will work. What will happen instead?

Purpose of the problem

To help you appreciate the disastrous effects that innocent-looking groups may have because of their weak acidity.

Suggested solution

The OH group is the Wicked Witch of the West in this problem. Whoever planned these syntheses expected it to lie quietly and do nothing. All chemists have to learn that things don't go our way just because we want them to do so. Here, although the OH group is only a weak acid (pK_a about 16) it will give up its proton to the very basic Grignard reagents. In the first case, one molecule of Grignard is destroyed but the reaction might succeed if two equivalents were used.

The second case is hopeless as the Grignard reagent destroys itself by intramolecular deprotonation. This synthesis could be rescued by putting a protecting group on the OH.

- 'Protecting groups' are discussed in chapter 23.

PROBLEM 11

Which of these bases would you choose to deprotonate the following molecules? Very strong bases are more challenging to handle so you should aim to use a base which is just strong enough for the job, but not unnecessarily strong.

Choice of bases:

KOH   NaH   BuLi   NaHCO₃

Purpose of the problem

To help you match basicity to acidity—an important part of choosing reagents for many reactions.

Suggested solution

We can start by estimating the pK_a of the most acidic proton in each of the substrates to be deprotonated, and likewise estimating the pK_a of the conjugate acids of the proposed bases. Most of these values were discussed in chapter 8, and you were encouraged to commit some of them to memory.

■ Bicarbonate may be new to you, but you might reasonably guess that it has basicity similar to a carboxylate anion. The base sodium hydride, and the pK_a of its conjugate acid, appears on p. 237 of the textbook.

most acidic

| pK_a carboxylic acid about 5 | pK_a protonated aromatic amine about 5 | pK_a alcohol about 15 | pK_a alkyne about 25 | pK_a amine about 35 |

| NaHCO$_3$ bicarbonate pK_a (H$_2$CO$_3$) about 5 | KOH hydroxide pK_a (H$_2$O) about 15 | | NaH hydride pK_a (H$_2$) about 35 | BuLi butyllithium pK_a (alkane) about 50 |

most basic

Any of the bases will deprotonate compounds above and to the left of them. So, to deprotonate the least acidic of these, the amine, you would choose to use butyllithium. To deprotonate the alkyne (a reaction which is commonly used to make C–C bonds) you could use BuLi, or alternatively sodium hydride (NaH). BuLi has to be handled under an inert atmosphere, while NaH, although it reacts with water, can be spooned out safely as a suspension in oil.

■ The lithium derivative that results from deprotonating this amine with BuLi is known as 'LDA' and features frequently in later chapters of the textbook.

The alcohol has a pK_a close to that of water, so hydroxide is not a good choice for complete deprotonation, and sodium hydride is commonly used for this purpose. Hydroxide will however deprotonate both the carboxylic acid, to make a carboxylate salt, or the ammonium ion, to make the free amine. Bicarbonate is also commonly used for this purpose: although it is only just basic enough to do the job, the deprotonation reaches completion because it is not an equilibrium: protonated bicarbonate forms carbonic acid which decomposes irreversibly to water and carbon dioxide.

Suggested solutions for Chapter 9

9

PROBLEM 1

Propose mechanisms for the first four reactions in the chapter.

Purpose of the problem

Rehearsal of the basic mechanisms from chapter 9.

Suggested solution

Each reaction involves nucleophilic attack of the organometallic reagent on the aldehyde or ketone followed by protonation. You may draw the intermediate as an anion or with an O–metal bond as you please. Note the atom-specific arrows to show which atom is the nucleophile. In the third reaction the allyl-Li might attack through either end.

PROBLEM 2

What products would be formed in these reactions?

Purpose of the problem

The toughest test: predicting the product. The sooner you get practice the better.

Suggested solution

Though prediction is harder than explanation, you should get these right the first time as only the last one has a hint of difficulty. In the first example, the ethyl Grignard reagent acts as a base to remove a proton from the alkyne. Whether you draw the intermediate as an alkyne anion or a Grignard reagent is up to you. Notice that sometimes we give the protonation step at the end and sometimes not. This is the general practice among organic chemists and you may or you may not bother to draw the mechanism of this step.

For the second example, just make the organometallic reagent and add it to the carbonyl group. Cyclobutanone is more electrophilic than many other ketones because of the strain of a carbonyl group in a four-membered ring. This time the protonation is shown.

The third example raises the question of which halogen is replaced. In fact bromine is more easily replaced than chlorine. Iodine is more easily replaced than either and fluorine usually does not react. Don't be disappointed if you failed to see this.

■ This synthesis of carboxylic acids is on pp. 190-191 of the textbook.

PROBLEM 3

Suggest alternative routes to fenarimol different from the one in the textbook on p. 192. Remind yourself of the answer to problem 2 above.

Purpose of the problem

Practice in choosing alternative routes.

Suggested solution

Three aromatic rings are joined to a tertiary alcohol in fenarimol, so the alternatives are to make organometallic reagents from different aromatic compounds. Two aromatic compounds must be joined to form a ketone and the third added as an organometallic reagent. You will meet ways to make ketones like these in chapter 21. You'll need the insight from problem 2 above to help you choose Br as the functional group to be lithiated or converted into a Grignard reagent. Here are two possible methods:

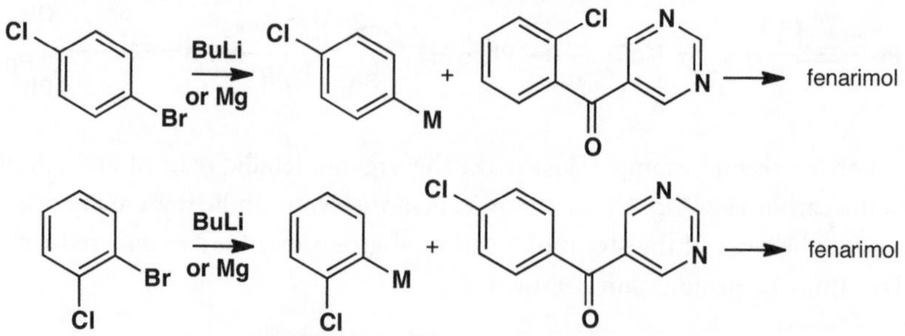

PROBLEM 4

Suggest two syntheses of the bee pheromone heptan-2-one.

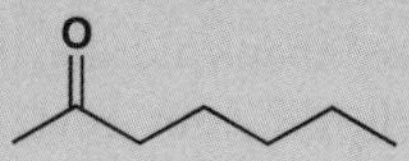

Purpose of the problem

Further exploration of the use of organometallic compounds. This time you'll probably need the oxidation of alcohols from p. 194 of the textbook.

Suggested solution

There are of course many different solutions but the most obvious are to make the corresponding secondary alcohol and oxidize it. Two alternatives are shown here.

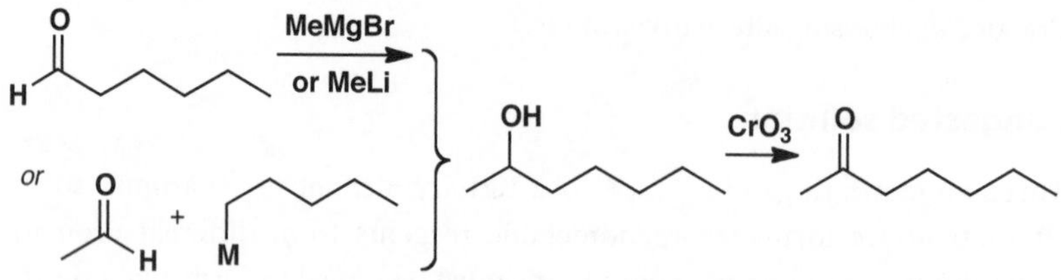

PROBLEM 5

The antispasmodic drug biperidin is made by the Grignard addition reaction shown here. What is the structure of the drug? Do not be put off by the apparent complexity of the structure: just use the chemistry of Chapter 9.

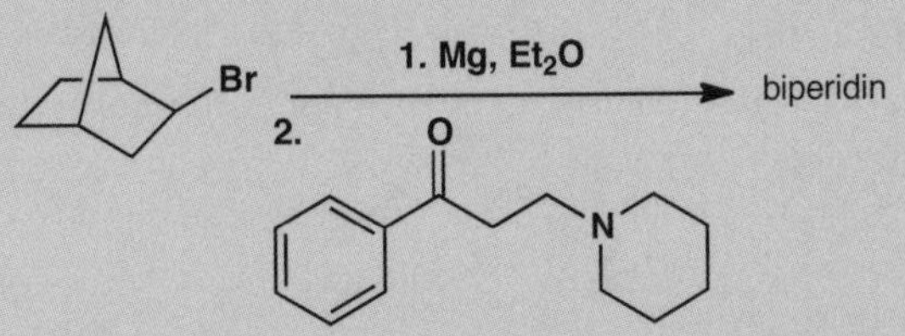

How would you suggest that the drug procyclidine should be made?

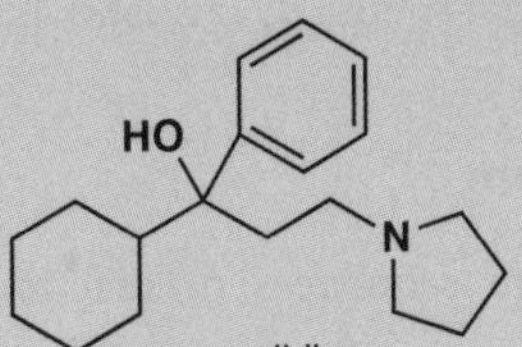

procyclidine

Purpose of the problem

Exercise in product prediction in a more complicated case and a logical extension to something new.

Suggested solution

A Grignard reagent must be formed from the alkyl bromide and this must add to the ketone. Aqueous acidic work up (not mentioned in the problem as is often the case) must give a tertiary alcohol and that is biperidin.

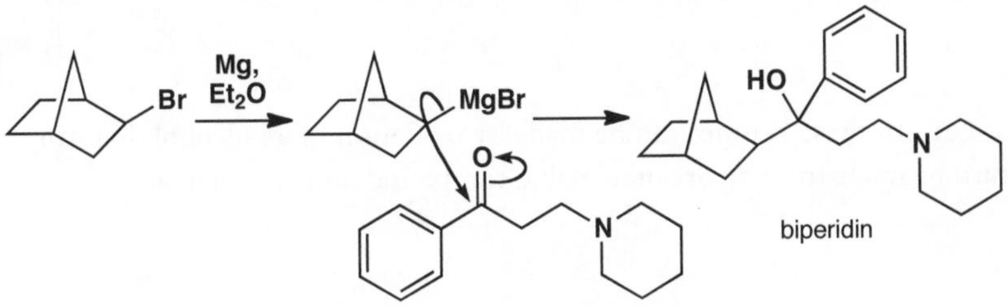

To get procyclidine, we must change both the alkyl halide and the ketone but the reaction is very similar.

procyclidine

PROBLEM 6

The synthesis of the gastric antisecretory drug rioprostil requires this alcohol.

(a) Suggest possible syntheses starting from ketones and organometallics.
(b) Suggest possible syntheses of the ketones in part (a) from aldehydes and organometallics (don't forget about CrO_3 oxidation).

Purpose of the problem

Your first introduction to sequences of reactions where more complex molecules are created.

Suggested solution

There are three one-step syntheses from ketones and organometallic compounds. We have used 'M' to indicate the metal—it might be Li or MgX (in other words, the organometallic could be an organolithium or a Grignard reagent).

Each of these ketones can be made by oxidation of an alcohol that can in turn be made from an organometallic compound and an aldehyde.

PROBLEM 7

Why is it possible to make the lithium derivative A by Br/Li exchange, but not the lithium derivative B?

Purpose of the problem

Revision of the stability of carbanions and its relevance to lithium/bromine exchange.

Suggested solution

The first example is a vinyl bromide and vinyl (sp^2) carbanions are more stable than saturated (sp^3) carbanions because of the greater s-character in the C–Li bond. The second example is saturated, like BuLi, but it is a *tertiary* alkyl bromide. The *t*-alkyl carbanion would be less stable than the primary one and its lithium derivative less stable than BuLi, so it is not formed.

PROBLEM 8

How could you use these four commercially available starting materials

PhCHO   EtI CO_2 ⬠—Br

to make the following three compounds?

Purpose of the problem

Thinking about synthesis: how to put molecules together

Suggested solution

The first compound contains a phenyl and an ethyl group, so you could convert the ethyl iodide to a Grignard reagent and add it to the aldehyde. The product is an alcohol, so you need to use CrO_3 to oxidize it to the ketone

The second compound is a carboxylic acid, which can come from addition of the Grignard reagent derived from cyclopentyl bromide to carbon dioxide.

The third compound is a tertiary alcohol, which you could make by addition of the same cyclopentyl Grignard reagent to a ketone. The ketone will also need to be made by oxidation of an alcohol, itself derived from benzaldehyde and the cyclopentyl Grignard reagent.

Suggested solutions for Chapter 10

10

PROBLEM 1

Suggest reagents to make the drug phenaglycodol by the route below.

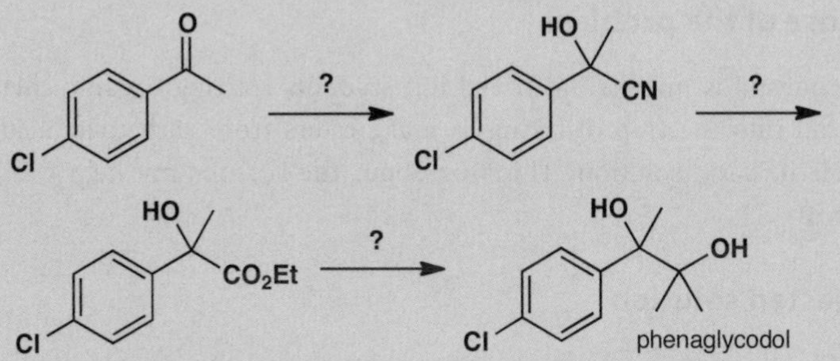

Purpose of the problem

Simple revision of addition to carbonyl groups from chapter 6.

Suggested solution

The first step is a simple addition of cyanide to a ketone (p. 127 of the textbook) usually carried out with NaCN and an acid, such as acetic acid. The second step is an acid-catalysed addition of an alcohol to a nitrile (p.-213 of the textbook). Finally there is a double addition of an organometallic reagent to an ester (p. 216 of the textbook). One way of doing all this is shown below.

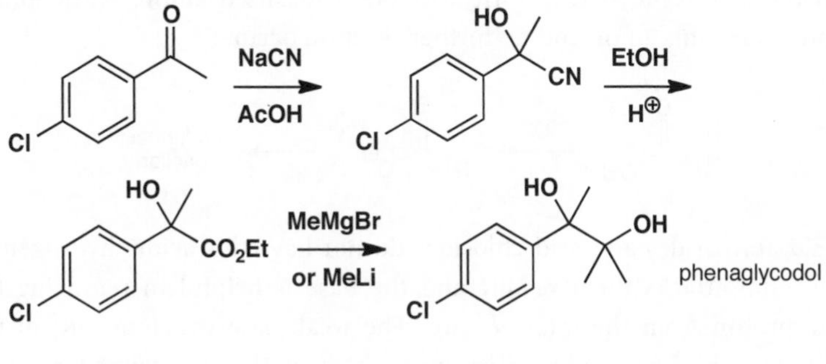

PROBLEM 2

Direct ester formation from carboxylic acids (R^1CO_2H) and alcohols (R^2OH) works in acid solution but not in basic solution. Why not? By contrast, ester formation from alcohols (R^2OH) and acid anhydrides [($R^1CO)_2O$)] or chlorides (R^1COCl) is commonly carried out in basic solution in the presence of bases such as pyridine. Why does this work?

Purpose of the problem

These questions may sound trivial but students starting organic chemistry often fall into the trap of trying to make esters from carboxylic acids and alcohols in basic solution. Thinking about the reasons my help you avoid this error.

Suggested solution

The direct reaction works in acid solution as the carboxylic acid is protonated (at the carbonyl group, note) and becomes a good electrophile. Later the tetrahedral intermediate is protonated and can lose a molecule of water.

■ This mechanism is described in detail on p. 208 of the textbook.

In basic solution, the first thing that happens is the removal of the proton from the carboxylic acid to form a stable delocalized anion. Nucleophiles cannot attack this anion and no further reaction occurs.

Acid anhydrides and acid chlorides do not have this acidic hydrogen so the alcohol attacks them readily and the base is helpful in removing the acidic proton from the intermediate. The weak base pyridine (pK_a of the conjugate acid 5.5) is ideal. The product from the uncatalysed reaction would be HCl from the acid chloride and the base also removes that.

PROBLEM 3

Predict the success or failure of these attempted substitutions at the carbonyl group. You should use estimated pK_a values in your answer and, of course, draw mechanisms.

Purpose of the problem

A chance to try out the correlation between leaving group ability and pK_a explained in the textbook (p. 205).

Suggested solution

You need to draw mechanisms for the formation of the tetrahedral intermediate and check that it is in the right protonated form. Then you need to check which potential leaving group is the best, using appropriate estimated pK_a values. The first and the last proposals will succeed but the second will not as chloride ion is a better leaving group than even a protonated amine and this reaction would go backwards.

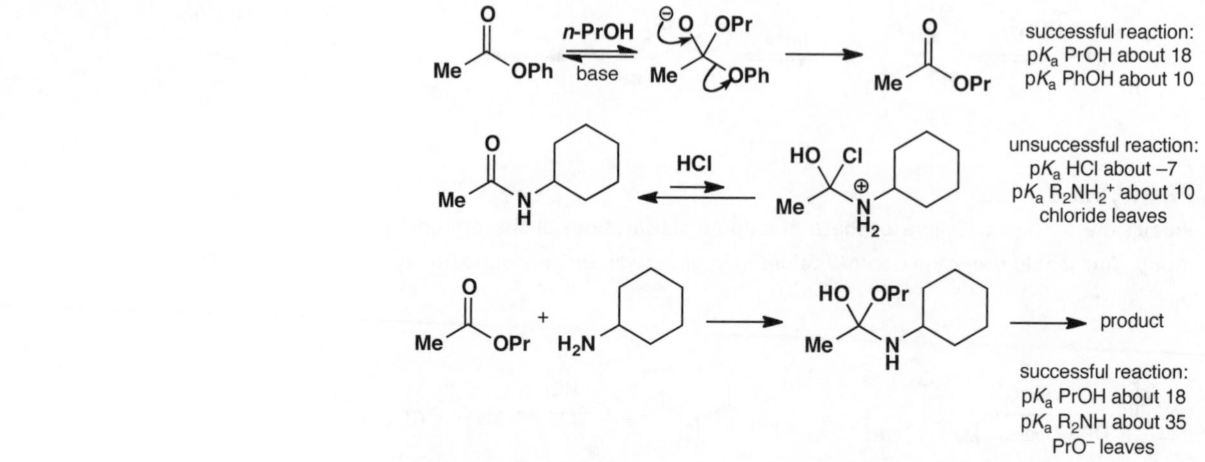
n-PrOH
base
successful reaction:
pKa PrOH about 18
pKa PhOH about 10
HCl
unsuccessful reaction:
pKa HCl about −7
pKa R2NH2+ about 10
chloride leaves
product
successful reaction:
pKa PrOH about 18
pKa R2NH about 35
PrO− leaves

PROBLEM 4

Suggest mechanisms for these reactions.

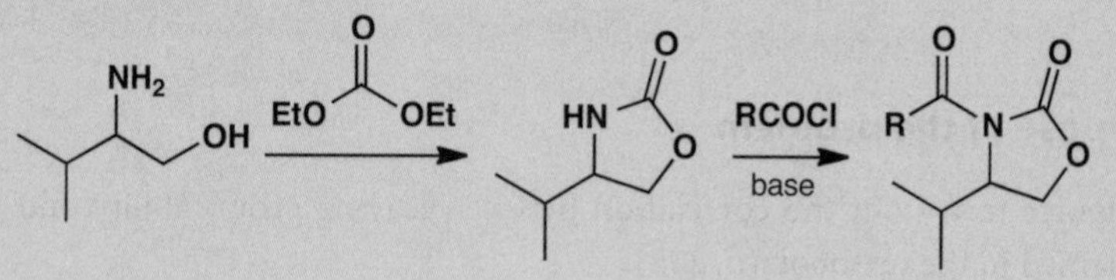
EtO
OEt
RCOCl
base

Purpose of the problem

Drawing mechanisms for nucleophilic substitution on important compounds including cyclic and dicarbonyl compounds.

Suggested solution

In the first reaction there are two nucleophilic substitutions and you must decide which nucleophile attacks first. The amine is a better nucleophile than the alcohol. The cyclization occurs because, in the intermediate for the second substitution, there are two alcohols as potential leaving groups. Either can leave but when the ring opens again, the alcohol is still part of the molecule and will re-cyclize, but if the ethoxide leaves it is lost into solution and does not come back.

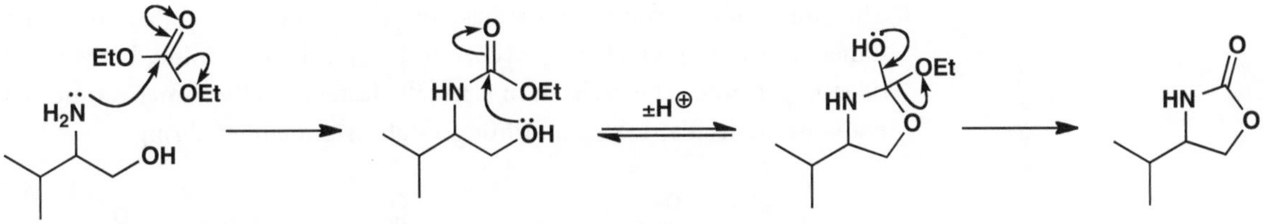

The second reaction is more straightforward. The amide proton is quite acidic and will be removed by the base making a better nucleophile. Notice that in these suggested solutions we are using the shorthand of the double-headed arrow on the carbonyl group.

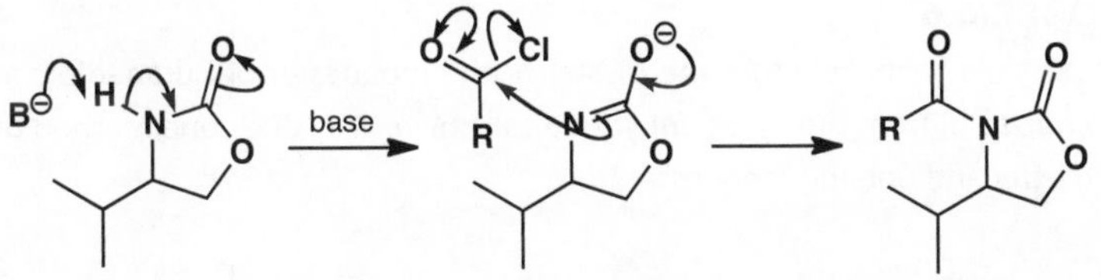

■ The use of the 'double-headed' arrow shorthand is explained on p. 217 of the textbook.

PROBLEM 5

In making esters of the naturally occurring amino acids (general structure below) it is important to keep them as their hydrochloride salts. What would happen to these compounds if they were neutralized?

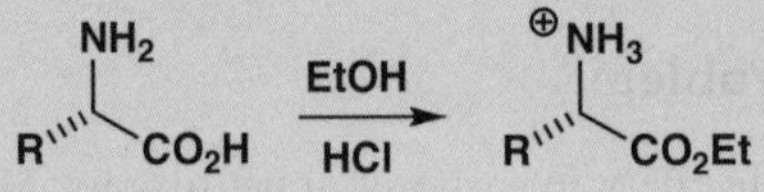

Purpose of the problem

Exploration of a simple reaction that can go seriously wrong if we do not think about what we are doing.

Suggested solution

The amino acids do not usually react with themselves as they exist mostly as the zwitterion. But after the acid is esterified it is much more electrophilic and the amino group is now nucleophilic.

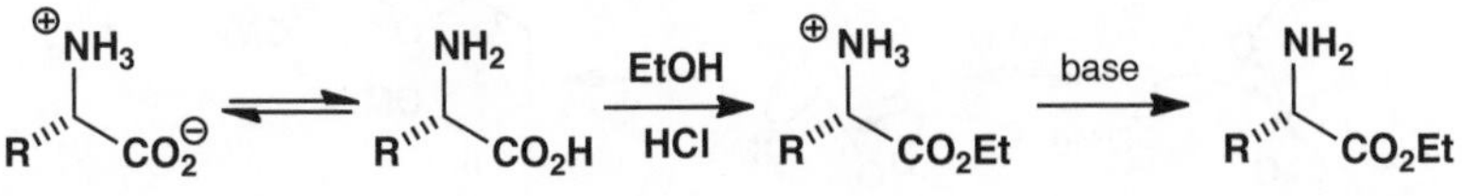

The amine of one compound attacks the ester group of another to form a dimer (a peptide) which may cyclize to form a double amide, known as a diketopiperazine. The cyclization is usually faster than the dimerization as it is an intramolecular reaction forming a stable six-membered ring.

PROBLEM 6

It is possible to make either the diester or the monoester of butanedioic acid (succinic acid) from the cyclic anhydride as shown. Why does one method give the diester and one the monoester?

Purpose of the problem

An exploration of selectivity in carbonyl substitutions. Mechanistic thinking allows you to say confidently whether a reaction will happen or not. This problem builds on problem 2.

Suggested solution

In basic solution the nucleophile is methoxide ion. This strong nucleophile attacks the carbonyl group to give a tetrahedral intermediate having two possible leaving groups. The ester anion is preferred (pK_a of RCO$_2$H about 5) to the alkoxide ion (pK_a of ROH about 15). This carboxylate anion cannot be protonated in basic solution and is not attacked by methoxide ion.

In acid solution the first reaction is similar, though the tetrahedral intermediate is neutral, and the carboxyl is still the better leaving group. The second esterification is now all right because methanol can attack the protonated carboxylic acid and water can be driven out after a second protonation. The second step is an equilibrium, with water and methanol about equal as leaving groups, but methanol is present in large excess as the solvent and drives the equilibrium across. We have omitted proton transfer steps.

PROBLEM 7

Suggest mechanisms for these reactions, explaining why these particular products are formed.

Purpose of the problem

A contrast between very reactive (acid chloride), less reactive (anhydride) and unreactive (amide) carbonyl compounds.

Suggested solution

The acid chloride reacts rapidly with the water and the carboxylic acid produced reacts rapidly with a second molecule of acid chloride. The anhydride reacts much more slowly (pK_a of HCl is –7 but the pK_a of RCO_2H is about 5) with water so there is a good chance of stopping the reaction there, especially when we use a low concentration of water in acetone solution. In this instance the chance is made a certainty because the anhydride precipitates from the solution and is no longer in equilibrium

■ The reactivity sequence of carboxylic acid derivatives is explained on pp. 202 and 206 of the textbook.

with the other reagents. It is usually possible to descend the reactivity sequence of acid derivatives.

■ The alkaline hydrolysis of amides is on p. 213 of the textbook.

The second reaction is an example of the alkaline hydrolysis of amides. Though the nitrogen atom is never a good leaving group, it will leave from the dianion and, once gone, it is quickly protonated and does not come back. This example also benefits from the release of the slight strain in the five-membered ring.

PROBLEM 8

Give mechanisms for these reactions, explaining the selectivity (or lack of it!) in each case.

Purpose of the problem

Analysis of a sequence of reactions where the first stops at the halfway stage but the second does not.

Suggested solution

One of the carbonyl groups of the anhydride must be attacked by LiAlH₄ and we need to follow that reaction through to see what happens next. The first addition of AlH₄⁻ produces a tetrahedral intermediate that decomposes with the loss of the only possible leaving group, the carboxylate ion, to give an aldehyde. That too is quickly reduced by AlH₄⁻ to give the hydroxy-acid as its anion, which is resistant to further reduction. In the acidic aqueous work-up, excess LiAlH₄ is instantly destroyed and the hydroxy-acid cyclizes to the lactone. The fact that the lactone is not formed under the reaction conditions is important: if it were, then it too would be reduced by the LiAlH₄.

The second reaction starts similarly with the Grignard reagent adding to the ester carbonyl group and the tetrahedral intermediate losing the only possible leaving group. Again, a reactive carbonyl compound is produced: a ketone that is more electrophilic than the ester, so it adds the Grignard reagent even faster. Work-up in aqueous acid gives the diol.

PROBLEM 9

This reaction goes in one direction in acid solution and in the other direction in basic solution. Draw mechanisms for the reactions and explain why the product depends on the conditions.

Purpose of the problem

A reminder that carbonyl substitutions are equilibria and that removal of a product from an equilibrium may decide which way the reaction goes. Practice at drawing mechanisms for intramolecular reactions.

Suggested solution

The equilibrium we are concerned with is that between the two products and we can draw what would happen in neutral solution.

The amine attacks the ester in the usual way to give the tetrahedral intermediate which decomposes with the loss of the better leaving group: phenols are reasonably acidic (pK_a PhOH = 10) so the phenoxy anion is a much better leaving group than ArNH⁻. In strongly basic solution, the phenol product is fully deprotonated, so again, the equilibrium lies to the right. In acidic solution the starting amine is fully protonated, pulling the equilibrium back over to the left.

PROBLEM 10

Amelfolide is a drug used to treat cardiac arrhythmia. Suggest how it could be made from 4-nitrobenzoic acid and 2,5-dimethylaniline.

Purpose of the problem

A reminder to avoid a common error in proposed reactions of carboxylic acids.

Suggested solution

It is tempting to try and react the amine directly with the acid, but unfortunately the only product this would give is the ammonium carboxylate salt: the amine deprotonates the acid, and the carboxylate anion that results is no longer electrophilic. With alcohols, esters can be formed from carboxylic acids under acid catalysis, but with amines the acid catalyst just protonates the amine, and it is no longer nucleophilic! The simplest solution is to convert the carboxylic acid to an acid chloride and allow that to react with the amine. Additional base will neutralize the HCl by-product.

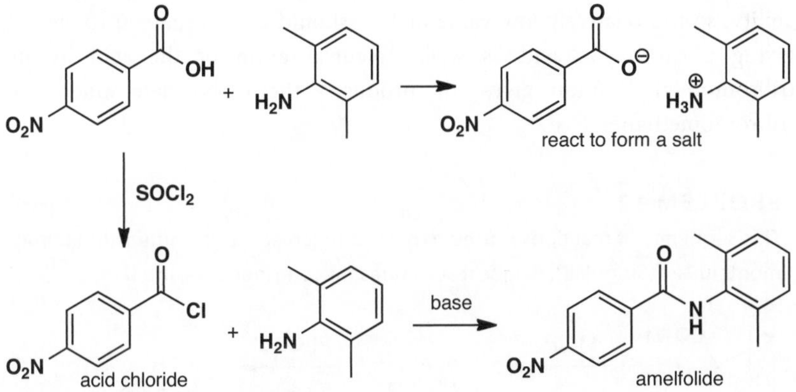

react to form a salt

acid chloride   base   amelfolide

PROBLEM 11

Given that the pK_a of tribromomethane, CHBr$_3$ (also known as bromoform) is 13.7, suggest what will happen when this ketone is treated with sodium hydroxide.

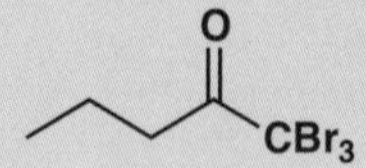

Purpose of the problem

Predicting reactivity with an unusual leaving group.

Suggested solution

The best approach to new reactions is to start drawing curly arrows for steps you know are reasonable, and to see where they take you. Here, we are treating a carbonyl compound, an electrophile, with hydroxide, a nucleophile, so the first step is likely to be addition of hydroxide to the C=O group. You have seen many, many reactions that start this way.

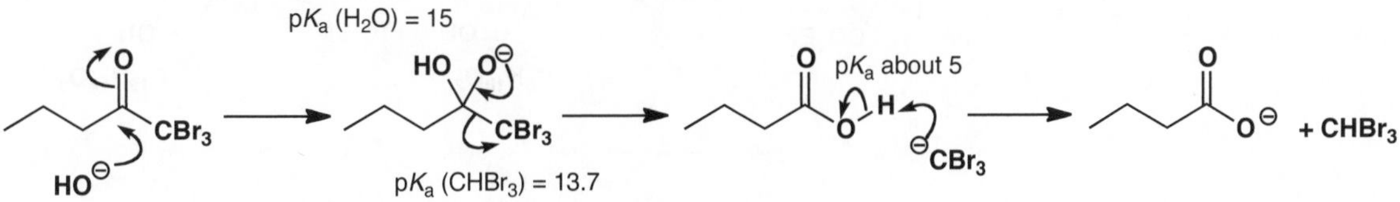

pK_a (H$_2$O) = 15

pK_a (CHBr$_3$) = 13.7

pK_a about 5

+ CHBr$_3$

The result looks like a tetrahedral intermediate: the only possible leaving groups are the hydroxide (which takes us back to starting materials) or the anion Br_3C^-. You learnt in chapter 10 to use pK_a to estimate leaving group ability, so the relatively low value of 13.7 should encourage you to eject it, giving a carboxylic acid as well. Neutralization of the acid by the tribromomethyl anion gives the products—the carboxylate anion and tribromomethane.

■ This reaction is known as the 'bromoform' reaction and is described on pp. 462-3 of the textbook.

PROBLEM 12

This sequence of reactions is used to make a precursor to the anti-asthma drug montelukast (Singulair). Suggest structures for compounds **A** and **B**.

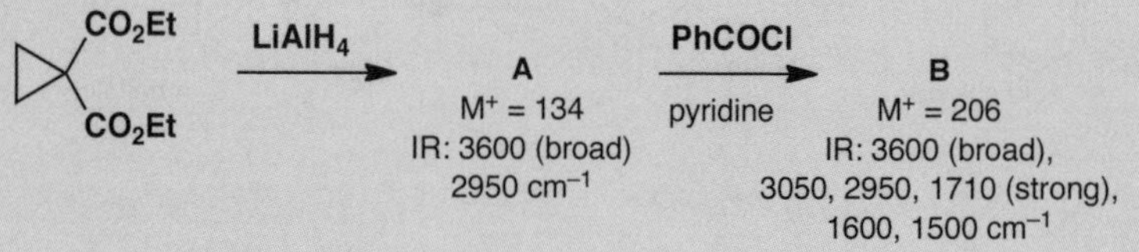

Purpose of the problem

Deducing the presence of functional groups from mass and infra-red spectra.

Suggested solution

Lithium aluminium hydride reduces esters to alcohols, so the only question here is whether it reduces one, or both esters. The IR tells us that there is an alcohol (3600 cm^{-1}) and no carbonyl group (which you would expect around 1700 cm^{-1}) so we can assume that both esters have been reduced. The diol structure below is consistent with the mass of the molecular ion.

Alcohols react with acid chlorides to form esters, so again we have the choice between a single or double ester formation. The IR tells us that one of the alcohols is still present, along with a carbonyl at 1710 cm^{-1}, and the mass of the product is consistent with the structure below.

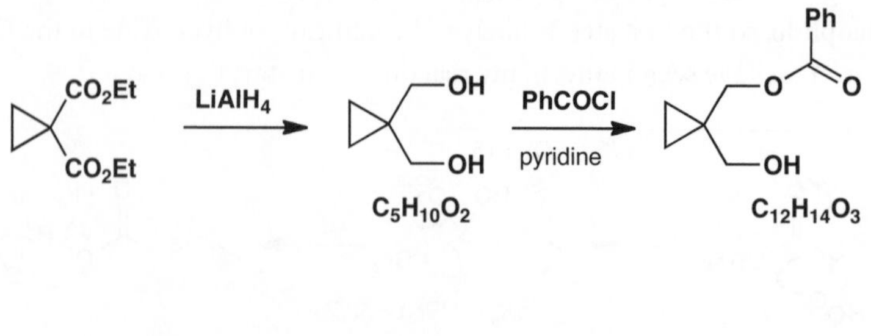

Suggested solutions for Chapter 11

Purpose of the problem

To see if you can draw mechanisms for two of the main classes of reactions in the chapter.

Suggested solution

As MeOH is present in large excess as the solvent, it probably adds first. This also makes the intermediate for the addition of chloride a stable oxonium ion. The mechanism is very like that for acetal formation and, if you added chloride first, that is also a reasonable mechanism.

The second example is imine formation—attack by an amine nucleophile and dehydration of the intermediate. Don't forget to protonate the OH group so that it can leave as a water molecule.

PROBLEM 2

Each of these compounds is an acetal, that is a molecule made from an aldehyde or ketone and two alcohol groups. Which compounds were used to make these acetals?

Purpose of the problem

Practice at the recognition of acetals and working out how to make them.

Suggested solution

All we have to do is to identify the hidden carbonyl group by finding the only carbon atom having two C–O bonds. This atom is marked with a grey circle. If you imagine breaking the two C–O bonds you will discover the carbonyl group and the alcohols.

PROBLEM 3

Suggest mechanisms for these two reactions of the smallest aldehyde, formaldehyde (methanal CH₂=O).

Purpose of the problem

Extension of simple acetal chemistry into related reactions with nitrogen.

Suggested solution

Both reactions start in the same way by attack of a nitrogen nucleophile on formaldehyde. Acid catalysis is not necessary for this step. The first reaction ends with the formation of the iminium ion by acid-catalysed dehydration.

In the other reaction a second amino group is waiting to capture the iminium ion by cyclization to form a stable five-membered ring.

PROBLEM 4

In the textbook (p. 104) we showed you a selective hydrolysis of an acetal. Why were the other acetals (one is a thioacetal) not affected by this treatment? How would you hydrolyse them? Chloroform ($CHCl_3$) is the solvent.

Purpose of the problem

Revision of the different types of acetal and their relative reactivity.

Suggested solution

Cyclic acetals are more stable than non-cyclic ones as we explain on p. 228 of the textbook. Hydrolysis needs more vigorous conditions. Thioacetals are much harder to hydrolyse because sulfides are even less basic than ethers. They can be hydrolysed using electrophiles that attack sulfur readily, such as Hg(II) or methylating agents. This is one possible solution:

PROBLEM 5

In the textbook (p. 228) we say that the Grignard reagent below is 'an unstable structure—impossible to make.' Why is this? What would happen if you tried to make it?

Purpose of the problem

Revision of the danger of mutually destructive functional groups.

Suggested solution

There are various possibilities that all arise from the presence of a carbonyl group and a Grignard in the same molecule. These two would react together. They might cyclize to form a four-membered ring or a bimolecular reaction might lead to a dimer and perhaps polymerization.

PROBLEM 6

Suggest mechanisms for these reactions.

Purpose of the problem

Extension of acetal and imine formation into examples where the intermediate is trapped by a different nucleophile.

Suggested solution

The first reaction starts with the usual attack of an alcohol on the aldehyde but the second nucleophile is the carboxylic acid. Though a poor nucleophile, it is good enough to react with an oxonium ion, particularly in a cyclization.

The second reaction starts with nucleophilic attack by the amine on the more electrophilic carbonyl group—the ketone. Imine formation is followed by cyclization and this second step is normal nucleophilic substitution of an ester (chapter 10). The imine double bond moves into the ring to secure conjugation with the ester.

The third example uses very simple molecules and again starts with imine formation. Cyanide is the nucleophile that captures the iminium ion and a second imine formation completes the mechanism.

PROBLEM 7

Don't forget the problem in the summary on p. 238 of the textbook: suggest a mechanism for the formation of this thioacetal.

Purpose of the problem

Extension of the mechanism for acetal formation to dithioacetal (dithiane) formation.

Suggested solution

The mechanism is a direct analogue of acetal formation. The dehydration step is more difficult: the C=S bond is less stable than the C=O bond because overlap of 2p and 3p orbitals is not as good as overlap of two 2p orbitals of similar size and energy.

PROBLEM 8

In chapter 6 we described how the anti-leprosy drug dapsone could be made soluble by the formation of a 'bisulfite adduct'. Now that you know about the reactions described in chapter 11, you should be able to draw a mechanism for this reaction. The adduct is described as a 'prodrug', meaning that it is not the drug but gives rise to the drug by chemistry within the body. How might this happen?

Purpose of the problem

Revision of chemistry from chapter 6 with a challenging mechanistic problem: did you avoid the trap?

Suggested solution

The trap is to go straight to the product by displacing hydroxide ion from the formaldehyde bisulfite adduct. Hydroxide is a very bad leaving group and reactions like this never occur.

To avoid this trap we must use carbonyl chemistry. First we must make formaldehyde from its adduct and add it to the amino group of dapsone.

Now we can form an iminium salt and add the bisulfite back into this reactive electrophile to give the final product. This is loss of carbonyl oxygen in an unusual setting as the carbonyl was not there at the start and is present only in the intermediates.

PROBLEM 9

This stable product can be isolated from the reaction between benzaldehyde and ammonia. Suggest a mechanism.

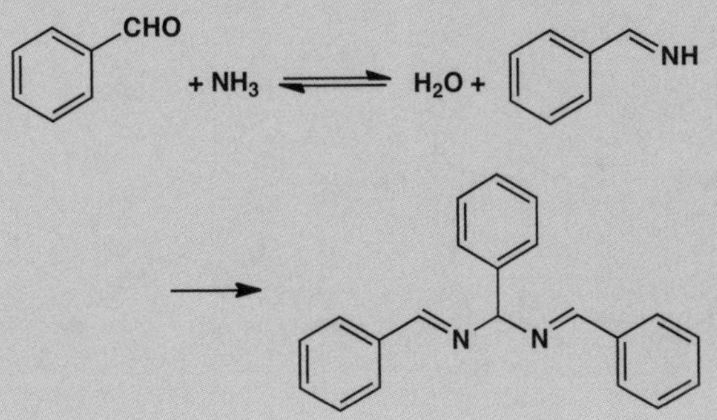

Purpose of the problem

Revision of aminal formation—the all-nitrogen version of acetal formation.

Suggested solution

Imine formation follows the usual pathway (pp. 230–32 of the textbook) but this imine is unstable, as are most primary imines, and it reacts with more benzaldehyde. This reaction starts normally enough but dehydration of the first intermediate produces a strange looking cation with two double bonds to the same nitrogen atom. Addition of another imine gives the final product. The benzene rings play no part in these reactions so we shall represent them as Ph, but they do stablize the final product by conjugation with the imines.

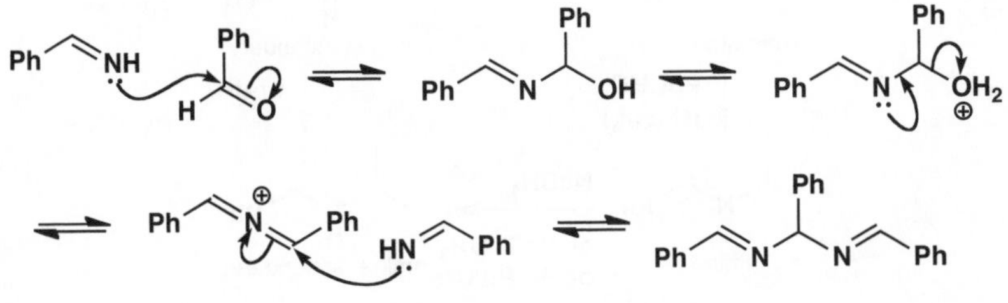

PROBLEM 10

In the following scheme

(a) Identify the functional group in each molecule, and

(b) Suggest a reagent or reagents for carrying out each transformation represented by an arrow.

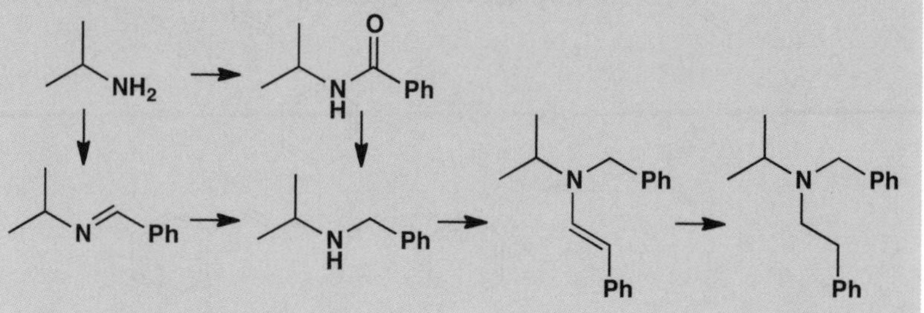

Purpose of the problem

Some important transformations of nitrogen containing functional groups.

Suggested solution

Primary amines are transformed into amides by substitution reactions of acid chlorides, and to imines by condensation with an aldehyde in the presence of an acid catalyst. Both amides and imines may be reduced to amines: amides need LiAlH₄, while imines may be reduced by sodium borohydride, sodium cyanoborohydride, or hydrogenation over a palladium catalyst.

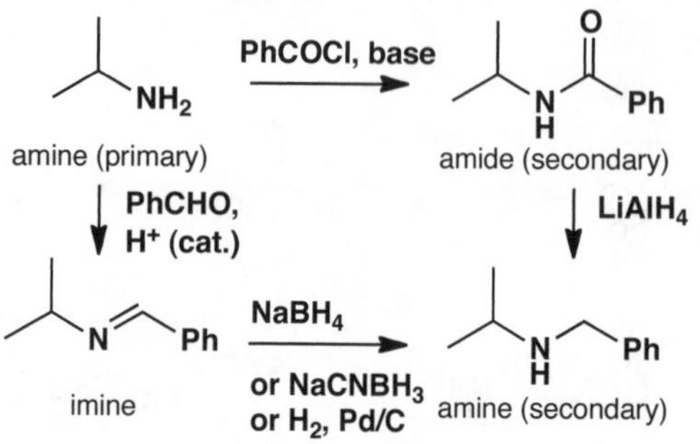

Secondary amines react with aldehydes to form enamines, which may be reduced to amines by hydrogenation, or (via their iminium ion tautomer) with sodium borohydride or sodium cyanoborohydride.

PROBLEM 11

Three chemical steps convert cyclohexane-1,4-dione into a compound which is used for the synthesis of the anti-migraine drug frovatriptan. Suggest how this transformation is carried out.

Purpose of the problem

Designing a route to a real pharmaceutical target.

Suggested solution

Both carbonyl groups have undergone substitution. One of them is converted to an acetal, so we must treat the ketone with a diol and an acid catalyst. Primary amines are transformed into amides by substitution reactions of acid chlorides, and to imines by condensation with an aldehyde in the presence of an acid catalyst.

The other ketone must be converted into an amine, so we can use reductive amination: we could first make the imine with methylamine, and reduce it; alternatively we can use NaCNBH$_3$ to reduce the imine as it forms.

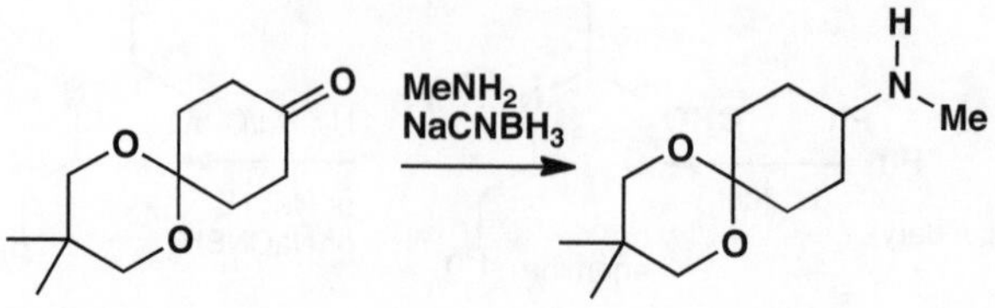
MeNH2
NaCNBH3
H
N
Me
O
O
O
O
O

Suggested solutions for Chapter 12

12

PROBLEM 1

In the comparison of stability of the last intermediates in the substitution at the carbonyl group of acid chlorides or anhydrides to make esters (chapter 10) we preferred one of these intermediates to the other:

Why is the one more stable than the other? If you were to treat an ester with acid, which of the two would be formed?

Purpose of the problem

Revision of contribution of delocalization to stability, particularly of cations.

Suggested solution

The positive charge on the more stable cation is delocalized over both oxygen atoms making it more stable than the other that has a localized cation on one oxygen atom. Protonation of the ester gives the more stable cation as both oxygens combine to make the carbonyl oxygen more nucleophilic.

PROBLEM 2

This reaction shows third-order kinetics as the rate expression is

$$\text{rate} = [\text{ketone}][\text{HO}^-]^2$$

Suggest a mechanism for the reaction.

Purpose of the problem

Interpretation of unexpected kinetics to find a mechanism

Suggested solution

The hydroxide ion must attack the ketone to form a tetrahedral intermediate. The best leaving group from this intermediate is the hydroxide ion that has just come in (pK_a of H_2O is about 15) rather than the alkyne anion. If we use the second hydroxide ion to deprotonate the intermediate, only one leaving group remains, though it is a poor one, and the decomposition of the dianion must be the rate-determining step. This mechanism is found for substitutions at the carbonyl group with very bad leaving groups, as in the hydrolysis of amides (p. 213 in the textbook).

PROBLEM 3

Draw an energy profile diagram for this reaction. You will of course need to draw the mechanism first. Suggest which step in this mechanism is likely to be the slow step and what kinetics would be observed

Purpose of the problem

Practice at drawing energy profile diagrams as one way to present the energetics of mechanisms.

Suggested solution

The first thing is to draw the mechanism of the reaction.

■ This mechanism is described in detail on p. 127 of the textbook.

The first step is bimolecular and forms a new C–C bond. The second step is just a proton transfer between oxygen atoms and is certainly fast. The first step must be the rate-determining step and the intermediate must have a higher energy than the starting material or the product. In this answer we have used the style of energy-profile diagrams used in the textbook (e.g. p. 252) but there is nothing sacred about this—any similar diagram is fine.

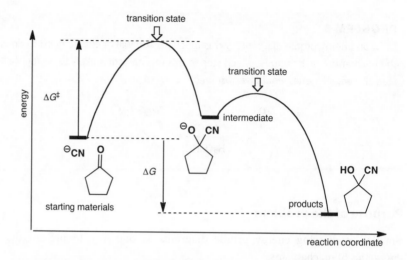

PROBLEM 4

What would be the effect of solvent changes on these reactions? Would the reactions be accelerated or retarded by a change from a polar to a non-polar solvent?

$$Ph_3P \xrightarrow[\text{solvent}]{Br_2} Ph_3\overset{\oplus}{P}\text{-Br}$$

$$\overset{\ominus}{O}\underset{\overset{\oplus}{N}Me_3}{\overset{O}{\parallel}}\xrightarrow[\text{solvent}]{\text{heat}} CO_2 + NMe_3$$

$$\underset{R}{\overset{O}{\parallel}}\underset{OMe}{} \xrightarrow[\text{solvent}]{NH_3} \underset{R}{\overset{O}{\parallel}}\underset{NH_2}{}$$

Purpose of the problem

Practice at assessing the likely effect of solvent polarity in terms of the mechanism of the reaction.

Suggested solution

It is essential to draw a mechanism for each reaction and to identify the rate-determining step in each case. The first two reactions are one-step processes so that makes life easier.

$$Ph_3P\colon \curvearrowright Br \curvearrowright Br \longrightarrow Ph_3\overset{\oplus}{P}\text{-Br} + Br^{\ominus} \qquad \overset{O}{\underset{\overset{\oplus}{N}Me_3}{\parallel}} \longrightarrow CO_2 + NMe_3$$

Now we need to draw the transition state for each reaction so that we can assess whether it is more or less polar than the starting materials. The way to do this is described on p. 251 of the textbook.

In the first reaction uncharged starting materials form a partly charged transition state. A polar solvent will stabilize the transition state and accelerate the reaction. In the second case a charged (zwitterionic) starting material gives a partly charged transition state. A polar solvent will stabilize both starting material and transition state but it will stabilize the starting material more. The energy gap will increase and the reaction go more slowly.

The third reaction is different as it has more than one step. It is a carbonyl substitution of the kind we met in chapter 10. The nucleophile (ammonia) attacks the carbonyl group to form a tetrahedral intermediate that decomposes with the loss of the better leaving group.

We have marked two steps 'fast' because they are just proton transfers between nitrogen and oxygen atoms. Either of the other two steps might be rate determining. In this substitution the leaving group is relatively good (compare problem 2) and the rate-determining step is the first: the usual one for carbonyl substitutions. In this step, neutral starting materials turn into a charged (zwitterionic) intermediate so the transition state is becoming charged and the reaction is accelerated by more polar solvents.

PROBLEM 5

Comment on the effect of acid and base on these equilibria.

Purpose of the problem

Practice at assessing how equilibrium constants respond to acid and base.

Suggested solution

■ The mechanisms in acid and in base are described on p. 208 and p. 210 of the textbook.

The first example is cyclic ester (lactone) formation that will go well in acid solution. In base the acidic proton will be removed and cyclization is no longer possible (see p. 210 in the textbook).

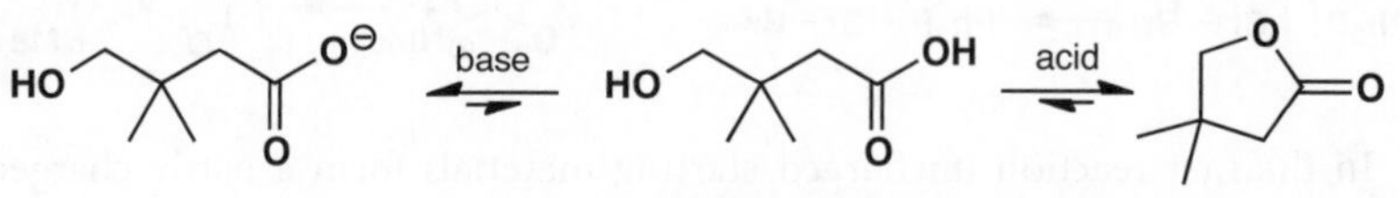

base   acid

The second example is cyanohydrin formation from a ketone (see p. 127 in the textbook). The reaction is reversible but in basic solution the cyanide anion is more stable than the oxyanion of the cyanohydrin and the carbonyl group is more stable than C–O plus C–C so the reaction runs backwards. In more acidic solution (pH less than about 12) the oxyanion will be protonated and the reaction driven towards the right.

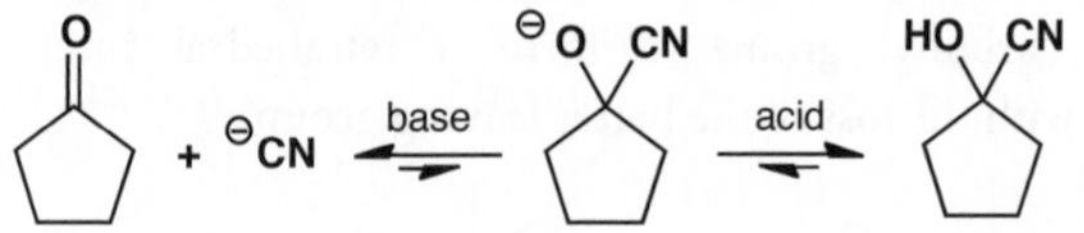

base   acid

Suggested solutions for Chapter 13

13

PROBLEM 1

How many signals will there be in the ^{1}H NMR spectrum of each of these compounds? Estimate the chemical shifts of the signals.

Purpose of the problem

Considering the effects of symmetry on proton (rather than carbon) NMR, and practice in estimating chemical shift.

Suggested solution

Considerations of symmetry apply equally to ^{1}H and to ^{13}C NMR. This is the answer, with different types of proton marked with different letters.

Estimating the chemical shift in ^{1}H NMR requires you to modify your experience of ^{13}C NMR to the narrower range of proton shifts and to consider that aromatic protons are in a distinct region from alkene protons. In each case we give a reasonable estimate and then the actual values. If your values agree with our estimates, you have done well. If you get something near the actual values, be very proud of yourself. The first compound has hydrogens on sp^2 carbon atoms bonded to two nitrogen atoms—hence the very large shift. The fourth molecule has two methyl groups directly bonded to electropositive silicon—hence the very small shift. The rest are more easily explained.

δ 2.29

estimate
δ 1–1.5 (a)
2.2–2.7 (b)

δ 1.24
estimate
δ 1–1.5 (a)
2.2–2.7 (b)

δ 3.20
MeO OMe
Me₂N　Me
δ 2.27　δ 1.20
estimate
δ 1.0–1.5 (a)
2.2–2.7 (b)
3–3.5 (c)

δ 0.45
Si
F₃C　N
Me
δ 2.27
estimate
δ 1.0–1.5 (a), 0–1 (b)
3.1 (c)
δ 1.05

δ 8.0
O H
O
δ 1.5
estimate
δ 1–1.5 (a)
8–10 (b)

PROBLEM 2

The following products might possibly be formed from the reaction of MeMgBr with the cyclic anhydride shown. How would you tell the difference between these compounds using IR and ¹³C NMR? With ¹H NMR available as well, how would your task be easier?

1. MeMgBr
2. H⊕, H₂O

Purpose of the problem

Further thinking the other way round—from structure to data. Contrasting the limitations of ¹³C NMR with data from ¹H NMR spectra.

Suggested solution

The molecular formula of the compounds varies so a mass spectrum would be useful. The compounds with an OH group would show a broad U-shaped band at above 3000 cm⁻¹. The cyclic ester would have a C=O stretch at about 1775 cm⁻¹, the ketones at about 1715 cm⁻¹, and the CO₂H group a band at about 1715 cm⁻¹ as well as a very broad band from 2500 to 3500 cm⁻¹. In the ¹³C NMR the acid and ester would have a carbonyl peak at about 170–180 ppm, but the ketones would have one at about 200 ppm. The number and position of the other signals would also vary.

¹³C NMR (ppm):	5 signals:	3 signals:	5 signals:	6 signals:
	1 at about 170	1 at about 200	1 at about 170	1 at about 200
	1 at 50–100	2 at 0–50	1 at 50–100	1 at about 170
	3 at 0–50		3 at 0–50	1 at 50–100
				3 at 0–50

In the proton NMR, all compounds would show two linked CH_2 groups as a pair of triplets except in the second compound as there the symmetry makes the two the same and would give a singlet. All except the second have a 6H singlet for the CMe_2 group. The second compound has two singlets because of the symmetry. The last has an isolated Me group. The OH and CO_2H protons might show up as broad signals at any chemical shift.

PROBLEM 3

One isomer of dimethoxybenzoic acid has the ¹H NMR spectrum δ_H (ppm) 3.85 (6H, s), 6.63 (1H, t, J 2 Hz), and 7.17 (2H, d, J 2 Hz). One isomer of coumalic acid has the ¹H NMR spectrum δ_H (ppm) 6.41 (1H, d, J 10 Hz), 7.82 (1H, dd, J 2, 10 Hz), and 8.51 (1H, d, J 2Hz). In each case, which isomer is it? The bonds sticking into the centre of the ring can be to any carbon atom.

Purpose of the problem

First steps in using coupling to decide structure.

Suggested solution

The coupling constants in the first spectrum are all too small to be between hydrogens on neighbouring carbon atoms, and there must be symmetry in the molecule. There is only one structure that answers these criteria: 3,5-dimethoxybenzoic acid.

The second compound has one coupling of 10 Hz, and this must be between protons on neighbouring carbon atoms. The other coupling is 2 Hz and this is too small to be anything but *meta* coupling. There are two structures that might be right. In fact the first one is correct and you might have worked this out from the very large chemical shift—almost in the aldehyde region—of the isolated proton with only a 2 Hz coupling. This proton is on an alkene carbon bonded to oxygen in the first structure, but on a simple alkene carbon in the second.

PROBLEM 4

Assign the NMR spectra of this compound and justify your assignments. 'Assign' means 'say which signal belongs to which atom'.

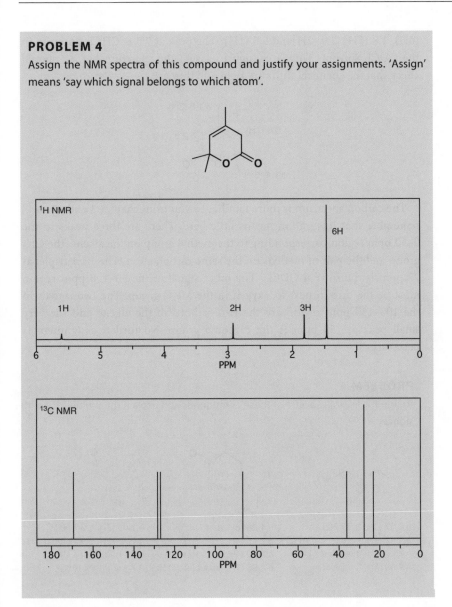

Purpose of the problem

Practice in the interpretation of real NMR spectra – this is harder than if the spectra have already been analysed and reported as a list of peaks.

Suggested solution

There is no coupling in this proton NMR spectrum which makes it much easier, but you should measure the chemical shifts and estimate the number of protons in each signal from the integration. Here we have: δ_H (ppm) 1.4

(6H), 1.8 (3H), 2.9 (2H)and 5.6 (1H). The peak at 7.5 is CHCl₃ impurity in the CDCl₃ solvent. This is enough to assign the spectrum but we should check that the chemical shifts are right and they are.

The carbon spectrum is more familiar to you from chapter 3 and you will remember that integration means little here. There are three peaks in the 0–50 ppm region corresponding to the methyl group on the alkene, the CH_2 group and the pair of methyls on the same carbon atom. The 1:1:1 triplet at 77 ppm is the solvent CDCl₃. The other signal in the 50–100 ppm region must be the carbon next to oxygen in the Me₂C group. The two signals in the 100–150 ppm region are the two carbons of the alkene and the very small peak at 150 ppm is the carbonyl group. No further assignment is necessary.

PROBLEM 5

Assign the ¹H NMR spectra of these compounds and explain the multiplicity of the signals

δ 0.97 (3H, t, *J* 7 Hz)
δ 1.42 (2H, sextuplet, *J* 7 Hz)
δ 2.00 (2H, quintet, *J* 7 Hz)
δ 4.40 (2H, t, *J* 7 Hz)

δ 1.08 (6H, d, *J* 7 Hz)
δ 2.45 (4H, t, *J* 5 Hz)
δ 2.80 (4H, t, *J* 5 Hz)
δ 2.93 (1H, sextuplet *J* 7 Hz)

δ 1.00 (3H, t, *J* 7 Hz)
δ 1.75 (2H, sextuplet, *J* 7 Hz)
δ 2.91 (2H, t, *J* 7 Hz)
δ 7.4–7.9 (5H, m)

Purpose of the problem

First serious practice in correlating splitting patterns and chemical shift.

Suggested solution

Redrawing the molecules with all the hydrogens showing probably helps at this stage, though you will not do this for long. The spectrum of 1-nitrobutane can be assigned by integration and splitting pattern without even looking at the chemical shifts! Just counting the number of neighbours and adding one gives the multiplicity and leads to the assignment. Alternatively you could inspect the chemical shifts which get smaller the further the protons are from the nitro group. Everything fits.

The next compound has an isopropyl group, typically a 6H doublet at about δ 1 ppm and a 1H septuplet with a larger chemical shift. Assigning the two triplets for the two CH_2 groups in the ring is not so easy as they are very similar. It doesn't really matter which is which as this uncertainty does not affect our identification of the compound.

The aromatic ketone happens to have all five aromatic protons overlapping so they cannot be sorted out. This is not unusual and a signal in the 6.5–8 region described as '5H, m' usually means a monosubstituted benzene ring. The side chain is straightforward with the CH_2 group next to the ketone having the largest shift. All the coupling constants happen to be the same (7 Hz) as is usual in an open-chain compound.

PROBLEM 6

The reaction below was expected to give the product **A** and did indeed give a compound with the correct molecular formula by its mass spectrum. However the NMR spectrum of this product was:

δ_H (ppm) 1.27 (6H, s), 1.70 (4H, m), 2.88 (2H, m), 5.4–6.1 (2H, broad s, exchanges with D_2O) and 7.0–7.5 (3H, m).

Though the detail is missing from this spectrum, how can you already tell that this is not the expected product?

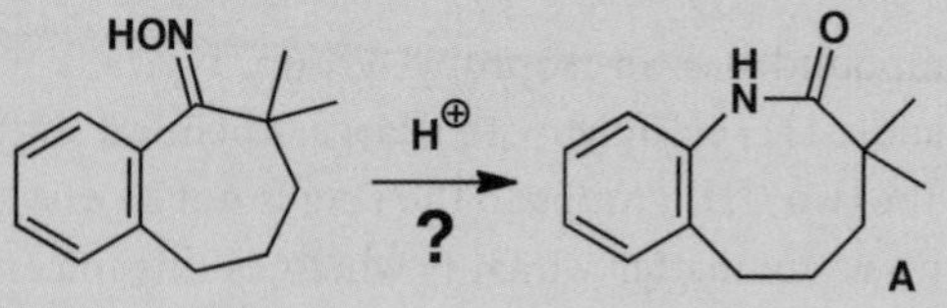

Purpose of the problem

To show that it is helpful to predict the NMR spectrum of an expected product provided that the structure is rejected if the NMR is 'wrong'.

Suggested solution

The spectrum is all wrong. There are only three aromatic Hs instead of the four expected. There are two exchanging hydrogens, presumably in NH_2 and not the one expected. The only thing that is expected is the chain of three CH_2 groups. If you managed to work out the product that was actually formed, you should be very pleased.

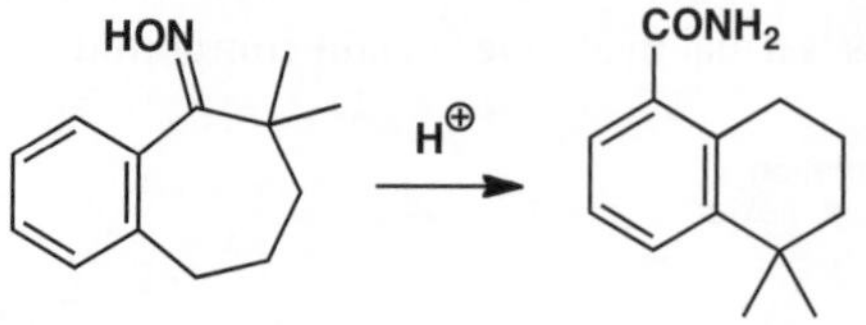

■ This surprising result was reported by B. Amit and A. Hassner, *Synthesis*, 1978, 932. The expected reaction was a Beckmann rearrangement (pp. 958–960) but what actually happened was a Beckmann fragmentation (pp. 959–960) followed by intramolecular Friedel-Crafts alkylation.

Now you know the structure of the product, you should be able to assign the spectrum and confirm the result.

PROBLEM 7

Assign the 400 MHz ¹H NMR spectrum of this enynone as far as possible, justifying both chemical shifts and coupling patterns.

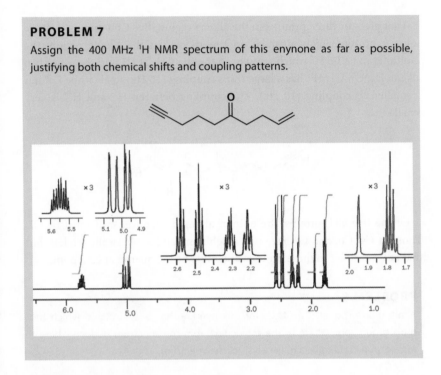

Purpose of the problem

Practice at interpretation of more complicated ¹H NMR spectra.

Suggested solution

First measure the spectrum and list the data. The expansions make it easier to see the coupling but even so we are going to have to call the signal at 5.6 ppm a multiplet. For the rest of the signals you should have measured the J values. Coupling is measured in Hz and at 400 MHz each chemical shift unit of 1 ppm is 400 Hz, so each subunit of 0.1 ppm is 40 Hz.

δ /ppm	integration	multiplicity	coupling, J / Hz	comments
5.6	1H	m	?	alkene region
5.05	1H	d with fine splitting	16.3	alkene region
4.97	1H	d with fine splitting	10.4	alkene region
2.58	2H	t with fine splitting	6.5	next to C=O or C=C
2.47	2H	t with fine splitting	6.5	next to C=O or C=C
2.32	2H	q with fine splitting	6.5	next to C=O or C=C
2.21	2H	t with fine splitting	6.5	next to C=O or C=C
1.95	1H	broad s	-	alkyne?
1.77	2H	q	6.5	not next to anything

That gives us three protons in the alkene region, five CH₂ groups and one solitary proton which must be on the alkyne. In the alkene region, the multiplet must be H^2 which couples to the the CH₂ at C3 and the other two alkene Hs. On C1, H^{1a} has a large *trans* coupling (16 Hz) to H^2 while H^{1b} has a smaller *cis* coupling (10 Hz). The coupling between H^{1a} and H^{1b} is very small.

Of the five CH₂ groups, the quintet at small chemical shift must be C7. Those at C4, C6, and C8 have two neighbours and are basically triplets, but that at C3 couples to three protons and must be the quartet at 2.32 ppm.

PROBLEM 8

A nitration product ($C_8H_{11}N_3O_2$) of this pyridine has been isolated which has a nitro group somewhere in the molecule. From the spectrum deduce where the nitro group is and give a full analysis of the spectrum.

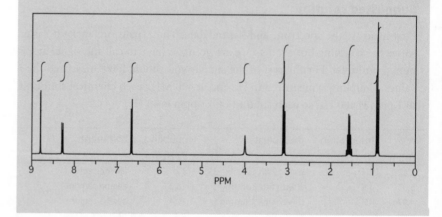

Purpose of the problem

Practice at working out the structure of a reaction product.

Suggested solution

The nitro group might go on the pyridine ring or on the aliphatic side chain or even, perhaps, on the nitrogen atom. Checking the integral shows that it must have gone on the pyridine: the propyl side chain is still there (CH$_3$ triplet, CH$_2$ quintet, and a CH$_2$ triplet with a large chemical shift). The NH proton is till there at 4.0 ppm. But there are now only three protons on the pyridine ring (at 6.7, 8.3, and 8.8 ppm).

There are four possible structures. The most significant feature of the aromatic ring is the proton at very large chemical shift (8.8) with only very small coupling. Protons next to nitrogen in pyridine rings have very large chemical shifts so this rules out all the structures except the second.

The nitro group also increases the shifts of neighbouring protons and so we can assign the spectrum. The rather high field of the proton on the pyridine ring at 6.6 ppm is explained by the electron-donating effect of the amino group.

PROBLEM 9

Interpret this ^{1}H NMR spectrum.

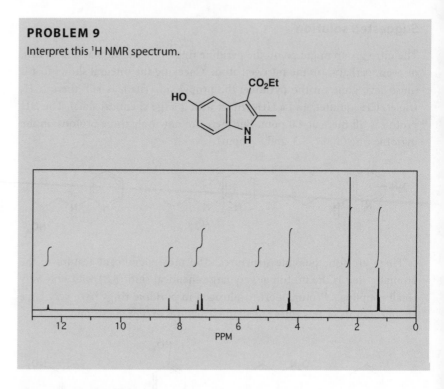

Purpose of the problem

Further correlation of chemical shift and coupling with interpretation of longer-range coupling.

Suggested solution

The ethyl group is easy to find—a typical 3H triplet at 1.2 ppm and a 2H quartet at 4.3 ppm. The large shift of the CH$_2$ group tells us it is next to O. The methyl group is also easy—a 3H singlet at 2.3 ppm, typical of a methyl group on an alkene. At the other end of the spectrum, the broad singlet at 12.5 ppm can be only the OH or the NH; the other is at 5.4 ppm. That leaves the three signals in the aromatic region. You may not be able to see clearly the couplings, but they are: δ_H (ppm) 7.2 (1H, dd, *J* 9, 2 Hz), 7.5 (1H, d, *J* 9 Hz), and 8.4 (1H, d, *J* 2 Hz). The larger coupling is typical *ortho* and the small coupling typically *meta* so we can assign the whole spectrum.

PROBLEM 10

Suggest structures for the products of these reactions, interpreting the spectroscopic data. Most of the reactions will be new to you, and you should aim to solve the structures from the data, not by guessing what might happen.

A, $C_{10}H_{14}O$
ν_{max} (cm^{-1}) C–H and fingerprint only
δ_C (ppm) 153, 141, 127, 115, 59, 33, 24
δ_H (ppm) 1.21 (6H, d, J 7 Hz), 2.83 (1H, septuplet, J 7 Hz), 3.72 (3H, s), 6.74 (2H, d, J 9 Hz) and 7.18 (2H, d, J 9 Hz)

B, $C_8H_{14}O_3$
ν_{max} (cm^{-1}) 1745, 1730
δ_C (ppm) 202, 176, 62, 48, 34, 22, 15
δ_H (ppm) 1.21 (6H, s), 1.8 (2H, t, J 7 Hz), 2.24 (2H, t, J 7 Hz), 4.3 (3H, s) and 10.01 (1H, s)

C, $C_{14}H_{15}NO_2$
ν_{max} (cm^{-1}) 1730
δ_C (ppm) 191, 164, 132, 130, 115, 64, 41, 29
δ_H (ppm) 2.32 (6H, s), 3.05 (2H, t, J 6 Hz) 4.20 (2H, t, J 6 Hz), 6.97 (2H, d, J 7 Hz), 7.82 (2H, d, J 7 Hz) and 9.97 (1H, s)

Purpose of the problem

Practice at determining the structures of reaction products of moderate complexity. This is a very common pastime of real chemists!

Suggested solution

Compound **A** contains the two reagents combined with the loss of HBr. The four Hs at 6.4 and 7.18 suggest the other reagent is attached to the benzene ring. The OMe group is still there (3H singlet at 3.72 ppm) and the new signals are a coupled 6H doublet and 1H septuplet—an isopropyl group. The compound is one of three isomers.

ortho   MeO   *meta*   MeO   *para*   MeO

The two 2H doublets coupled with *J* 9 Hz show that the product has symmetry and only the *para* isomer will fit, as both the *ortho* and *meta* compounds have four different protons. We have used the proton NMR alone but we could have mentioned that there are no functional groups other than the ether and that the four aromatic signals in the ^{13}C NMR reflect the symmetry of the product.

MeO $\xrightarrow{\text{AlCl}_3}$ Br

141 / 24 / 33 / 24 / 153 / MeO / 59 / 127/115

7.18 2H, d   1.21 6H, d   6.74 2H, d   3.72 MeO 3H, s   2.83, 1H septuplet

Compound **B** combines the two reagents with the loss of Me_3Si and the gain of H. Both the IR and the ^{13}C NMR show the appearance of a second carbonyl group—the ester (1745 cm^{-1} and 176 ppm) has been joined by an aldehyde or ketone (1730 cm^{-1} and 202 ppm). The proton NMR shows it is an aldehyde (10.01, 1H, s). There is also a CMe_2 group but it is no longer part of an alkene (proton and carbon NMR show that the alkene has gone). The OMe of the ester has survived. Finally, and very helpfully, there are two open chain CH_2 groups linked together (the two triplets with *J* 7 Hz). One of them (2.24 ppm) has to be next to something and that can only be a carbonyl group as there is nothing else. So we have:

Me Me   ?—CO_2Me ?—CHO   H H / ? / H H O / ?

Though saying what 'ought to happen' is not always helpful, it obviously makes much more sense to consider first a solution in which the ester group stays where it is, on a chain of two carbon atoms, than one in which it moves mysteriously to the other end of the molecule. We prefer the first of these two possibilities:

OHC CO_2Me MeO_2C CHO

Real evidence comes from the lack of coupling of the aldehyde proton which would surely be a triplet in the second structure. The first structure is indeed correct.

Adding up the atoms for compound **C** reveals that the two reagents have joined together with the loss of HF. The 1,4-disubstituted benzene ring is still there (same pattern as compound **A**) as is the aldehyde (1730 cm⁻¹, 191 ppm and 9.97 ppm). The NMe₂ group and the CH₂–CH₂ chain from the other reagent have also survived. It looks as though the fluoride has been displaced by the oxygen of the alcohol and this is indeed what has happened.

Suggested solutions for Chapter 14

14

PROBLEM 1

Are these molecules chiral? Draw diagrams to justify your answer.

Purpose of the problem

Reinforcement of the very important criterion for chirality. Make sure you understand the answer.

Suggested solution

Only one thing matters: does the molecule have a plane of symmetry? We need to redraw some of them to see if they do. On no account look for chiral centres or carbon atoms with four different groups or anything else. *Just look for a plane of symmetry.* If the molecule has one, it isn't chiral. The first compound has been drawn with carboxylic acids represented in two different ways. The two CO₂H groups are in fact the same and the molecule has a plane of symmetry (shown by the dashed lines). It isn't chiral.

The second compound is chiral but if you got this wrong don't be dismayed. Making a model would help but there are only two plausible candidate planes of symmetry: the ring itself, in the plane of the page, and a plane at right angles to the ring. The molecule redrawn below with the tetrahedral centre displayed shows that the plane of the page isn't a plane of symmetry as the CO_2H is on one side and the H on the other, and neither is the plane perpendicular to the ring, as Ph is on one side and H on the other. No plane of symmetry: molecule is chiral.

The third compound is not chiral because of its high symmetry. All the CH_2 groups are identical so the alcohol can be attached to any of them. The plane of symmetry (shown by the dotted lines) may be easier to see after redrawing, and will certainly be much easier to see if you make a model.

The fourth compound needs only the slightest redrawing to make it very clear that it is not chiral. The dashed line shows the plane of symmetry at right angles to the paper.

■ Spiro compounds, which contain two rings joined at a single atom, are discussed on p. 653 of the textbook.

The final acetal (which is a spiro compound) is drawn flat but the central carbon atom must in fact be tetrahedral so that the two rings are orthogonal. By drawing first one and then the other ring in the plane of the page it is easy to see that neither ring is a plane of symmetry for the other because of the oxygen atoms.

PROBLEM 2

If a solution of a compound has an optical rotation of +12, how could you tell if this was actually +12 or really −348 or +372?

Purpose of the problem

Revision of the meaning of optical rotation and what it depends on.

Suggested solution

Check the equation (p. 310 of the textbook) that states that rotation depends on three things: the rotating power of the molecule, the length of the cell used in the polarimeter, and the concentration of the solution. We can't change the first, we may be able to change the second, but the third is easiest to change. If we halve the concentration, the rotation will change to +6, −174, or +186. That is not quite good enough as the last two figures are the same, but any other change of concentration will distinguish them.

PROBLEM 3

Cinderella's glass slipper was undoubtedly a chiral object. But would it have rotated the plane of polarized light?

Purpose of the problem

Revision of cause of rotation and optical activity.

Suggested solution

No. The macroscopic shape of an object is irrelevant. Only the molecular structure matters as light interacts with electrons in the molecules. Glass is not chiral (it is usually made up of inorganic borosilicates). Only if the slipper had been made of single enantiomers of a transparent substance would it have rotated the plane of polarized light. The molecules of Cinderella's left foot are the same as those in her right foot, despite both feet being macroscopically enantiomeric.

PROBLEM 4

Discuss the stereochemistry of these compounds. *Hint:* this means saying how many diastereoisomers there are, drawing clear diagrams of each, and stating whether they are chiral or not.

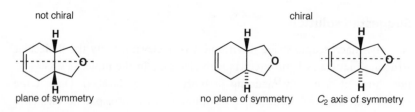

Purpose of the problem

Making sure you can handle this important approach to the stereochemistry of molecules.

Suggested solution

Just follow the hint in the question! Diastereoisomers are different compounds so they must be distinguished first. Then it is easy to say if each diastereoisomer is chiral or not. The first two are simple:

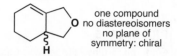

The third structure could exist as two diastereoisomers. The one with the *cis* ring junction has a plane of symmetry and is not chiral. The one with the *trans* ring junction has no plane of symmetry and is chiral (it has C_2 symmetry). Only one enantiomer is shown here.

not chiral   chiral

plane of symmetry   no plane of symmetry C_2 axis of symmetry

The last compound is most complicated as it has no symmetry at all. We can have two diastereoisomers and neither has a plane of symmetry. Both the *cis* compound and the *trans* compound can exist as two enantiomers.

enantiomers of the *cis* compound

enantiomers of the *trans* compound

PROBLEM 5

In each case state, with explanations, whether the products of these reactions are chiral and/or enantiomericaly pure.

biological
reduction
enzyme

heat

S-(+)-glutamic acid

LiAlH$_4$

aqueous
work-up

(±)

Purpose of the problem

Combining mechanism and stereochemical analysis for the first time.

Suggested solution

We need a mechanism for each reaction, a stereochemical description for each starting material (achiral, chiral? enantiomerically enriched?) and an analysis of what happens to the stereochemistry in each reaction. Don't forget: you can't get single enantiomers out of nothing—if everything that goes into a reaction is racemic or achiral, so is the product.

In the first reaction the starting material is achiral as the two CH$_2$OH side chains are identical. The product is chiral as it has no plane of symmetry but it cannot be one enantiomer as that would require one of the CH$_2$OH side chains to cyclize rather than the other. It must be racemic.

The starting material for the second reaction is planar and achiral. If the reagent had been sodium borohydride, the product would be chiral but racemic. But an enzyme, because it is made up of enantiomerically pure components (amino acids), can deliver hydride to one side of the ketone only. We expect the product to be enantiomerically enriched.

In the third reaction, the starting material is one enantiomer of a chiral compound. So we need to ask what happens to the chiral centre during the reaction. The answer is nothing as the reaction takes place between the amine and the carboxylic acid. The product is a single enantiomer too.

■ Amides don't usually form well from amines and carboxylic acids (see p. 207 of the textbook). But in this case the reaction is intramolecular, and with a fair degree of heating (the product is known trivially as 'pyroglutamic acid') the amide-forming reaction is all right.

The final problem is a bit of a trick. The starting material is chiral, but racemic while the product is achiral as the two CH_2CH_2OH side chains are identical so there can be a plane of symmetry between them. The mechanism doesn't really matter but we might as well draw it.

PROBLEM 6
This compound racemizes in base. Why is that?

Purpose of the problem

To draw your attention to the dangers in working with nearly symmetrical molecules and revision of ester exchange (textbook p. 209).

Suggested solution

Ester exchange in base goes in this case through a symmetrical (achiral) tetrahedral intermediate with a plane of symmetry. Loss of the right hand leaving group gives one enantiomer of the ester and loss of the left hand leaving group gives the other.

PROBLEM 7

Assign a configuration (*R* or *S*) to each of these compounds.

Purpose of the problem

Nomenclature may be the least important of the organic chemist's necessary skills, but giving *R* or *S* designation to simple compounds is an essential skill. These three examples check your basic knowledge of the rules.

Suggested solution

Carrying out the procedure given in the chapter (pp. 308–9 of the textbook) we prioritize the substituents 1–4 and deduce the configuration. In all these cases '4' is H and goes at the back when we work out the configuration. The first compound is Pirkle's chiral solvating agent, used to check the purity of enantiomerically enriched samples. The next is the amino acid cysteine and, despite being the natural enantiomer, is *R* because S ranks higher than O (all other natural amino acids are *S*). The third is natural citronellol having three carbon atoms on the chiral centre. They are easily ranked by the next atom along the chain or the atom beyond that if necessary.

(R)-2,2,2-trifluoro-1-(9-anthryl)ethanol   (R)-cysteine   (R)-citronellol

PROBLEM 8

Just for fun, you might try and work out just how many diastereoisomers there are of inositol and how many of them are chiral.

inositol

Purpose of the problem

Fun, it says! There is a more serious purpose in that the relationship between symmetry and stereochemistry is interesting, and, for this human brain chemical, important to understand.

Suggested solution

If we start with all the OH groups on one side and gradually move them over, we should get the answer. If you got too many diastereoisomers, check that some are not the same as others. There are eight diastereoisomers altogether and, remarkably, only one is chiral. All the others have at least one plane of symmetry (shown as dotted lines).

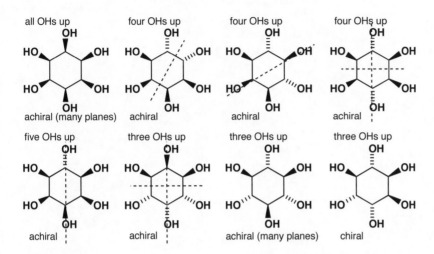

all OHs up
achiral (many planes)

four OHs up
achiral

four OHs up
achiral

four OHs up
achiral

five OHs up
achiral

three OHs up
achiral

three OHs up
achiral (many planes)

three OHs up
chiral

15

Suggested solutions for Chapter 15

PROBLEM 1

Suggest mechanisms for the following reactions, commenting on your choice of S_N1 or S_N2.

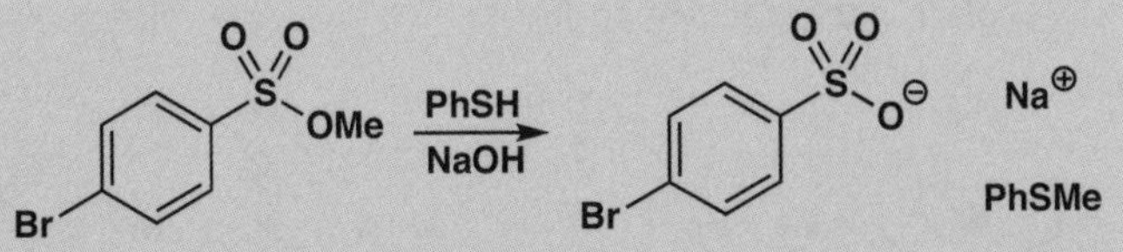

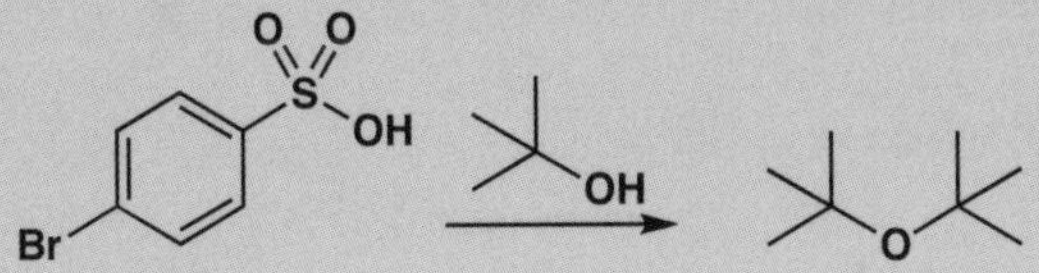

Purpose of the problem

Simple example of the two important mechanisms of chapter 15: S_N1 and S_N2.

Suggested solution

NaOH (pK_a of water about 16) removes the proton from PhSH (pK_a about 7) rapidly as this is a proton transfer between electronegative atoms. Clearly the methyl group must be transferred from O to S and this must be an S_N2 reaction.

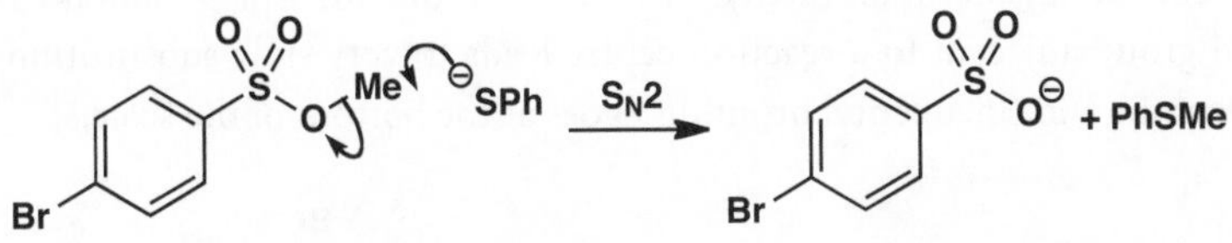

The first reagent in the second reaction resembles the reagent in the first reaction but it is the free sulfonic acid and not the ester. The ether product must come from the displacement of OH from one molecule of t-BuOH by the OH group of the other and this can only be an S_N1 reaction. The OH group leaves as H_2O after being protonated by the sulfonic acid.

PROBLEM 2

Arrange the following in order of reactivity towards the nucleophile sodium azide. Give a brief comment for each compound to explain what factor influences its place in the reactivity scale.

Purpose of the problem

Revision of the factors affecting reactivity in a series of S_N2-reactive molecules.

Suggested solution

None of these compounds has structural features necessary to promote S_N1 (not even the third: notice that the bromine is attached to a primary carbon, even though there is a *tert*-butyl group in the molecule), so we need to think about S_N2 reactivity only. In general, steric hindrance slows down S_N2 reactions, so we can start by saying that methyl bromide > *n*-butyl bromide > cyclohexyl bromide. But how do the other two fit into the scale? An adjacent carbonyl group accelerates S_N2 reactions enormously, so the ketone will react even faster than methyl bromide. On the other hand, a bulky *tert*-butyl group adjacent to a reaction centre leads to very slow substitution, so this compound ('neopentyl bromide') goes at the bottom of the scale.

■ The summary table on p. 347 of the textbook might be useful revision here.

PROBLEM 3

Draw mechanisms for these reactions, explaining why these particular products are formed.

Purpose of the problem

How to choose between S_N1 and S_N2 when the choice is more subtle.

Suggested solution

The first compound has two leaving groups—both secondary chlorides. The one that leaves is next to oxygen so that suggests S_N1 (the oxygen lone pair can stabilize the cation) as does the reagent: surely MeO⁻ would be used for S_N2.

The second compound has only one leaving group and that must be protonated before it can leave. It has two possible sites for attack by the nucleophile (Cl⁻), one primary and one secondary. As the primary is chosen, this must be S_N2.

PROBLEM 4

Suggest how to carry out the following transformations.

Purpose of the problem

Choosing reagents for substitution reactions of alcohols.

Suggested solution

The first compound can react by S_N1, so we have an opportunity to use an excellent reaction: just treating the alcohol with HBr will give the bromide. The second compound is primary, so we have to make it react by S_N2. But we can't simply try to get OH⁻ to leave (hydroxide is never a leaving group!). We have to convert the alcohol into a better leaving group, and we can do this just by treating with PCl_3 (or we could make a sulfonate leaving group and displace with chloride).

The final example must be an S_N2 reaction because it involves an inversion of configuration. Again, hydroxide cannot be a leaving group, so we have to make the alcohol into a tosylate first. We could then use hydroxide as a nucleophile, but to avoid competing elimination reactions the next step is usually done in two stages using acetate as the nucleophile and then hydrolysing to the alcohol.

PROBLEM 5

Draw mechanisms for these reactions and give the stereochemistry of the product.

Purpose of the problem

Drawing mechanisms for two types of nucleophilic substitution in the same sequence to make a β-lactam antibiotic.

Suggested solution

We need an S_N2 reaction at a primary carbon and a nucleophilic substitution at the carbonyl group with the amino group as the nucleophile in both cases. The carbonyl group reaction probably happens first. Don't worry if you didn't deprotonate the amide before the S_N2 reaction. The stereochemistry is the same as that of the starting material (CO_2Et up as drawn) as no change has occurred at the chiral centre.

PROBLEM 6

Suggest a mechanism for this reaction. You will find it helpful first of all to draw good diagrams of reagents and products.

$$t\text{-BuNMe}_2 + (\text{MeCO})_2\text{O} \longrightarrow \text{Me}_2\text{NCOMe} + t\text{-BuO}_2\text{CMe}$$

Purpose of the problem

Revision of chapter 2 and practice at drawing mechanisms of unusual reactions.

Suggested solution

First draw good diagrams of the molecules as the question suggests.

With an unfamiliar reaction, it is best to identify the nucleophile and the electrophile and see what happens when we unite them. The nitrogen atom is obviously the nucleophile and one of the carbonyl groups must be the electrophile.

We must lose a *t*-butyl group from this intermediate to give one of the products and unite it with the acetate ion to give the other. This must be an S_N1 rather than an S_N2 at a *t*-butyl group.

PROBLEM 7

Predict the stereochemistry of these products. Are they diastereoisomers, enantiomers, racemic or what?

Purpose of the problem

Revision of stereochemistry from chapter 14 and practice at applying it to substitution reactions.

Suggested solution

The starting material in the first reaction has a plane of symmetry so it is achiral: the stereochemistry shows only which diastereoisomer we have. Attack by the amine nucleophile at either end of the epoxide (the two ends are the same) must take place from underneath for inversion to occur. The product is a single diastereoisomer but cannot, of course, be a single enantiomer so it doesn't matter which enantiomer you have drawn. The stereochemistry of the Ph group cannot change—it is just a spectator.

The starting material for the second reaction is also achiral as it too has a plane of symmetry. The stereochemistry merely shows that the two OTs groups are on the same side of the molecule as drawn. Displacement with sulfur will occur with inversion and it is wise to redraw the intermediate before the cyclization. This 'inverts' the chiral centre so that we can see that the stereochemistry of the product has the methyl groups *cis*. There are various ways to draw this.

PROBLEM 8

What are the mechanisms of these reactions, and what is the role of the $ZnCl_2$ in the first step and the NaI in the second?

Purpose of the problem

Exploration of two different kinds of catalysis in substitution reactions.

Suggested solution

The $ZnCl_2$ acts as a Lewis acid and can be used either to remove chloride from MeCOCl or to complex with its carbonyl oxygen atom, in either case making it a better electrophile so that it can react with the unreactive oxygen atom of the cyclic ether. Ring cleavage by chloride follows.

The second reaction is an S_N2 displacement of a reasonable leaving group (chloride) by a rather weak nucleophile (acetate). The reaction is very slow unless catalysed by iodide ion—a better nucleophile than acetate and a better leaving group than chloride.

■ Iodide behaves as a nucleophilic catalyst: see p. 358 of the textbook.

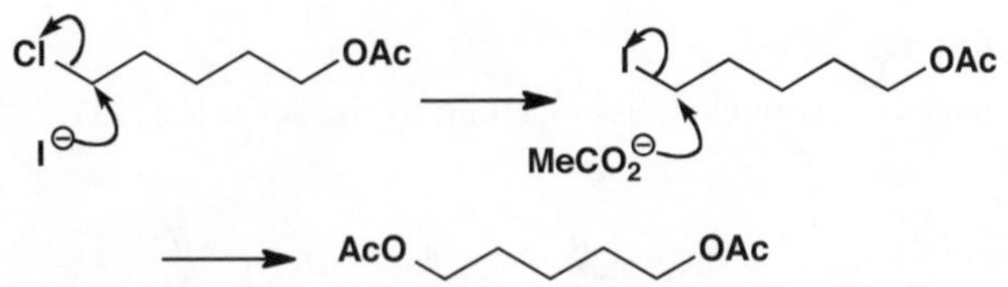

■ You can read more about this in B. S. Furniss *et al.*, *Vogel's textbook of organic chemistry* (5th edn), Longmans, Harlow, 1989, p. 492.

PROBLEM 9

Describe the stereochemistry of the products of these reactions.

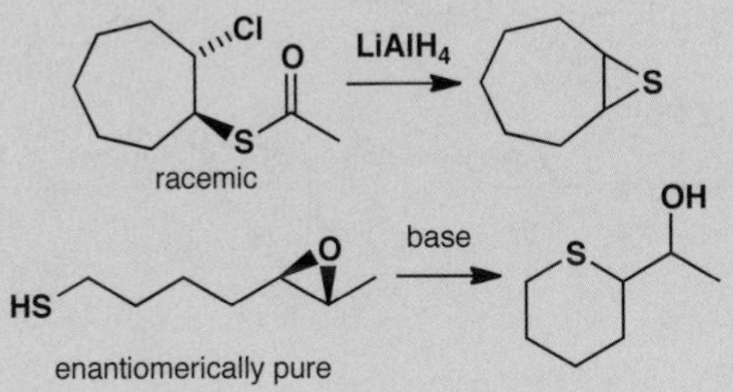

racemic

enantiomerically pure

Purpose of the problem

Nucleophilic substitution and stereochemistry, with a few extra twists.

Suggested solution

The ester in the first example is removed by reduction leaving an oxyanion that cyclizes by intramolecular S_N2 reaction with inversion giving one diastereoisomer (*cis*) of the product. The product is achiral.

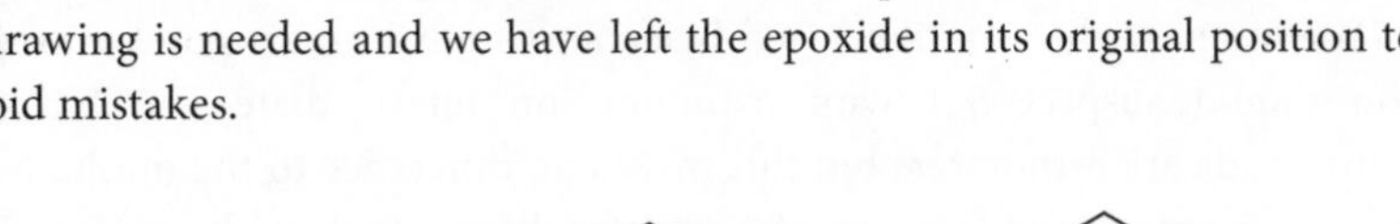

The second case involves an intramolecular S_N2 reaction on one end of the epoxide. The reaction occurs stereospecifically with inversion and so one enantiomer of one diastereoisomer of the product is formed. Some redrawing is needed and we have left the epoxide in its original position to avoid mistakes.

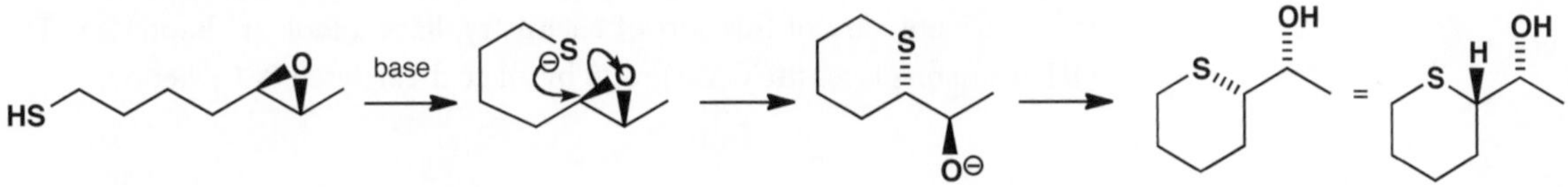

PROBLEM 10

State, with reasons, whether these reactions will be S_N1 or S_N2.

Purpose of the problem

Taxing examples of the choice between our two main mechanisms. The last two differ only in reaction conditions.

Suggested solution

The first reaction offers a choice between an S_N2 reaction at a tertiary carbon or an S_N1 reaction next to a carbonyl group. Neither looks very good but experiments have shown that these reactions go with inversion of configuration and they are about the only examples of S_N2 reactions at tertiary carbon. They work because the p orbital in the transition state is stabilized by conjugation with the carbonyl group: S_N2 reactions adjacent to C=O groups are usually fast.

The moment that you see acetal-like compounds in the second example, you should suspect S_N1 with oxonium ion intermediates. In fact the compounds are orthoesters but this makes no difference to the mechanism. If you are not sure of this sort of chemistry, have a look at chapter 11. The OH group displaces the OMe group by an acid-catalysed S_N1 reaction.

The last two examples add the same group (OPr) to the same compound (an epoxide) to give different products. We can tell that the first is S_N1 as PrOH adds to the more substituted (tertiary and benzylic) position. Inversion occurs because the nucleophile prefers to add to the less hindered face opposite the OH group. If you said that it is an S_N2 reaction at a benzylic centre with a loose cationic transition state, you may well be right.

stable cation   addition to less hindered side

The second is easier as the more reactive anion adds to the less hindered centre with inversion and this must be S_N2.

PROBLEM 11

The pharmaceutical company Pfizer made the antidepressant reboxetine by the following sequence of reactions. Suggest a reagent for each step, commenting on aspects of stereochemistry or reactivity.

reboxetine
(Prolift®, Vestra®)

Purpose of the problem

Thinking about substitution reactions in a real synthesis. It might look challenging, but each step uses a reaction you have already met.

Suggested solution

The first step is the attack of a nucleophile on an epoxide. It's an S_N2 reaction, because it goes with inversion of configuration, and we need a phenol as the nucleophile. To make the phenol more reactive, we probably want to deprotonate it to make the phenoxide, and NaOH will do this. Why does this end of the epoxide react? Well, it is next to a phenyl ring, and benzylic S_N2 reactions are faster than reactions at 'normal' secondary carbons. Next the end hydroxyl group is made into a leaving group (a 'mesylate'), for which we need methanesulfonyl chloride (mesyl chloride) and triethylamine. The primary hydroxyl group must react faster than the secondary one because it is less hindered.

The next stage is an intramolecular subtitution leading to formation of a new epoxide. The hydroxyl group is the nucleophile and the methanesulfonate (MsO⁻) group the leaving group. We need base to do this, and sodium hydroxide is a good choice. Now the epoxide can be opened (at its more reactive, less hindered end) with a nitrogen nucleophile: ammonia might be a possible choice, but often better is azide, followed by reduction by hydrogenation or LiAlH₄. The amine product is converted into an amide, so we need an acid chloride and base.

■ The use of azide as a nitogen nucleophile and as an alternative to ammonia is described on pp. 353-4 of the textbook.

Another intramolecular substitution follows, this time with an alcohol nucleophile displacing a chloride leaving group to make a new ring. A strong base will make the alcohol nucleophilic by deprotonating it to form the epoxide, and KOt-Bu works here (though if you just suggested 'base' that is fine: only experimentation will show which works best). Finally, the amide is reduced to an amine, for which we need LiAlH₄.

Suggested solutions for Chapter 16

16

PROBLEM 1

Identify the chair or boat rings in the following structures and say why this particular structure is adopted.

Purpose of the problem

Exploration of simple examples of chair and boat forms.

Suggested solution

The first three are relatively simple. The first has a chair with all substituents equatorial. The second is forced to have a boat as no chair is possible but the third has a normal chair with a 1,3-diaxial bridge.

Two have nothing but boats as chairs are impossible. One has three boats and the other just the one as there is only one six-membered ring.

The remaining molecule is adamantane – a tiny fragment of a diamond. It is more symmetrical than a paper diagram can show but a model reveals a beautifully symmetrical structure. All the rings are chairs, though some don't look very chair-like in our diagrams.

PROBLEM 2

Draw clear conformational drawings of these molecules, labelling each substituent as axial or equatorial.

Purpose of the problem

Simple practice at drawing chair cyclohexanes with axial and equatorial substituents.

Suggested solution

Your drawings may look different from ours but make sure the rings have parallel sides and don't 'climb upstairs' (pp. 371–372 in the textbook). Make sure that the axial bonds are vertical and the equatorial bonds parallel to the next ring bond but one. The easiest strategy with this question is to draw a ring accurately, and then to add the substituents. The first molecule is simple: there is only one substituent, a large bromine atom, so it goes equatorial. In the second molecule, the two substituents have to be on opposite sides of the ring: this allows them both to be equatorial, which of course they prefer. The last two molecules are dominated by the large *t*-butyl group which insists on being equatorial. Once you have put an equatorial *t*-butyl group on the last molecule you find that there is no choice but to put the Me and OH groups axial.

PROBLEM 3

Would the substituents in these molecules be axial, equatorial, or a mixture between the two?

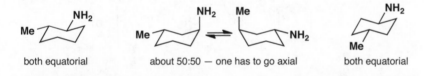

Purpose of the problem

Simple practice at drawing chair cyclohexanes and deciding whether the substituents are axial or equatorial. Remember to decide by drawing and not by trying to remember rules.

Suggested solution

All three molecules have a free choice as the substituents aren't large and are about the same size. Note that all three molecules have their substituents '*trans*' but in two they are both equatorial and in one they are axial/equatorial.

both equatorial   about 50:50 — one has to go axial   both equatorial

PROBLEM 4

Which of these two compounds would form an epoxide on treatment with base?

Purpose of the problem

Exploration of the relationship between conformation and mechanism.

Suggested solution

The mechanism is easy (intramolecular S$_N$2) and the conformation of a *trans* decalin is fixed so we can start with the conformational drawings.

The first compound can get the necessary 'attack from the back' angle of 180° between the nucleophile (O$^-$) and the leaving group (Br) for the intramolecular reaction. But this is impossible for the second compound.

PROBLEM 5

It is more difficult to form an acetal from the first of these compounds than from the second. Why is this?

■ Acetal formation is on p. 224 in the textbook; the importance of thermodynamic control in acetal formation is discussed on p. 226.

Purpose of the problem

Exploration of the effect of conformation on equilibria in thermodynamically controlled reactions.

Suggested solution

The mechanism of the reaction is normal acetal formation and is irrelevant to the question as acetal formation is thermodynamically controlled: it is only the structure and stability of the product that matters. We need to look at the conformations of the molecules to find out which is the more stable.

Axial groups 1,3-related to the ketone are not important as there is no axial group on the ketone. But one of the oxygen atoms in the acetal must be axial and there is now a bad 1,3-diaxial interaction with the methyl group in the first but not in the second acetal. Though the first ketone is slightly less stable than the second, the first acetal is markedly less stable than the second.

PROBLEM 6

Hydrolysis of the tricyclic bromide below in water gives an alcohol. What is the conformation of the bromide and what will be the stereochemistry of the alcohol?

Purpose of the problem

Exploring the conformation of a tricyclic system and discovering that conformation can control stereochemistry even of S_N1 reactions. (Yes, you do need to work out for yourself that this is an S_N1 reaction.)

Suggested solution

The mechanism of the reaction is obviously S_N1 as it is a tertiary bromide and the reagent is water, a weak nucleophile. The water molecule can approach from either side of the planar cation intermediate. There is a unique conformation of the starting material with all three rings in chair conformations and this will be much preferred in the product. The reaction goes with retention as it is under thermodynamic control.

PROBLEM 7

Treatment of the triol below with benzaldehyde in acid solution produces one diastereoisomer of of an acetal but none of the alternative acetal. Why is one acetal preferred? (*Hint*: what controls acetal formation?) What is the stereochemistry of the undefined centre in the acetal that is formed?

Purpose of the problem

Exploration of conformational control in acetal formation.

Suggested solution

Acetal formation is thermodynamically controlled (p. 226 in the textbook) so we need look only for the most stable possible product. The one that is *not* formed is a *cis*-decalin as that would be significantly less stable than the *trans*-decalin that is formed. The phenyl group prefers to adopt an equatorial position and that will decide the stereochemistry as all the acetals are in equilibrium. The remaining OH has to be axial because of its configuration in the starting material.

PROBLEM 8

The compound below is the painkiller tramadol. Draw the most likely conformation of its six-membered ring.

Purpose of the problem

Conformation of a drug molecule.

Suggested solution

As before, just draw a chair, then add substituents. Here we have one carbon with two substituents, but the large aryl ring will prefer to be equatorial. This also allows the amine substituent to be equatorial.

Suggested solutions for Chapter 17

17

PROBLEM 1

Draw mechanisms for these elimination reactions.

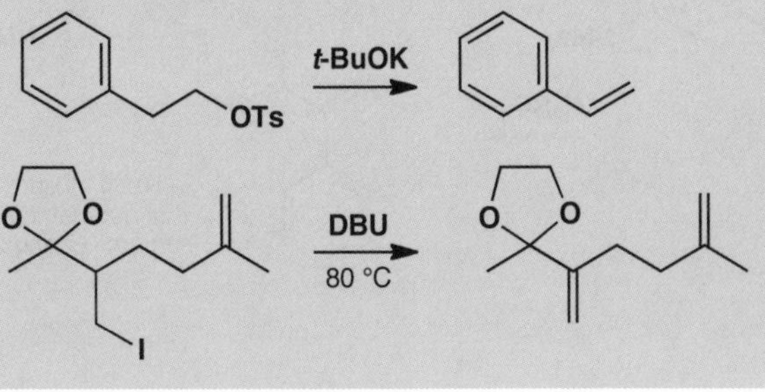

Purpose of the problem

Exercise in drawing simple eliminations.

Suggested solution

These are both E2 reactions as the leaving groups are on primary carbons. In fact both of these reaction are in the textbook (pp. 387 and 391 of the textbook).

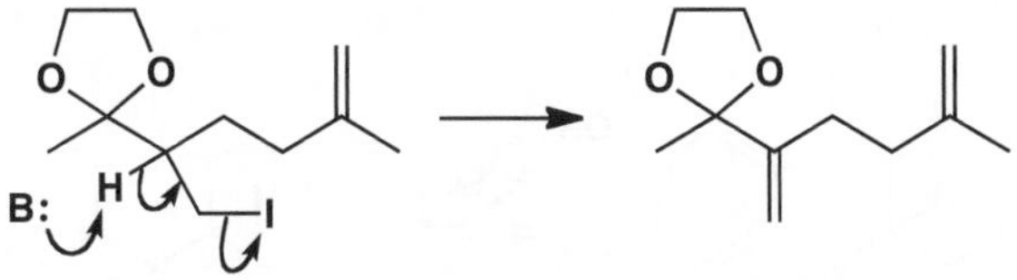

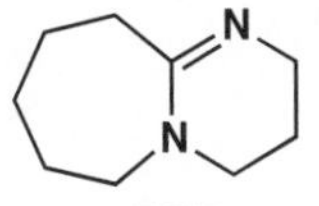

DBU

■ The structure of the amidine base, DBU, and why it is used in elimination reactions is discussed in the textbook on p. 387.

PROBLEM 2

Give a mechanism for the elimination reaction in the formation of tamoxifen, a breast cancer drug, and comment on the roughly 50:50 mixture of geometrical isomers (*cis-* and *trans-*alkenes)

50:50 mixture of geometrical isomers (*E-* and *Z-*)

Purpose of the problem

Thinking about the stereochemical consequences of E1.

Suggested solution

■ The fact that equilibration of the products of E1 elimination gives the most stable possible alkene is discussed in the textbook on p. 394.

The tertiary alcohol leaving group, the acid catalyst, and the 50:50 mixture all suggest E1 rather than E2. There is only one proton that can be lost and, as there is very little difference between the isomeric alkenes, equilibration probably gives the 50:50 mixture.

PROBLEM 3

Suggest mechanisms for these eliminations. Why does the first give a mixture and the second a single product?

64% yield, 4:1 ratio

Purpose of the problem

Regioselectivity of eliminations.

Suggested solution

Whether the first reaction is E1 or E2, there are two sets of hydrogen atoms that could be lost in the elimination. The conditions suggest E1 and the major product may be so because of equilibration.

■ The fact that equilibration of the products of E1 elimination gives the most stable possible alkene is discussed in the textbook on p. 394.

The second reaction produces a more stable tertiary cation from which any of six protons could be lost, but all give the same product. Repetition gives the diene.

PROBLEM 4

Explain the position of the alkene in the products of these reactions. The starting materials are enantiomerically pure. Are the products also enantiomerically pure?

Purpose of the problem

Examples of E1cB in the context of absolute stereochemistry.

Suggested solution

■ E1cB reactions are on p. 399 in the textbook

The first reaction is an E1cB elimination of a β-hydroxy-ketone. The product is still chiral although it has lost one stereogenic centre. The other (quaternary) centre is not affected by the reaction so the product is enantiomerically pure.

The second example already has an electron-rich alkene (an enol ether) present in the starting material so this is more of an E1 than an E1cB mechanism. The intermediate is a hemiacetal that hydrolyses to a ketone (p. 224 in the textbook). The product has two chiral centres unaffected by the reaction and is still chiral so it is also enantiomerically pure.

PROBLEM 5

Explain the stereochemistry of the alkenes in the products of these reactions.

Purpose of the problem

Display your skill in a deceptive example of control of alkene geometry by elimination.

Suggested solution

The first reaction is stereospecific *cis* addition of hydrogen to an alkyne to give the *cis*-alkene. The intermediate is therefore a *cis,cis*-diene and it may seem remarkable that it should become a *trans,trans*-diene on elimination. However, when we draw the mechanism for the elimination, we see that there need be no relationship between the stereochemistry of the intermediate and the product as this is an E1 reaction and the cationic intermediate can rotate into the most stable shape before conversion to the aldehyde.

■ The hydrogenation of alkynes to give *cis* alkenes is described on p. 537 of the textbook.

PROBLEM 6

Suggest a mechanism for this reaction and explain why the product is so stable.

Purpose of the problem

Exploring what might happen on the way to an elimination and explaining special stability.

Suggested solution

■ If you have already read chapter 20 you may have preferred to form the enol of the remaining ketone and eliminate directly.

The obvious place to start is cyclization of the phenol onto a ketone to form a six-membered ring. The product is a hemiacetal that will surely eliminate by a combination of hemiacetal hydrolysis and the E1cB mechanism.

The final product is particularly stable as the right hand ring is aromatic. It has two alkenes and a lone pair on oxygen, making six electrons in all. If you prefer you can show the delocalization to make the ring more benzene-like.

PROBLEM 7

Comment on the position taken by the alkene in these eliminations.

Purpose of the problem

Further exploration of the site occupied by the alkene after an elimination.

Suggested solution

The first is an E1cB reaction after methylation makes the amine into a leaving group. The alkene has to go where the amine was (and in conjugation with the ketone).

The second is also E1cB and so the alkene must end up conjugated with the ketone. But this time the leaving group is on the ring so that is where the alkene goes. The stereochemistry is irrelevant as the enolate has lost one chiral centre and there is no requirement in E1cB for H and OH to be antiperiplanar.

The third is an E2 reaction so there *is* now a requirement for H and Br to be *anti*-periplanar. This means that the Br must be axial and only one hydrogen is then in the right place.

PROBLEM 8

Why is it difficult (though not impossible) for cyclohexyl bromide to undergo an E2 reaction? What conformational changes must occur during this reaction?

Purpose of the problem

Simple exploration of the relationship between conformation and mechanism.

Suggested solution

Cyclohexyl bromide prefers the chair conformation with the bromine equatorial. It cannot do an E2 reaction in this conformation as E2 requires the reacting C–H and C–Br bonds to be anti-periplanar. This can be achieved if the molecule first flips to put the C–Br bond in an unfavourable axial conformation.

favourable conformation:
Br equatorial

unfavourable conformation:
Br axial

PROBLEM 9

Only one of these bromides eliminates to give alkene A. Why? Neither alkene eliminates to give alkene B. Why not?

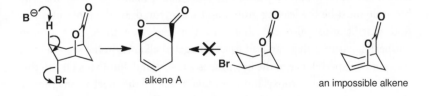

Purpose of the problem

Helping you to understand that cage molecules often have restricted opportunity for elimination.

Suggested solution

The first molecule has one H antiperiplanar to the Br atom so elimination can occur. The second has no hydrogens antiperiplanar to Br. Alkene B is a bridgehead alkene and cannot exist (see the textbook, pp. 389–390).

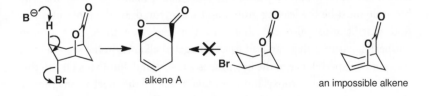

PROBLEM 10

Account for the constrasting results of these two reactions.

Purpose of the problem

How configuration controls mechanism, and an alternative type of elimination.

Suggested solution

The two compounds differ only in their configuration, and as they both have a *tert*-butyl group they have no choice about their conformation. The bromide must be the leaving group, and when you draw the molecules you find that it must also be axial. In the first case there is a proton antiperiplanar to it that can lead to a conjugated alkene. In the second case, the bond antiperiplanar to the bromine is a C–C bond, but that's OK on this occasion because decarboxylation can take place by the mechanism shown. There is an antiperiplanar C–H bond on the other side of course, but the decarboxylation must be faster than simple E2 elimination.

Suggested solutions for Chapter 18

18

PROBLEM 1

A compound C_6H_5FO has a broad peak in the infrared at about 3100-3400 cm^{-1} and the following signals in its (proton decoupled) ^{13}C NMR spectrum. Suggest a structure for the compound and interpret the spectra.

δ_C (ppm) 157.4 (d, J 229 Hz), 151.2 (s), 116.3 (d J 7.5 Hz), and 116.0 (d, J 23.2 Hz).

Purpose of the problem

A reminder that coupling may occur in ^{13}C NMR spectra too and can be useful.

Suggested solution

All the signals are in the sp^2 region and two (at >150 ppm) are of carbons attached to electronegative elements. As the formula contains C_6, a benzene ring is strongly suggested. The IR spectrum tells us that we have an OH group, so the compound is one of these three:

The symmetry of the spectrum suggests the *para* disubstituted compound as there are only four types of carbon atom. We can assign the spectrum by noting that the very large coupling (J 229) must be a $^2J_{CF}$ and the zero coupling must be the carbon furthest from F, i.e. the *para* carbon. The intermediate couplings are for the other two carbons and the CF coupling diminishes with distance.

116.3 (d, J 7.5 Hz) - - - - - → 157.4 (d, J 229 Hz)
151.2 (s) - - - - - → ← - - - - - 116.0 (d, J 23.2 Hz)

PROBLEM 2

The natural product bullatenone was isolated in the 1950s from a New Zealand myrtle and assigned the structure **A**. Then authentic compound **A** was synthesized and found not to be identical to natural bullatenone. Predict the expected ^{1}H NMR spectrum of **A**. Given the full spectroscopic data, not available in the 1950s, say why **A** is definitely wrong and suggest a better structure for bullatenone.

A: alleged bullatenone

Spectra of isolated bullatenone:

Mass spectrum: m/z 188 (10%) (high resolution confirms $C_{12}H_{12}O_2$), 105 (20%), 102 (100%), and 77 (20%)

Infrared: 1604 and 1705 cm^{-1}

^{1}H NMR: δ_H (ppm) 1.43 (6H, s), 5.82 (1H, s), 7.35 (3H, m), and 7.68 (2H, m).

Purpose of the problem

Detecting wrong structures teaches us to be alert to what the spectra are telling us rather than what we expect or want.

Suggested solution

The mass spectrum and IR are all right for **A** but the NMR shows at once that the structure is wrong. There is a monosubstituted benzene ring all right, but the aliphatic protons are a 6H singlet, presumably a CMe$_2$ group, and a 1H singlet in the alkene region at 5.82 ppm.

The fragments we have are Ph, carbonyl, a CMe$_2$ group, and an alkene with one proton on it. That adds up to $C_{12}H_{12}O$ leaving only one oxygen to fit in somewhere. There must still be a ring or there would not be enough hydrogen atoms and the ring must be five-membered (just try other possibilities yourself). There are three ring systems we can choose and each can have the Ph group at either end of the alkene, making six possibilities in all.

The last four are esters (cyclic esters or lactones) and they would have a C=O frequency at 1745–1780 cm^{-1} so **D–G** are all wrong. The hydrogen on the alkene cannot be next to oxygen as it would have a very large chemical shift indeed whereas it is close to the 'normal' alkene shift of 5.25 ruling out structure **C**. Structure **B** is correct and the spectrum can be assigned. Compound **B** has now been synthesized and proved identical to natural bullatenone.

■ You can read the full story in W. Parker *et al.*, *J. Chem. Soc.* 1958, 3871. See also T. Reffstrup and P. M. Boll, *Acta Chem. Scand.*, 1977, **31B**, 727; *Tetrahedron Lett.*, 1971, 4891, and R. F. W. Jackson and R. A. Raphael, *J. Chem. Soc., Perkin Trans. 1*, 1984, 535.

B:
bullatenone

PROBLEM 3

Suggest structures for each of these reaction products, interpreting the spectroscopic data. You are *not* expected to give mechanisms for the reactions and you must resist the temptation to say what 'ought to happen'. These are all unexpected products.

A, $C_6H_{12}O_2$
ν_{max} (cm^{-1}) 1745
δ_C (ppm) 179, 52, 39, 27
δ_H (ppm) 1.20 (9H, s) and 3.67 (3H, s)

B, $C_6H_{10}O_3$
ν_{max} (cm^{-1}) 1745, 1710
δ_C (ppm) 203, 170, 62, 39, 22, 15
δ_H (ppm) 1.28 (3H, t, J 7 Hz), 2.21 (3H, s)
3.24 (2H, s) and 4.2 (2H, q, J 7 Hz)

C, m/z 118
ν_{max} (cm^{-1}) 1730
δ_C (ppm) 202, 45, 22, 15
δ_H (ppm) 1.12 (6H, s), 2.28 (3H, s)
and 9.8 (1H, s)

Purpose of the problem

A common situation in real life—you carry out a reaction, isolate the product, and it's something quite different from what you were expecting. What is it?

Suggested solution

Compound **A** has a carbonyl group (IR) that is an acid derivative (179 ppm in the ¹³C NMR). The 9H singlet in the proton NMR must be a *t*-Bu group and the 3H singlet at 3.67 ppm must be an OMe group. Putting these four fragments together we get a structure immediately. The IR is typical for an ester (1715 + 30 = 1745 cm⁻¹).

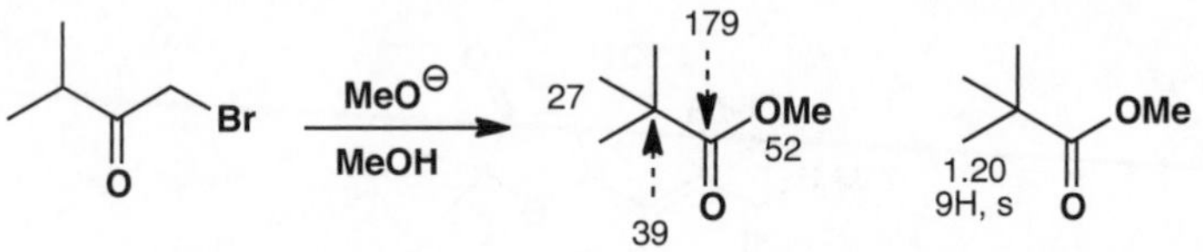

Compound **B** again has an ester (1745 cm⁻¹ and 170 ppm) but it also has a ketone (1710 cm⁻¹ and 203 ppm). The proton NMR shows an OEt group (3H triplet and 2H quartet) together with another methyl group next to something electron-withdrawing (that can only be a C=O as there isn't anything else), and a CH₂ group with no coupling at 3.24 ppm. This is 2 ppm away from a 'normal' CH₂ but it can't be next to O as we've used up all the O atoms already. It must be between two electron-withdrawing groups. These can only be carbonyls so this CH₂ is isolated between the two carbonyl groups and we have the structure.

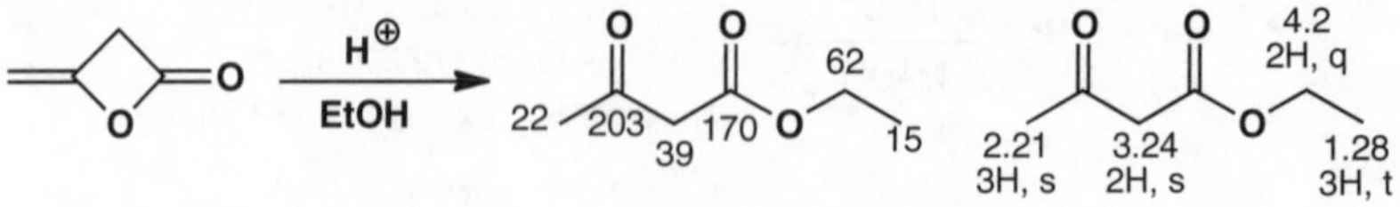

Compound **C** has no formula given, just a molecular ion in the mass spectrum. The most obvious formula is C₅H₁₀O₃ but S is 32 while O is 16 so it might be C₅H₁₀OS. We must look at the rest of the spectra for clarification. There is a carbonyl group (1730 cm⁻¹) that is an aldehyde or ketone (202 ppm). The proton NMR shows a CMe₂ group (6H, s), a methyl group at 2.8 ppm that doesn't look like an OMe (expected 3-3.5 ppm), but might be an SMe. The carbon spectrum also suggests SMe rather than OMe at 45 ppm, and one hydrogen atom at 9.8 ppm that looks like an aldehyde. We know we have these fragments:

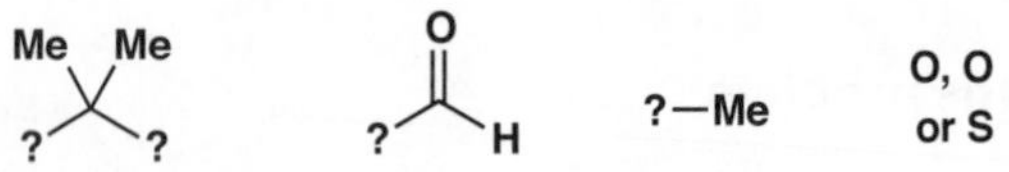

It is not possible to construct a molecule with two extra oxygen atoms but without an OMe, and those we could propose look rather unstable, such as:

Only one compound is possible if we have an S atom—this fits the data very much better and indeed is the correct structure. It has a genuine aldehyde (not a formate ester) and SMe fits better than OMe the signal at δ_H 2.8 ppm and δ_C 45 ppm.

PROBLEM 4

Suggest structures for the products of these reactions.

Compound **A**: $C_7H_{12}O_2$; IR 1725 cm^{-1}; δ_H (ppm) 1.02 (6H, s), 1.66 (2H, t, J 7 Hz), 2.51 (2H, t, J 7 Hz), and 4.6 (2H, s).

Compound **B**: m/z 149/151 (M$^+$ ratio 1:3); IR 2250 cm^{-1}; δ_H (ppm) 2.0 (2H, quintet, J 7 Hz), 2.5 (2H, t, J 7 Hz), 2.9 (2H, t, J 7 Hz), and 4.6 (2H, s).

Purpose of the problem

More practice at the important skill of total structure determination.

Suggested solution

The starting material for **A** is $C_7H_{12}O_3$ and appears just to have lost an oxygen atom. As the reagent is NaBH$_4$, the chances are that two hydrogens have been added and the oxygen lost as a water molecule. The IR spectrum shows a carbonyl group and the frequency suggests an ester or a strained ketone. The NMR shows two joined CH$_2$ groups, one at 2.51 being next to a functional group, not O and so it must be C=O. There is also an unchanged CMe$_2$ group and an isolated CH$_2$ group next to oxygen at 3.9 ppm. There is only one reasonable structure.

The mass spectrum of compound B shows that it has chlorine in it, the IR shows a CN group and the proton NMR shows eight Hs. If we assume that no carbons have been lost, the most reasonable formula is C_5H_8ClNS. The compound has lost a water molecule. The NMR shows three linked CH_2 groups with triplets at the ends and a quintet in the middle. The shifts of the terminal CH_2s show that they are next to functional groups but not Cl. This means we must have a unit $-SCH_2CH_2CH_2CN$. All that remains is the isolated CH_2 group with a large chemical shift evidently joined to both the S and Cl. The large shift comes from 1.5 + 1 (S) + 2 (Cl) = 4.5 ppm. Again only one structure emerges.

PROBLEM 5

Two alternative structures are shown for the products of these reactions. Explain in each case how you would decide which product is actually formed. Several pieces of evidence will be required and estimated values are better than general statements.

Purpose of the problem

To get you thinking the other way round: from structure to data. What are the important pieces of evidence?

Suggested solution

There are many acceptable ways in which you could answer this question ranging from choosing just one vital statistic for each pair to analysing all the data. We'll adopt a middle way and point out several important distinctions. In the first example, one main difference is the ring size, seen mainly in the IR. Both are esters (about 1745 cm^{-1}) but we should add 30 cm^{-1} for the five-membered ring. The functional group next to OCH_2 is also different—an OH in one case and an ester in another. There will be other differences too of course.

In the second case there are also differences in the IR C=O stretch between the aldehyde (about 1730 cm^{-1}) and the conjugated ketone (about

1680 cm^{-1}). The aldehyde proton and the number of protons next to oxygen make a clear distinction. There will also be differences in the ^{1}H and ^{13}C NMR signals of the benzene rings as one is conjugated to a C=O group and the other is not. This reaction actually gave a mixture of both compounds.

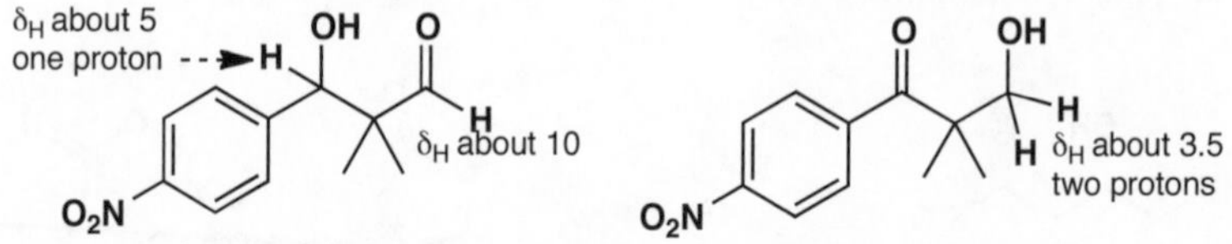

PROBLEM 6

The NMR spectra of sodium fluoropyruvate in D$_2$O are given below. Are these data compatible with the structure shown? If not, suggest how the compound might exist in this solution.

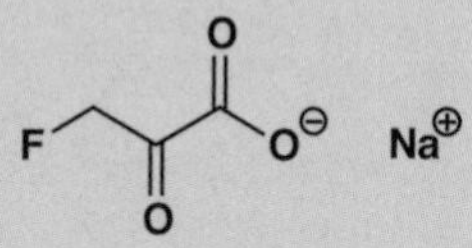

δ_H (ppm) 4.43 (2H, d, *J* 47 Hz);

δ_C (ppm) 83.5 (d, *J* 22 Hz), 86.1 (d, *J* 171 Hz), and 176.1 (d, *J* 2 Hz).

Purpose of the problem

To show how NMR spectra can reveal more than just the identity of a compound.

Suggested solution

The proton NMR spectrum is all right as we expect a large shift: from the chart on p. 276 of the textbook, we can predict 1.3 + 1 (C=O) + 2 (F) = 4.3 ppm and the coupling to fluorine is fine. The carbon NMR shows the carboxylate carbon at 176 ppm with a small coupling to F as it is so far away. The CH$_2$ carbon is at 86.1 ppm with a huge coupling as it is joined directly to F. So far, so good. But what about the C=O group itself? We should expect it at about 200 ppm but it is at 83.5 with the expected intermediate coupling. It cannot be a carbonyl group at all. So what could have happened in D$_2$O? The obvious answer is that a hydrate is formed from this very electrophilic carbonyl group.

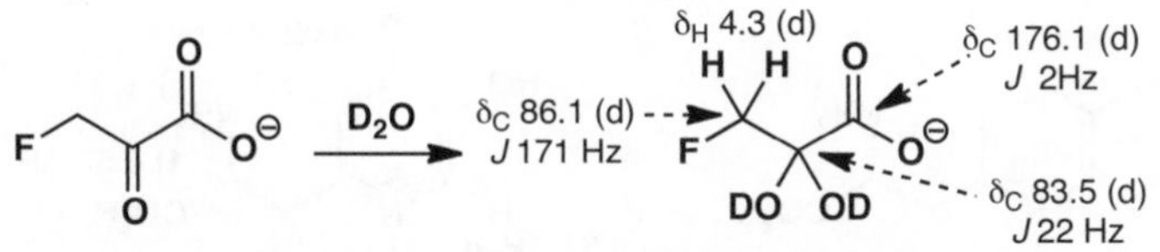

PROBLEM 7

An antibiotic isolated from a microorganism was crystallized from water and formed different crystalline salts in either acid or base. The spectroscopic data were:

Mass spectrum 182 (M^+, 9%), 109 (100%), and 74 (15%).

δ_H (ppm in D_2O at pH<1) 3.67 (2H, d, J 7), 4.57 (1H, t, J 7), 8.02 (2H, m), and 8.37 (1H, m).

δ_C (ppm in D_2O at pH<1) 33.5, 52.8, 130.1, 130.6, 130.9, 141.3, 155.9, and 170.2. Suggest a structure for the antibiotic.

Purpose of the problem

Structure determination of a compound with biological activity from a natural source.

Suggested solution

The solubility and salt formation suggest the presence of both acidic and basic groups, perhaps CO_2H and NH_2 as this is a natural compound. If so, the ^{13}C peak at 170.2 ppm is the CO_2H group. The five carbons in the sp^2 region and protons at 8.0 and 8.4 suggest an aromatic ring, probably a pyridine. The mass spectrum gives an even molecular ion (182) so there must be another nitrogen atom beyond the one in the pyridine. The two sets of aliphatic protons are coupled and the large shift of the ^1H signal at 4.57 ppm suggests a proton between CO_2H and NH_3^+ (pH <1). We have these fragments:

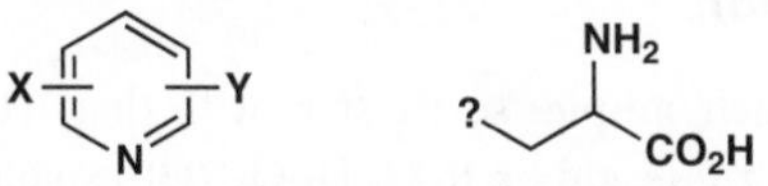

Presumably the aliphatic part must be X or Y, and that leaves just one oxygen atom for a formula of $C_8H_{10}N_2O_3 = 182$. Only six of the ten H atoms show up in the NMR because the OH, NH_3^+, and CO_2H protons all exchange rapidly at pH <1.

■ The details of the structure and spectra are in S. Inouye *et al.*, *Chem. Pharm. Bull.*, 1975, **23**, 2669; S. R. Schow *et al.*, *J. Org. Chem.*, 1994, **59**, 6850 and B. Ye and T. R. Burke, *J. Org. Chem.*, 1995, **60**, 2640.

azatyrosine

H 8.02, 2H, m
HO
H
NH₃⊕
H 4.57, 1H, t
8.37, 1H, m
CO₂H
H H
3.67, 2H, d

PROBLEM 8

Suggest structures for the products of these two reactions.

HO_2C CO_2H + ⟶ **A** $\xrightarrow{PhNH_2}$ **B**

Compound **A**:

m/z 170 (M⁺, 1%), 84 (77%), and 66 (100%);

IR 1773, 1754 cm⁻¹;

δ_H (ppm, CDCl₃) 1.82 (6H, s) and 1.97 (4H, s);

δ_C (ppm, CDCl₃) 22, 23, 28, 105, and 169.

Compound **B**:

m/z 205 (M+, 40%), 161 (50%), 160 (35%), 105 (100%), and 77 (42%);

IR 1670, 1720 cm⁻¹;

δ_H (ppm, CDCl₃) 2.55 (2H, m), 3.71 (1H, t, J 6 Hz), 3.92 (2H, m), 7.21 (2H, d, J 8 Hz), 7.35 (1H, t, J 8 Hz), and 7.62 (2H, d, J 8 Hz);

δ_C (ppm, CDCl₃) 21, 47, 48, 121, 127, 130, 138, 170, and 172.

Purpose of the problem

The other important kind of structure determination: compounds isolated from a chemical reaction.

Suggested solution

Compound **A** is much simpler so we start with that. The two reagents are $C_5H_6O_4$ and $C_5H_8O_2$ these add up to $C_{10}H_{14}O_6$ (230) so 60 has been lost. This looks like $C_2H_4O_2$ or, less likely (because it must be saturated—it has no double bond equivalents), C_3H_8O. If the first is right, **A** is $C_8H_{10}O_4$ which at least fits the proton NMR. The IR suggests two carbonyl groups. The ¹³C NMR shows only one, but there must be some symmetry as there are only five signals for eight carbon atoms. The only unsaturation we have identified is the two carbonyl groups so the signal at 105 ppm is very strange. It must be next to two oxygen atoms to have such a large shift. Either 22 or 105 must be the C of CMe$_2$. C$_8$H$_{10}$O$_4$ would have four double bond equivalents, so the last two degrees of unsaturation must be rings. The cyclopropane provides one and the other must link the two oxygen atoms in the second part-structure. So we have:

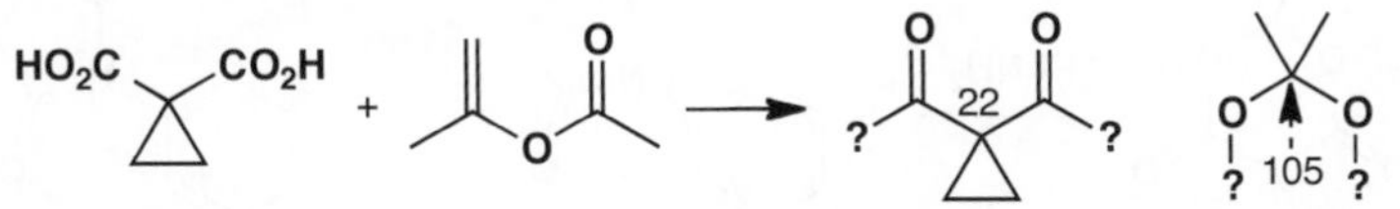

This accounts for all the atoms in **A** so all we need to do is join these two fragments together! The carbonyls are arranged rather like those in cyclic anhydrides and the two carbonyl peaks must be the symmetric and antisymmetric stretches.

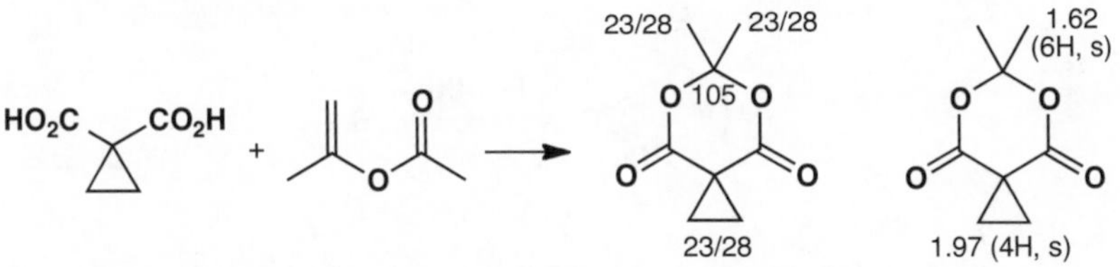

Compound **B** has nitrogen in it (it has an odd molecular weight) and clearly has a benzene ring from the NMR spectra, so we can put down PhN (= 91) as part of the structure. It also has two carbonyl groups (in the IR the one at 1670 cm^{-1} looks like an amide) and they are both acid derivatives (you can see that in the ^{13}C NMR). There are three aliphatic carbons, two CH$_2$s and one CH. Adding that together gives C$_{11}$H$_{10}$NO$_3$ = 188 so there is 17 missing that looks like OH. Since we need a second acid derivative and the OH is the only remaining heteroatom, it must be a carboxylic acid. Given that the CH is a triplet, it must be joined to one of the CH$_2$ groups and, as they are both multiplets, they must be joined to each other. There is one double bond equivalent to account for and that must be a ring. So we have:

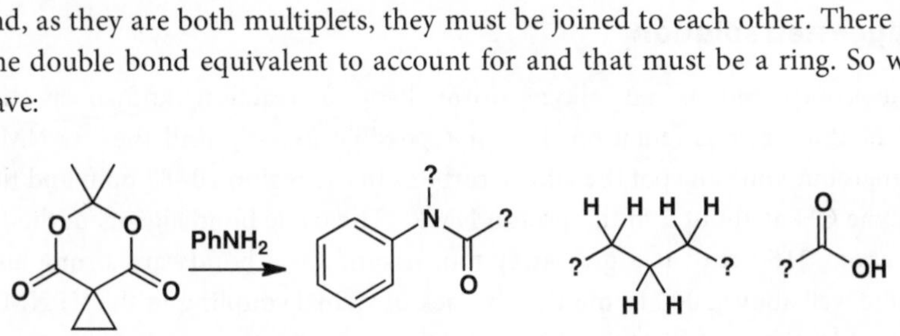

To assemble these three fragments into a molecule we must plug the amide into the C$_3$ fragment and put the CO$_2$H group in the last free position. We can do this in two ways. Proton NMR distinguishes them. The end CH$_2$ is attached either to the nitrogen atom (which would give an estimated shift

■ See S. Danishefsky and R. R. Singh, *J. Am. Chem. Soc.*, 1975, **97**, 3239.

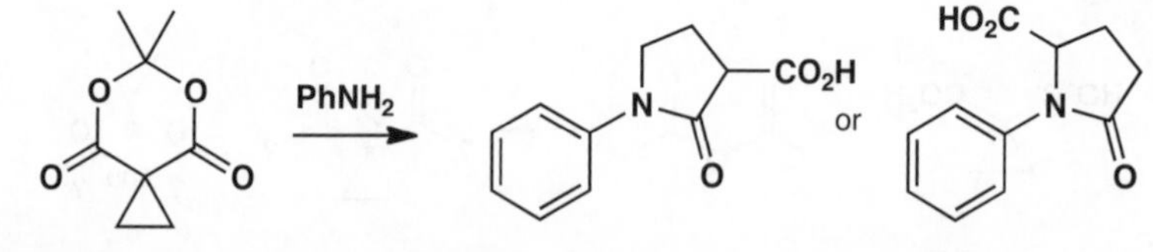

of 3.2 ppm) or to the carbonyl group (estimated shift 2.2 ppm) of the amide. The observed value (3.92) fits the first better. A similar estimate for the CH gives the same answer and the first structure is indeed correct.

PROBLEM 9

Treatment of this epoxy-ketone with tosyl hydrazine gives a compound with the spectra shown below. What is its structure?

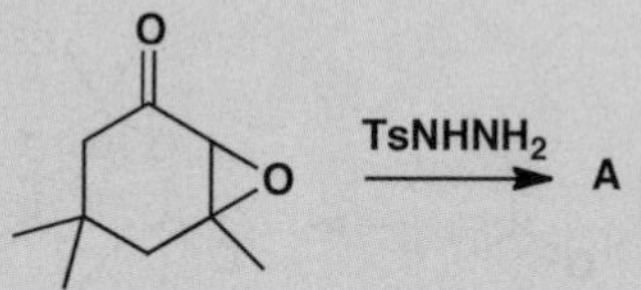

m/z 138 (M+, 12%), 109 (56%), 95 (100%), 81 (83%), 82 (64%), and 79 (74%);

IR 3290, 2115, 1710 cm^{-1};

δ_H (ppm in CDCl$_3$) 1.12 (6H, s), 2.02 (1H, t, *J* 3 Hz), 2.15 (3H, s), 2.28 (2H, d, *J* 3 Hz), and 2.50 (2H, s);

δ_C (ppm in CDCl$_3$) 26, 31, 32, 33, 52, 71, 82, 208.

Purpose of the problem

Further practice at structure determination, adding a curious chemical shift.

Suggested solution

■ See A. Eschenmoser *et al.*, *Helv. Chim. Acta*, 1971, **54**, 2896.

The compound is an alkyne formed by a reaction known as the Eschenmoser fragmentation. It is not possible to assign all the ^{13}C NMR signals but you can spot the alkyne carbons in the region 70–85 ppm and the alkyne CH at about 2 in the proton NMR. The triple bond signals in the IR at about 2150 cm^{-1} is a give-away too. Alkyne C–H bonds are strong and come well above 3000 in the IR. The lack of vicinal coupling in the ^{1}H NMR helps identify the rest of the skeleton of the molecule.

PROBLEM 10

Reaction of the epoxy-alcohol below with LiBr in toluene gave a 92% yield of compound **A**. Suggest a structure for this compound from the data:

mass spectrum gives $C_8H_{12}O$;

ν_{max} (cm^{-1}) 1685, 1618;

δ_H (ppm) 1.26 (6H, s), 1.83, 2H, t, J 7 Hz), 2.50 (2H, dt, J 2.6, 7 Hz), 6.78 (1H, t (J 2.6 Hz), and 9.82 (1H, s);

δ_C (ppm) 189.2, 153.4, 152.7, 43.6, 40.8, 30.3, and 25.9.

Purpose of the problem

Further practice at structure determination including a change in the carbon skeleton—a ring contraction.

Suggested solution

The compound **A** is a simple cyclopentenal. The ^{13}C NMR assignment is not at all certain.

■ G. Magnusson and S. Thorén, *J. Org. Chem.*, 1973, **38**, 1380.

PROBLEM 11

Female boll weevils (a cotton pest) produce two isomeric compounds that aggregate the males for food and sex. A few mg of two isomeric active compounds, grandisol and Z-ochtodenol, were isolated from 4.5 million insects. Suggest structures for these compounds from the data below. Signals marked * exchange with D_2O.

Z-ochtodenol:

m/z 154 ($C_{10}H_{18}O$), 139, 136, 121, 107, 69 (100%);

v_{max} (cm^{-1}) 3350, and 1660;

δ_H (ppm) 0.89 (6H, s), 1.35-1.70 (1H, broad m), 1.41* (1H, s), 1.96 (2H, s), 2.06 (2H, t, J 6 Hz), 4.11 (2H, d, J 7 Hz), and 5.48 (1H, t, J 7 Hz).

Grandisol:

m/z 154 ($C_{10}H_{18}O$), 139, 136, 121, 109, 68 (100%);

v_{max} (cm^{-1}) 3630, 3520, 3550, and 1642;

δ_H (ppm) 1.15 (3H, s), 1.42 (1H, dddd, J 1.2, 6.2, 9.4, 13.4 Hz), 1.35-1.45 (1H, m), 1.55-1.67 (2H, m), 1.65 (3H, s), 1.70-1.81 (2H, m), 1.91-1.99 (1H, m), 2.52* (1H, broad t, J 9.0 Hz), 3.63 (1H, ddd, J 5.6, 9.4, 10.2 Hz), 3.66 (1H, ddd, J 6.2, 9.4, 10.2 Hz), 4.62 (1H, broad s), and 4.81 (1H, broad s);

δ_C (ppm) 19.1, 23.1, 28.3, 29.2, 38.8, 41.2, 52.4, 59.8, 109.6, and 145.1.

Purpose of the problem

Further practice at structure determination of natural products.

Suggested solution

These are the structures. If you have other answers, check that these structures fit the data better.

■ J. M. Tumlinson *et al., Science,* 1969, **166**, 1010; but see K. Mori *et al., Liebigs Annalen,* 1989, 969 for the spectra of ochtodenol and K. Narasaka, *et al., Bull. Chem. Soc. Jap.,* 1991, **64**, 1471 for the spectra of grandisol.

Z-ochtodenol

grandisol

PROBLEM 12

Suggest structures for the products of these reactions.

Compound **A**:

$C_{10}H_{13}OP$, IR (cm^{-1}) 1610, 1235;

δ_H (ppm) 6.5-7.5 (5H, m), 6.42 (1H, t, J 17 Hz), 7.47 (1H, dd, J 17, 23 Hz), and 2.43 (6H, d, J 25 Hz).

Compound **B**:

$C_{12}H_{17}O_2$, IR (cm^{-1}) C-H and fingerprint only;

δ_H (ppm) 7.25 (5H, s), 4.28 (1H, d, J 4.8 Hz), 3.91 (1H, d, J 4.8 Hz), 2.96 (3H, s), 1.26 (3H, s) and 0.76 (3H, s).

Purpose of the problem

Structure determination of reaction products with extra twists: a nucleus with spin (P) and protons on the same carbon atom that are different in the NMR.

Suggested solution

The coupling constants $^3J_{PH}$ across the alkene are very large. Typically *cis* $^3J_{PH}$ is about 20 and *trans* $^3J_{PH}$ about 40. Geminal ($^2J_{PH}$) are also large but more variable. In **B** there is a stereogenic centre, meaning that the hydrogen atoms and methyl groups in the ring are different: they are either on the same side as MeO or the same side as Ph. (The term we will introduce in chapter 31 to describe such groups is 'diastereotopic'). We cannot say which H gives which signal.

■ F. Nerdel *et al., Tetrahedron Lett.,* 1968, 5751.

A

B

PROBLEM 13

Identify the compounds produced in these reactions. Warning! Do not attempt to deduce the structures from the starting materials, but use the data. These molecules are so small that you can identify them from ¹H NMR alone.

Data for **A**: C_4H_6; δ_H (ppm) 5.35 (2H, s) and 1.00 (4H, s)

Data for **B**: C_4H_6O; δ_H (ppm) 3.00 (2H, s), 0.90 (2H, d, J 3 Hz) and 0.80 (2H, d, J 3 Hz)

Data for **C**: C_4H_6O; δ_H (ppm) 3.02 (4H, t, J 5 Hz) and 1.00 (2H, quintet, J 5 Hz).

Purpose of the problem

Structure determination of reaction products by ¹H NMR alone.

Suggested solution

The very small shifts of cyclopropane protons may have worried you but they often have shifts of less than 1 ppm. Compounds **A** and **C** are simple enough but **B** may have amazed you. It is unstable but can be isolated and the two three-membered rings sit at right angles to each other, so as in problem 12 the protons on each side of the cyclopropane ring are different.

PROBLEM 14

The yellow crystalline antibiotic frustulosin was isolated from a fungus in 1978 and it was suggested the structure was an equilibrium mixture of **A** and **B**. Apart from the difficulty that the NMR spectrum clearly shows one compound and not an equilibrium mixture of two compounds, what else makes you unsure of this assignment? Suggest a better structure. Signals marked * exchange with D_2O.

Frustulosin:

m/z 202 (100%), 174 (20%);

v_{max} (cm^{-1}) 3279, 1645, 1613, and 1522;

δ_H (ppm) 2.06 (3H, dd, J 1.0, 1.6 Hz), 5.44 (1H, dq, J 2.0, 1.6 Hz), 5.52 (1H, dq, J 2.0, 1.0 Hz), 4.5* (1H, broad s), 7.16 (1H, d, J 9.0 Hz), 6.88 (1H, dd, J 9.0, 0.4 Hz), 10.31 (1H, d, J 0.4 Hz), and 11.22* (1H, broad s);

δ_C (ppm) 22.8, 80.8, 100.6, 110.6, 118.4, 118.7, 112.6, 125.2, 129.1, 151.8, 154.5, and 196.6.

Warning! This is difficult—after all, the original authors got it wrong initially.

Hint: How might the DBEs be achieved without a second ring?

Purpose of the problem

A serious and difficult determination of a natural product as a final challenge.

Suggested solution

Structure **B** is definitely wrong because the NMR shows only one methyl group, not two, and only one carbonyl group, not two. Structure **A** looks unlikely because it appears to be unstable, but that is not evidence. The NMR shows two protons on the same end of a double bond (at 5.44 and 5.52 ppm) with the characteristic small coupling, but they are coupled to a methyl group, presumably by allylic coupling, and the methyl group is too far away in **B**. But what is the signal at 80.8 in the ^{13}C NMR? The 'hint' was meant to guide you towards suggesting an alkyne. That solves many of the problems even though the carbons of the alkene and the aromatic ring cannot be assigned with confidence. At least the revised structure is one compound and not two.

■ The true structure was later described with the help of NMR as you can read in R. C. Ronald *et al.*, *J. Org. Chem.*, 1982, **47**, 2541 and M. S. Nair and M. Anchel, *Phytochemistry*, 1977, **16**, 390, revised from M. S. Nair and M. Anchel, *Tetrahedron Lett.*, 1975, 2641.

Suggested solutions for Chapter 19

Purpose of the problem

Simple examples of addition with regioselectivity.

Suggested solution

The first and last alkenes have different numbers of substituents at each end of the alkene and will give the more stable, more highly substituted cation on protonation. The middle one has the same number of substituents (one) at each end but they are very different in kind. The secondary benzylic cation is preferred to the non-conjugated alternative.

PROBLEM 2

Suggest mechanism and products for these reactions.

Purpose of the problem

Checking that you understand the bromination mechanism.

Suggested solution

The question of what product is formed is easily answered as we know bromine adds *trans* to alkenes. The products are both racemic, of course, as all reagents are achiral and only the relative stereochemistry is shown.

The mechanism is bromonium ion formation by electrophilic attack of bromine on the alkene and *trans* opening of the bromonium ion by bromide ion.

PROBLEM 3

What will be the products of the addition of bromine water to these alkenes?

Purpose of the problem

Checking that you understand the bromonium ion mechanism with an external nucleophile.

Suggested solution

The bromonium ion is formed again but now water attacks as the nucleophile as it is in large excess as the solvent. If the alkene is unsymmetrical, water attacks the more substituted end of the bromonium ion (p. 441 of the textbook). In any case, it does so with inversion.

PROBLEM 4

By working at low temperature with one equivalent of buffered solution of a peroxy-acid, it is possible to prepare the monoepoxide of cyclopentadiene. Why are these precautions necessary and why does a second epoxidation not occur under these conditions?

Purpose of the problem

A more complicated electrophilic addition with questions of stability and selectivity to consider.

Suggested solution

One of the alkenes in the diene reacts in the usual way to give, first of all, the monoepoxide. The reaction can be stopped there only if the remaining alkene is less nucleophilic than the alkenes in cyclopentadiene. This is indeed the case because the HOMO of a diene is higher in energy than the HOMO of a simple alkene. The HOMO of the diene (Ψ_2) results from antibonding addition of the two separate π-orbitals, making the diene more reactive than an isolated alkene.

■ More details on this explanation of the reactivity of dienes can be found on pp. 146–8 of the textbook.

HOMO of the diene

The other questions concern the low temperature, which favours the kinetic product and encourages epoxide formation on the remaining alkene. A by-product from the reaction is RCO_2H which could catalyse the opening of the epoxide to give a stable allyl cation (p. 336 in the textbook). The buffer prevents the mixture becoming too acidic.

danger of decomposition by allyl cation formation

PROBLEM 5

The synthesis of a tranquilizer uses this step. Give mechanisms for the reactions.

Purpose of the problem

An electrophilic addition followed by a substitution: revision of chapter 17.

Suggested solution

HBr adds to the alkene to form the tertiary cation that captures bromide ion.

The bromide is hydrolysed by water. This must be an S_N1 reaction as the bromide is tertiary and the nucleophile is water. The same cation is an intermediate in both reactions.

PROBLEM 6

Explain this result:

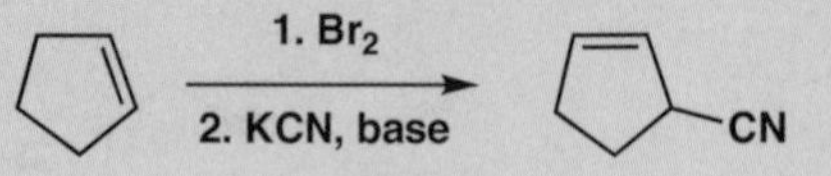

Purpose of the problem

An electrophilic addition followed by an elimination (revision of chapter 17) and a substitution (revision of chapter 15).

Suggested solution

Addition of bromine occurs first to give the *trans* dibromide in the usual way. Base then eliminates one of the bromides in an E2 reaction using the only available *trans* hydrogen atom. This gives a reactive allylic bromide (p. 336 in the textbook) that reacts with cyanide ion by a favourable S_N2 reaction to give the product.

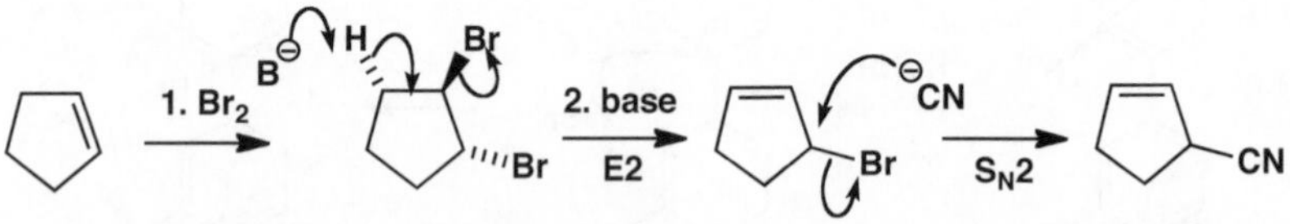

PROBLEM 7

Suggest a mechanism for the following reaction. What is the stereochemistry and conformation of the product? The product has these signals in its ^1H NMR spectrum: δ_H 3.9 (1H, ddq, J 12, 4, 7) and δ_H 4.3 (1H, dd, J 11, 3).

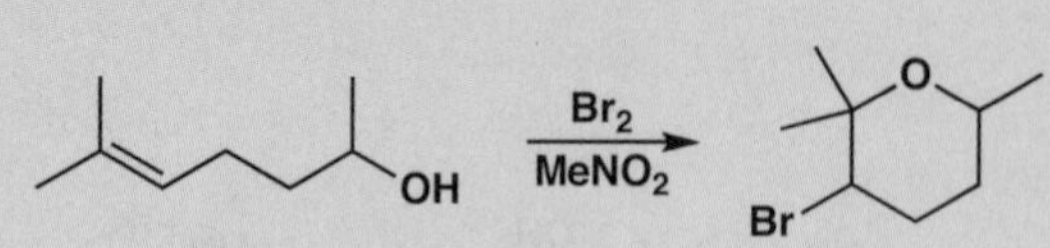

Purpose of the problem

Drawing a mechanism for bromination of an alkene with an internal nucleophile, and revision of NMR.

Suggested solution

The mechanism is formation of the bromonium ion and nucleophilic attack by the OH group at the more substituted carbon. The product has one carbon with two methyl groups, of which one must be axial and one equatorial. The remaining substituted carbons, with a Br and a Me substituent, could be axial or equatorial: you might expect them to prefer to be equatorial but is there any evidence? The NMR spectrum shows that the two protons listed, whose large shift indicates that they are next to Br and O, both have large coupling constants. You will see details of the values of coupling constants in six-membered rings in chapter 31, but you already know that large J values indicate parallel bonds (see pp. 293–4 of the textbook). This suggests that these C–H bonds are parallel (antiperiplanar) to their neighbours, something that is possible only if the C–H bonds are axial. The Br and the methyl group must therefore be equatorial.

PROBLEM 8

The two alkenes below can be converted into two regioisomers or two diastereoisomers as shown. Suggest reagents to achieve these transformations. What alternative starting material could you use to make the *trans* diol (bottom right)?

Purpose of the problem

Overview of methods for controlling selectivity in the reactions of alkenes.

Suggested solution

Hydration of the top alkene with aqueous acid goes via a carbocation intermediate, and would therefore give the product on the left. To get the product on the right we need a more roundabout method, involving hydroboration to place boron on the less hindered carbon and then oxidation to the alcohol (p. 446 of the textbook).

Using osmium tetroxide to add two hydroxyl groups to the second alkene will give the diastereoisomer on the left, because the mechanism ensures both O atoms add to the same side of the alkene (p. 442 of the textbook). To get the *trans* diol, we can make an epoxide with *m*-CPBA and then open it with water, an S_N2 reaction that proceeds with inversion of configuration.

A possible alternative would be to start with the *cis* alkene: the product will be the diol shown, which, if we redraw it in an open chain conformation, is the same diol as the one that comes from the *trans* alkene vis the epoxide.

20

Suggested solutions for Chapter 20

PROBLEM 1

Draw all possible enol forms of these carbonyl compounds and comment on their stability.

Purpose of the problem

Simple exercise in drawing enols with an extra twist.

Suggested solution

There is only one enol for the first compound and it might be stable because it is aromatic (two alkenes and one oxygen giving two electrons each, making six in all).

The second compound has more possibilities, one of which is very stable indeed as it has two benzene rings. We haven't drawn the mechanism for each enolization this time but note the different reaction arrows for tautomerism (equilibrium) and delocalization (two ways of drawing the same compound).

■ Aromaticity is discussed in chapter 7 of the textbook, pages 157–162. Compounds in which lone pairs contribute to the aromatic sextet are introduced on page 162, but also have two whole chapters devoted to them (chapters 29 and 30). These chapters are more advanced than you need at this stage.

PROBLEM 2

The proportions of enol in a neat sample of the two ketones below are rather different. Why is this?

0.0004% enol   62% enol

Purpose of the problem

Thinking about the stability of enols bearing stabilizing substituents.

Suggested solution

The first compound is an ordinary ketone with its strong C=O bond and so the less stable enol with its weaker C=C bond is present in only tiny amounts. The second compound has the special 1,3-relationship between two carbonyl groups that means a very stable enol. The stability comes from a major reason—conjugation of C=C and C=O—and a minor reason—intramolecular hydrogen bonding.

PROBLEM 3

The NMR spectrum of this dimethyl ether is complicated: the two MeO groups are different as are all the hydrogen atoms on the rings. However the diphenol has a very simple NMR spectrum—there are only two types of proton on the rings marked 'a' and 'b' on the diagram. Explain.

dimethyl ether

diphenol

Purpose of the problem

Exploring the way that tautomerism leads to equivalence.

Suggested solution

The protons in the ether are obviously all different as it has no symmetry. Tautomerization interconverts carbonyl groups and enols, and can make either of the enols in the diphenol into a carbonyl group and can make the carbonyl group into an enol, so all structures are equivalent. If this proton transfer (note that it is *not* a delocalization) is fast on the NMR time scale, all the H^{a}s will appear in one signal and all the H^{b}s will appear in another signal.

■ The idea that some interconversions take place too fast to be detected by NMR (they are fast on the NMR timescale) is covered in the blue boxes on p. 363 and p. 374 of the textbook.

PROBLEM 4

Suggest mechanisms for these reactions:

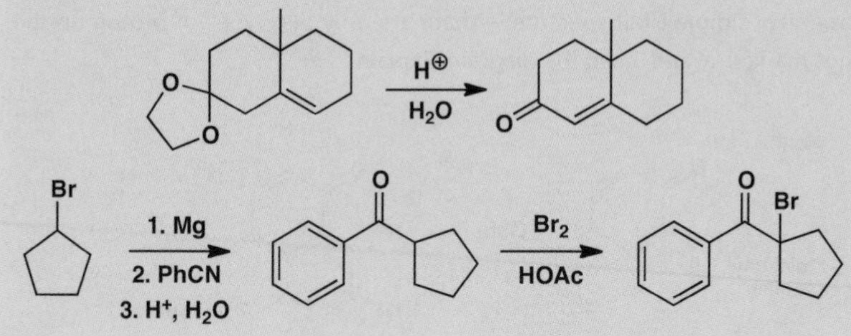

Purpose of the problem

Revision of mechanisms plus exploration of reactions involving enols.

Suggested solution

The first reaction starts with acetal hydrolysis and the product is an enone that is not conjugated. It can become conjugated in acid solution *via* the enol.

■ The mechanism of acetal hydrolysis is on p. 226 of the textbook.

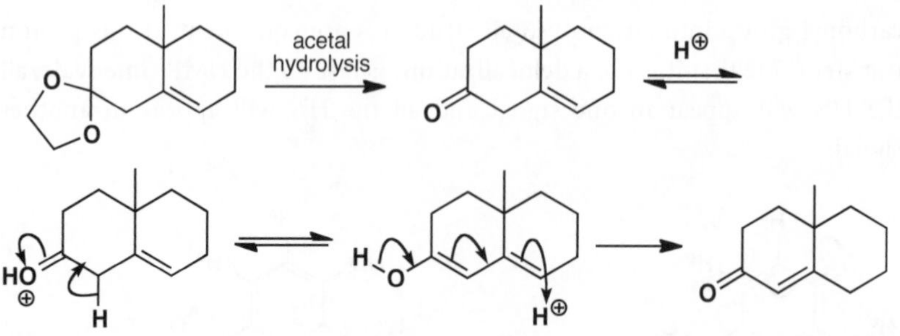

The second sequence starts with the reaction of a nitrile with a Grignard reagent and the hydrolysis of the product to a ketone: this is mentioned on pp. 220 and 231 of the textbook. The bromination in acid solution makes use of the only enol possible.

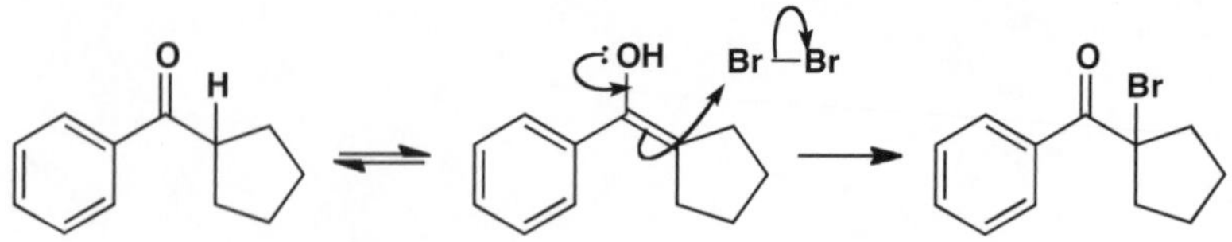

PROBLEM 5

Suggest mechanisms for these reactions and explain why these products are formed.

Purpose of the problem

Making sure that you understand why enols usually react through carbon but sometimes react through oxygen.

Suggested solution

The same ketone reacts in similar acidic conditions with different selectivity according to the electrophile. The bromination occurs at carbon as we should expect but the anhydride reacts at oxygen. So we can draw mechanisms to suit the product.

■ This 'double-headed curly arrow' short-hand is explained on p. 217 of the textbook.

Acid anhydrides, being carbonyl electrophiles, respond to charge density (they are 'hard' electrophiles) and react well with oxygen nucleophiles. Bromine, by contrast, is uncharged and unpolarized (it is a 'soft' electrophile) and reacts well with neutral nucleophiles such as alkenes.

■ 'Hard' and 'soft' reagents are explained on p. 357 of the textbook and revisited on p. 467.

PROBLEM 6

1,3-Dicarbonyl compounds such as **A** are usually mostly enolized. Why is this? Draw the enols available to compounds **B–E** and explain why **B** is 100% enol but **C, D,** and **E** are 100% ketone.

Purpose of the problem

Exploring enols of different kinds of 1,3-dicarbonyl compounds—an important class of enolizable compounds.

Suggested solution

Compound **A** is mostly enol because only the enol is delocalized over five atoms. A minor reason for this particular compound is the intramolecular hydrogen bond in the enol.

You might also have pointed out that there is another equally good enol that has the other carbonyl group enolized. The two enols are tautomers of each other and of the keto-ester.

That compound **B** is completely enolized shows that conjugation is much more important than hydrogen bonding, which is impossible with **B**. However, **B** has extra conjugation from the lone pairs on the extra oxygen atoms.

The remaining compounds have problems with enolization. Compound **C** can form an enol on the side away from the other carbonyl group, but cannot form an enol between the two ketones as it would be a 'bridgehead alkene'. These do not generally exist as the four substituents around the alkene cannot become planar.

■ See pp. 389–90 and p. 914 of the textbook for more on why bridgehead alkenes are usually impossible.

Compound **D** seems to have a perfectly reasonable enol. But the very large *tert*-butyl group, which would be out of the plane in the diketone, would have to lie in the plane if the enol were formed. The four-membered ring in **E** is already strained enough with two sp^2 atoms having 90° bond angles. The enol would have three such atoms and this is too much strain.

PROBLEM 7

Attempted nitrosation of this carboxylic acid leads to formation of an oxime with loss of CO_2. Why?

Purpose of the problem

Nitrosation of an enol, followed by an unusual step.

Suggested solution

Sodium nitrite in acid generates NO^+, which usually reacts with enols to form nitroso compounds and then oximes (p. 464 of the textbook). Let's follow the mechanism through with this acid. Enolization gives the nucleophile, which picks up NO⁺.

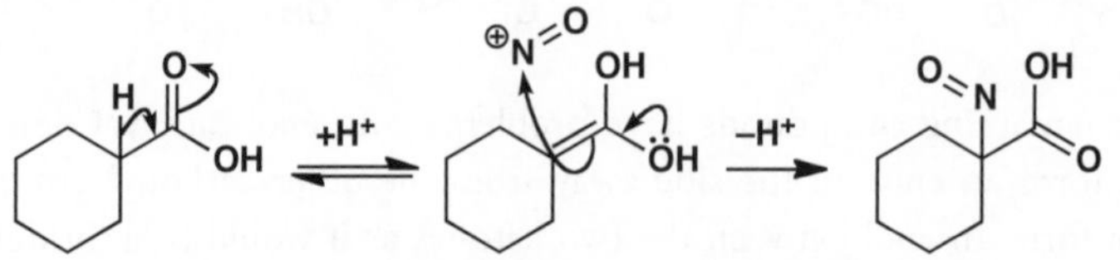

This nitroso compound can't tautomerize because there are no adjacent protons, so instead the compound forms an oxime by losing CO_2. We can draw a cyclic mechanism for this:

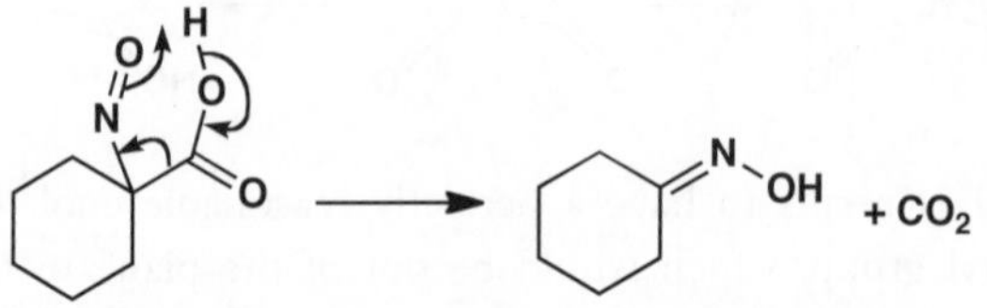

PROBLEM 8

This molecule is a perfumery compound with an intense flowery odour, but it isomerizes rapidly in base to its odourless diastereoisomer. Why?

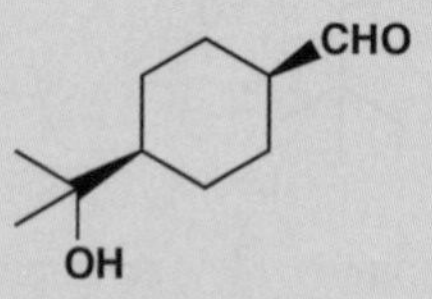

Purpose of the problem

The stereochemical consequences of enolization in a cyclohexane.

Suggested solution

Base will catalyse the enolization of our aldehyde, and when the carbonyl compound reforms, the proton could return to either the face it left from or the other side, allowing the compound to intercovert with its diastereoisomer. If we draw the compounds in their chair conformation, you can see that while the starting aldehyde has an axial substituent (which must be the less bulky CHO group), the diastereoisomer formed through enolization has two equatorial substituents and is more stable. In fact, the equilibrated mixture contains 92% of the equatorial product.

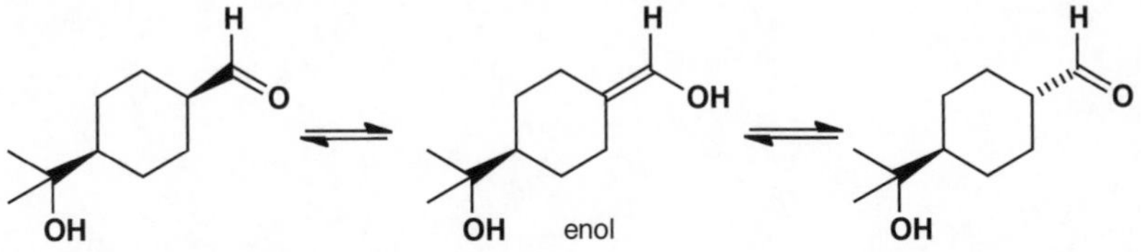

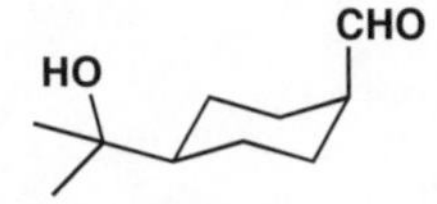

CHO group axial

both equatorial: more stable

Suggested solutions for Chapter 21

21

PROBLEM 1

All you have to do is to spot the aromatic rings in these compounds. It may not be as easy as you think and you should give some reasons for questionable decisions.

thyroxine: human hormone regulating metabolic rate

aklavinone: tetracycline antibiotic

colchicine: anti-cancer agent from the autumn crocus

calistephin: natural red flower pigment

methoxatin: coenzyme from bacteria living on methane

Purpose of the problem

Simple exercise in counting electrons with a few hidden tricks.

Suggested solution

Truly aromatic rings are marked with bold lines. Thyroxine has two benzene rings—obviously aromatic—and that's that. Aklavinone also has two aromatic benzene rings and we might argue about ring 2. It has four electrons as drawn, and you might think that you could push electrons round from the OH groups to give ring 2 six electrons as well. But if you try it, you'll find you can't.

Colchicine has one benzene ring and a seven-membered conjugated ring with six electrons in double bonds (don't count the carbonyl electrons as they are out of the ring). It perhaps looks more aromatic if you delocalize the electrons and represent it as a zwitterion. Either representation is fine.

Methoxatin has one benzene ring and one pyrrole ring—an example of an aromatic compound with a five-membered ring. The six electrons come from two double bonds and the lone pair on the nitrogen atom. The middle ring is not aromatic—even if you try drawing other delocalized structures, you can never get six electrons into this ring.

PROBLEM 2

First, as some revision, write out the detailed mechanism for these steps.

$$HNO_3 + H_2SO_4 \longrightarrow {}^{\oplus}NO_2$$

In a standard nitration reaction with, say, HNO_3 and H_2SO_4, each of these compounds forms a single nitration product. What is its structure? Explain your answer with at least a partial mechanism.

Purpose of the problem

Revision of the basic nitration mechanism and extension to compounds where selectivity is an issue.

Suggested solution

The basic mechanisms for the formation of NO_2^+ and its reaction with benzene appear on p. 476 of the textbook. Benzoic acid has an electron-withdrawing substituent so it reacts in the *meta* position. The second compound is activated in all positions by the weakly electron-donating alkyl groups (all positions are either *ortho* or *para* to one of these groups) but will react at one of the positions more remote from the alkyl groups because of steric hindrance.

The remaining two compounds have competing *ortho,para*-directing substituents but in each case the one with the lone pair of electrons (N or O) is a more powerful director than the simple alkyl group. In the first case nitrogen directs *ortho* but in the second oxygen activates both *ortho* and *para* and steric hindrance makes the *para* position marginally more reactive.

PROBLEM 3

How reactive are the different sites in toluene? Nitration of toluene produces the three possible products in the ratios shown. What would be the ratios if all the sites were equally reactive? What is the actual relative reactivity of the three sites? You could express this as x:y:1 or as a:b:c where a+b+c = 100. Comment on the ratio you deduce.

Purpose of the problem

A more quantitative assessment of relative reactivities.

Suggested solution

As there are two *ortho* and two *meta* sites, the ratio if all were equally reactive would be 2:2:1 *o:m:p*. The observed reactivity is 30:2:37 or 15:1:18 or 43:3:54 depending on how you expressed it. The *ortho* and *para* positions are roughly equally reactive because the methyl group is electron-donating. The *para* is slightly more reactive than the *ortho* because of steric hindrance. The *meta* position is an order of magnitude less reactive because the intermediate is not stabilized by electron-donation (σ-conjugation) from the methyl group.

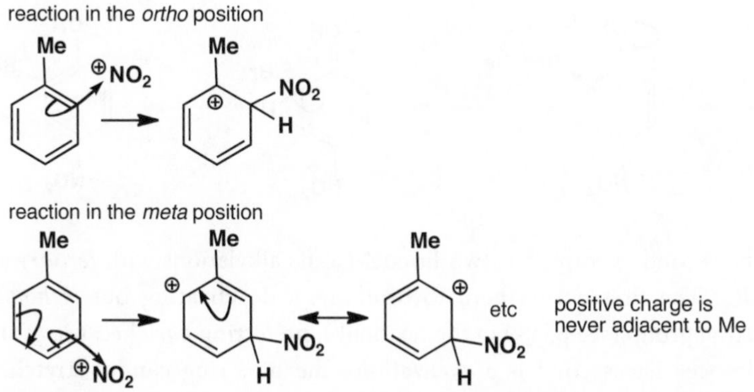

reaction in the *ortho* position

Me
$\oplus$NO$_2$

Me
NO$_2$
$\oplus$
H

reaction in the *meta* position

Me

Me
$\oplus$
$\oplus$NO$_2$

Me
$\oplus$
—NO$_2$
H

Me
$\oplus$
—NO$_2$
H

etc

positive charge is
never adjacent to Me

PROBLEM 4

Draw mechanisms for these reactions and explain the positions of substitution.

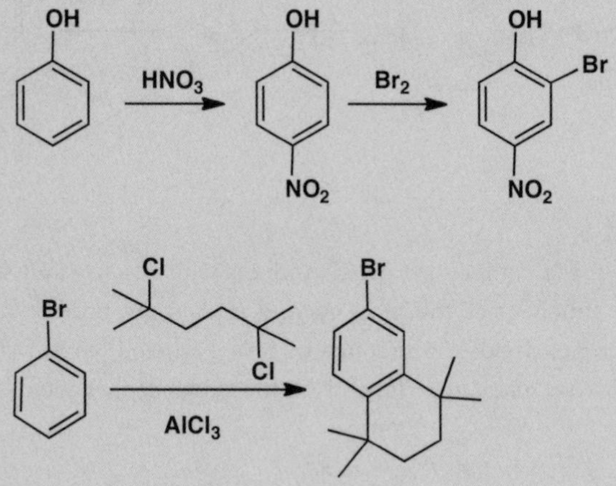

OH
HNO$_3$
OH
NO$_2$
Br$_2$
OH
Br
NO$_2$

Br
Cl
Cl
AlCl$_3$
Br

Purpose of the problem

More advanced questions of orientation with more powerful electron-donating groups.

Suggested solution

The OH group has a lone pair of electrons and dominates reactivity and selectivity. Steric hindrance favours the *para* product in the first reaction. The bromination has to occur *ortho* to the phenol as the *para* position is blocked.

:OH → ⊕OH → :OH → ⊕OH → OH

⊕NO₂   H NO₂   NO₂ Br—Br   Br H NO₂   Br NO₂

The second example has two Friedel-Crafts alkylations with *tertiary* alkyl halides. The first occurs *para* to bromine, a deactivating but *ortho,para-*directing group (see p. 489 in the textbook), preferring *para* because of steric hindrance. The second is a cyclization—the new ring cannot stretch any further than the next atom.

Cl ... AlCl₃ → Br ... Cl → Br ... and repeat → Br ...

PROBLEM 5

Nitration of these compounds gives products with the ¹H NMR spectra shown. Deduce the structures of the products and explain the position of substitution. WARNING: do not decide the structure by saying where the nitro group 'ought to go'! Chemistry has many surprises and it is the evidence that counts.

$NO_2^{\oplus}$ → ?

δ_H
7.77 (4H, d, J 10)
8.26 (4H, d, J 10)

Cl Cl $NO_2^{\oplus}$ → ?

δ_H
7.6 (1H, d, J 10)
8.1 (1H, dd, J 10,2)
8.3 (1H, d, J 2)

F $NO_2^{\oplus}$ → ?

δ_H
7.15 (2H, dd, J 7,8)
8.19 (2H, dd, J 6,8)

Purpose of the problem

Revision of the relationship between NMR and substitution pattern.

Suggested solution

The first product has only eight hydrogens so two nitro groups must have been added. The molecule is clearly symmetrical and the coupling constant is right for neighbouring hydrogens so a substitution on each ring must have occurred in the *para* position. Note that the hydrogen next to the nitro group has the larger shift. We can deduce that each benzene ring is an *ortho,para*-directing group on the other because the intermediate cation is stabilized by conjugation.

7.77 H   H8.26
(4H, d, *J* 10)   (4H, d, *J* 10)

The hydrogen count reveals that the next two products are mono-nitro compounds. There are two hydrogens *ortho* to nitro in the second compound and one of them also has a typical *ortho* coupling to a neighbouring hydrogen while the other has only a small coupling (2 Hz) which must be a *meta* coupling. Substitution has occurred *para* to one of the chlorines and *ortho* to the other. The chlorines are *ortho,para*-directing thus activating all remaining positions so steric hindrance must explain the site of nitration.

■ Vicinal (*ortho*) coupling constants in benzene rings are typically 8–10 Hz; *meta* coupling constants are typically <2 Hz: see pp. 295-6 of the textbook.

H 1H, d, *J* 2

1H, d, *J* 10 H   1H, dd, *J* 10,2

The third compound has the extra complication of couplings to fluorine. The coupling of 7 Hz shown by one hydrogen and 6 Hz shown by the other must be to fluorine as they occur once only. The symmetry of the compound and the typical *ortho* coupling between the hydrogens (8 Hz) shows that *para* substitution must have occurred.

■ The idea that heteronuclear couplings leave 'unpaired' coupling constants in the ¹H NMR spectrum is explained in the green box on p. 416 of the textbook.

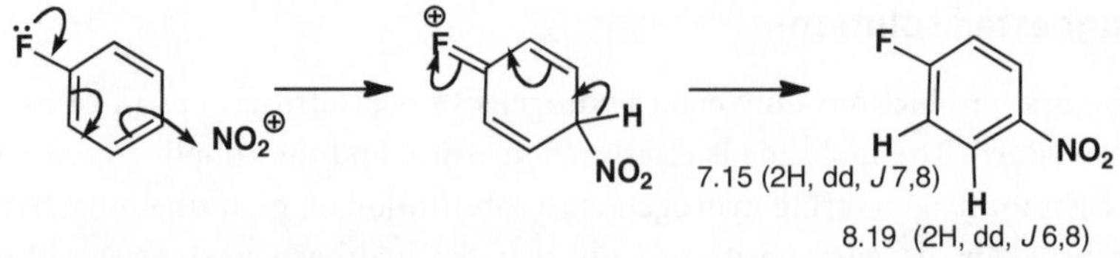

7.15 (2H, dd, *J* 7,8)

8.19 (2H, dd, *J* 6,8)

PROBLEM 6

Attempted Friedel-Crafts acylation of benzene with *t*-BuCOCl gives some of the expected ketone **A** as a minor product, as well as some *t*-butylbenzene **B**, but the major product is the substituted ketone **C**. Explain how these compounds are formed and suggest the order in which the two substituents are added to form compound **C**.

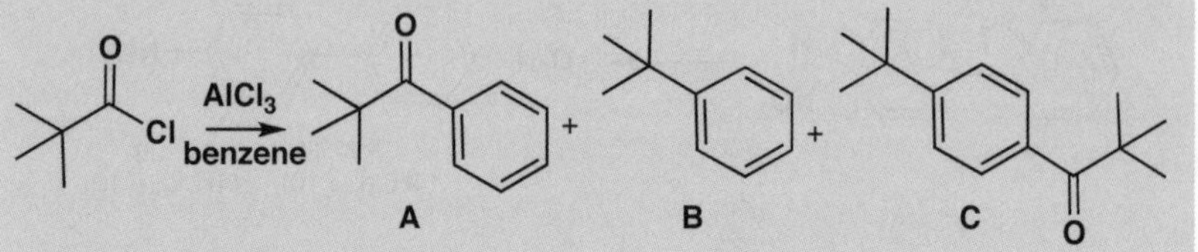

Purpose of the problem

Detailed analysis of a revealing example of the Friedel-Crafts reaction.

Suggested solution

■ Friedel-Crafts acylation is on p. 477 of the textbook.

The expected reaction to give **A** is a simple Friedel-Crafts acylation with the usual acylium ion intermediate.

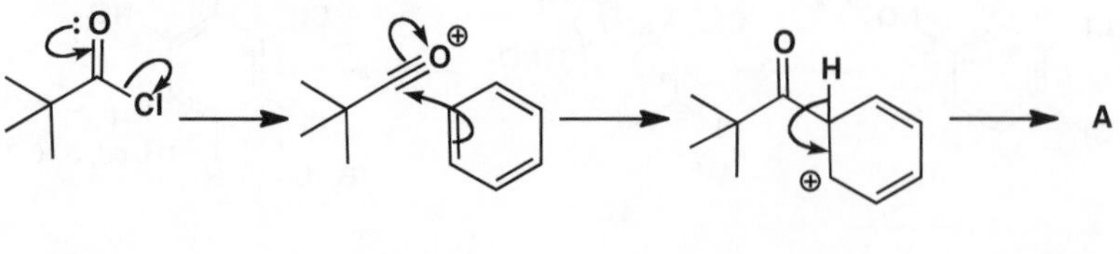

Product **B** must arise from a *t*-butyl cation and the only way that might be formed is by loss of carbon monoxide from the original acylium ion. Such a reaction happens only when the resulting carbocation is reasonably stable.

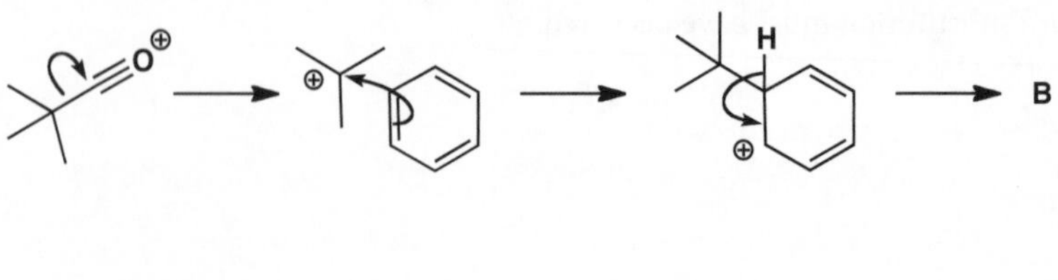

The main product **C** comes from the addition of both these electrophiles, but which adds first? The ketone in **A** is deactivating and *meta* directing but the *t*-butyl group in **B** is activating and *para*-directing so it must be added first.

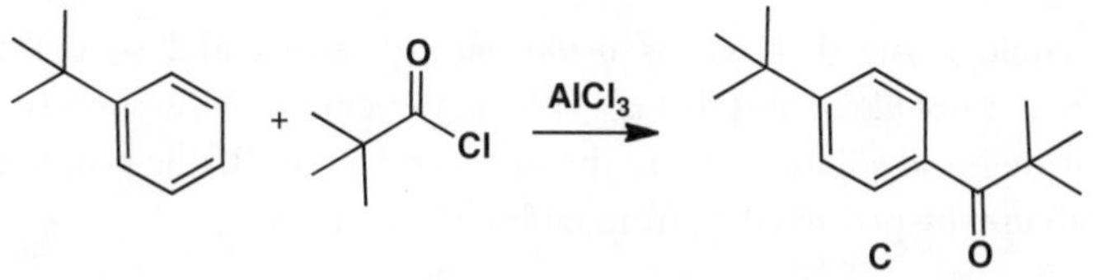

That answers the question but you might like to go further. Both **A** and **C** are formed by the alkylation of benzene as the first step. The decomposition of the acylium ion is evidently *faster* than the acylation of benzene. However, when **B** reacts further, it is mainly by acylation as only a small amount of di-*t*-butyl benzene is formed. Evidently the decomposition of the acylium ion is *slower* than the acylation of **B**! This is not unreasonable as the *t*-butyl group accelerates electrophilic attack—but it is a dramatic demonstration of that acceleration.

PROBLEM 7

Nitration of this heterocyclic compound with the usual HNO_3/H_2SO_4 mixture gives a single nitration product with the ^1H NMR spectrum shown below. Suggest which product is formed and why.

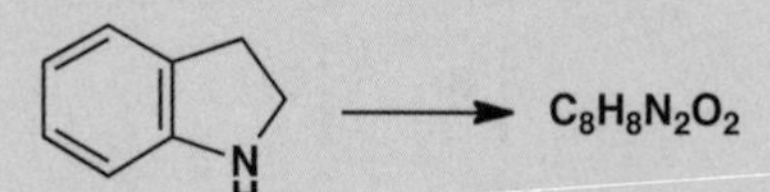

δ_H
3.04 (2H, t, J 7 Hz)
3.68 (2H, t, J 7 Hz)
6.45 (1H, d, J 8 Hz)
7.28 (1H, broad s)
7.81 (1H, d, J 1 Hz)
7.90 (1H, dd, J 8, 1 Hz)

Purpose of the problem

Revision of NMR and an attempt to convince you that the methods of chapter 21 can be applied to molecules you've not met before.

Suggested solution

The two 2H triplets and the broad NH signal show that the heterocyclic ring is intact. One nitro group has been added to the benzene ring. The proton at 7.81 with only one small (*meta*) coupling must be between the nitro group and the other ring and is marked on the two possible structures.

You could argue that NH is *ortho, para*-directing and so the second structure is more likely. But this is a risky argument as the reaction is carried out in strong acid solution where the nitrogen will mostly be protonated. It is safer to use the predicted δ_H from tables. Here we get:

Proton	*ortho*	*meta*	*para*	predicted δ_H
H^a	$NO_2 = +0.95$	$CH_2 = -0.14$	$NH = -0.25$	7.73
H^b	$NO_2 = +0.95$	$NH = -0.75$	$CH_2 = -0.06$	7.31

There's not much difference but H^a at 7.73 is closer to the observed 7.81, so it looks as though the small amount of unprotonated amine directs the reaction.

PROBLEM 8

What are the two possible isomeric products of this reaction? Which structure do you expect to predominate? What would be the bromination product from each?

Purpose of the problem

Getting you to think about alternative products and possible reactions on compounds that haven't been made (yet).

Suggested solution

The reaction is a Friedel-Crafts cyclization, as you could have deduced by the simple loss of water. The resulting cation could cyclize in two ways, arbitrarily called **A** and **B**. Steric hindrance suggests that **A** would be the more likely product.

■ A Top Tip: when you have a formula for a product, but no structure, is to compare it with the formula for the starting material—in this case, $C_{12}H_{18}O_2$.

Bromination will go either *ortho* or *para* to the methoxy group: **A** has two different positions *ortho* to the OMe, but the *para* position is blocked. The least sterically hindered position gives a 1,2,4,5-tetrasubstituted ring. **B** might give a mixture of *ortho* and *para* substitution.

PROBLEM 9

On p. 479 of the textbook we explain the formation of 2,4,6-tribromophenol by bromination of phenol in water. It looks as though we can go no further as all the *ortho* and *para* positions are brominated. But we can if we treat the tribromo-compound with bromine in an organic solvent. Account for the formation of the tetrabromo-compound.

The product is useful in brominations as it avoids using unpleasant Br_2. Suggest a mechanism for the following bromination and account for the selectivity.

Purpose of the problem

Exploration of interesting chemistry associated with electrophilic substitution on benzene rings.

Suggested solution

Phenol is so reactive that the fourth bromine adds in the *para* position. Now the molecule has a problem as there is no hydrogen on that carbon to be lost. So the phenolic hydrogen is lost instead. It is surprising but revealing that this loss of aromaticity is preferred to the alternative bromination at the *meta* position.

In the second reaction, one of the reactive bromines in the *para* position is transferred to the amine. It could have added *ortho* or *para* to the NMe_2 group but CF_3 is small and NMe_2 is large, because the two methyl groups lie in the plane of the ring, so steric hindrance rules. The other product is recovered tribromophenol.

■ Note that the *meta* directing effect of the deactivating CF_3 group is irrelevant (see p. 491 of the textbook).

PROBLEM 10

How would you make each of the following compounds from benzene?

Purpose of the problem

Choosing a synthetic route, taking into account the directing effects of the substituents involved.

Suggested solution

The first compound has a ketone substituent, which is electron-withdrawing and therefore *meta*-directing, and an amino group, which is electron-donating and therefore *ortho,para*-directing. Aromatic amino groups are best made by reduction of nitro groups, which are also meta directing, so there are two possibilities. We can either start with a Friedel-Crafts acylation of benzene to give the ketone, which we can nitrate in the *meta* position and then reduce, or we can start by nitrating benzene, then do the acylation and then reduce. Either is a reasonable solution.

The second compound has a bromo substituent, which is *ortho,para*-directing, and a *meta*-directing nitro group. We need the *para* relationship, so we must put the bromine in first, then nitrate.

Finally, a compound with two *para*-directors arranged *meta* to one another. This may seem a problem, but we must introduce the alkyl group by Friedel-Crafts acylation and reduction, since primary alkyl groups cannot be introduced by Friedel-Crafts alkylation. The acyl group will be *meta* directing, so that solves both problems. First acylate, then brominate, then reduce.

PROBLEM 1

Draw a mechanism for this reaction. Why is base unnecessary?

PhPH$_2$ + ⟍⟍CN (excess) ⟶ Ph—P(CH$_2$CH$_2$CN)$_2$

Purpose of the problem

Simple example of conjugate addition with a nucleophile from the second row of the periodic table.

Suggested solution

The phosphine is a good soft nucleophile with a high energy lone pair, well able to add in a conjugate fashion without help. In particular, the neutral phosphine does not need to be converted into its anion. The intermediate is a good base and removes a proton from itself, not necessarily intramolecularly.

PROBLEM 2

Which of the two routes suggested here would actually lead to the product?

Purpose of the problem

Do you understand the essentials of conjugate addition? Can you say when it *won't* happen?

Suggested solution

To get the product, the chloride must add in a conjugate fashion and ethyl Grignard in a direct fashion that removes the carbonyl group. Conjugate addition can happen only if the carbonyl group is intact so HCl must be added first.

In the other sequence, EtMgBr is likely to add to the carbonyl group direct and further addition of HCl may either substitute on the allylic alcohol or add the 'wrong way round' to the alkene.

PROBLEM 3

Suggest reasons for the different outcome of each of these reactions. Your answer must of course be mechanistically based.

Purpose of the problem

A reminder of the reactions possible with enones.

Suggested solution

The three reactions are: enolization and trapping with silicon, direct addition with a hard irreversible nucleophile, and conjugate addition with a softer reversible nucleophile.

PROBLEM 4

Suggest a mechanism for this reaction.

Purpose of the problem

Combination of conjugate addition and electrophilic aromatic substitution.

Suggested solution

The weakly nucleophilic benzene has evidently added in conjugate fashion to the enone in a kind of Friedel-Crafts reaction and we can use the Lewis acid to make the enone into the necessary cation.

PROBLEM 5

What is the structure of the product of this reaction and how is it formed? It has δ_C 191, 164, 132, 130, 115, 64, 41, 29 and δ_H 2.32 (6H, s), 3.05 (2H, t, *J* 6 Hz), 4.20 (2H, t, *J* 6 Hz), 6.97 (2H, d, *J* 7 Hz), 7.82 (2H, d, *J* 7 Hz), 9.97 (1H, s). You should obviously interpret the spectra to get the structure.

Purpose of the problem

Revision of NMR with an exercise in nucleophilic aromatic substitution.

Suggested solution

Summing the formulae of the two starting materials shows that this is a substitution of fluoride (the product is the sum of the starting materials less

HF). The aldehyde is still there (from the IR and the proton at 10 ppm) so the spectra are best interpreted by this structure:

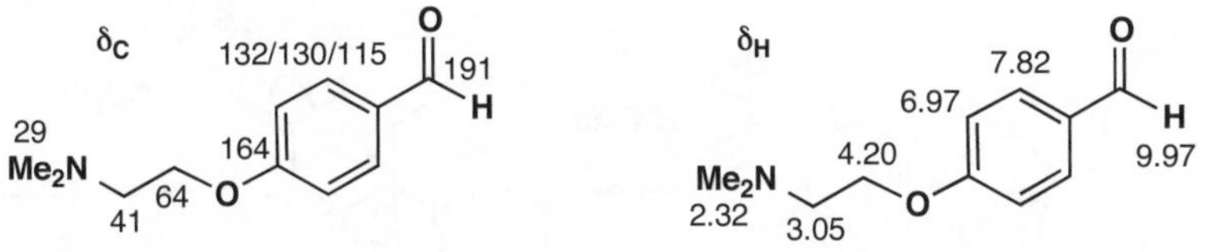

That suggests a simple nucleophilic aromatic substitution by the addition-elimination mechanism with both F and CHO assisting the first step.

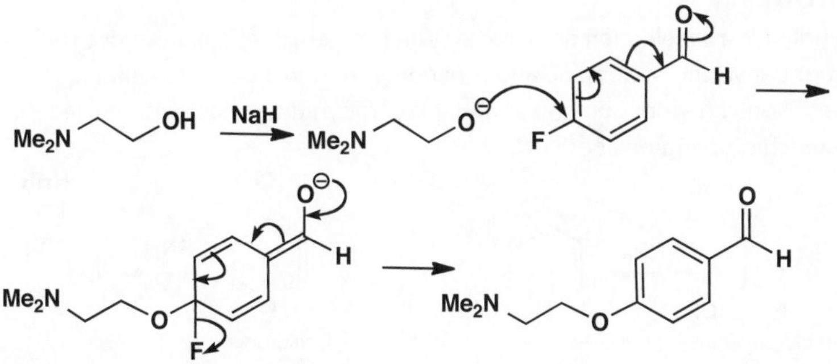

PROBLEM 6

Suggest a mechanism for this reaction, explaining the selectivity.

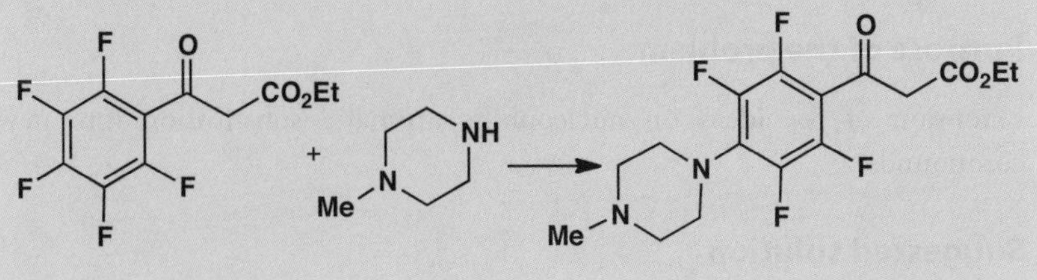

Purpose of the problem

Introduction to the mechanism and selectivity of nucleophilic aromatic substitution.

Suggested solution

Both *ortho* and *para* positions are activated by the ketone towards nucleophilic attack by the amine, but the *para* position is preferred because of steric hindrance between the large heterocyclic ring and the ketone. The substitution works because those five fluorine atoms make the ring very electron-deficient.

PROBLEM 7

Pyridine is a six-electron aromatic system like benzene. You have not yet been taught anything systematic about pyridine (that will come in chapter 29) but see if you can work out why 2- and 4-chloropyridines react with nucleophiles but 3-chloropyridine does not.

Purpose of the problem

Extension of the ideas on nucleophilic aromatic substitution into new compounds.

Suggested solution

The problem is to find somewhere to park the negative charge in the intermediate and the only possible place is on the pyridine nitrogen atom. This is easy with 2- and 4-choropyridine but impossible with 3-chloropyridine. Using a general nucleophile:

Amine formation by this reaction is particularly important as you will see in chapters 29 and 30. The mechanism is the same with a few proton transfers.

PROBLEM 8

How would you carry out these two conversions?

Purpose of the problem

Application of nucleophilic aromatic substitution in synthesis.

Suggested solution

Usually you would think of introducing NH_2 by nitration and reduction (chapter 21), but the regioselectivity is wrong for the first reaction: the methoxy group will direct nitration *ortho* to itself. An alternative is to introduce both NH_2 and CN as nucleophiles, but the ring is unactivated so we can't use the addition-elimination mechanism (there is nowhere for the negative charge to go). The successful alternatives are electrophilic aromatic substitution followed by diazonium salt formation and the benzyne method. Here are two possible routes. Nitration will insert the nitro group *ortho* to the more strongly electron-donating MeO group. Reduction, diazotization and substitution with copper cyanide by the S_N1 mechanism gives one product.

The other product could come from chlorination, elimination to give a benzyne, addition of amide anion to put the anion *ortho* to MeO (p. 524 in the textbook) and protonation.

PROBLEM 9

Suggest mechanisms for these reactions, pointing out why you chose the pathways.

Purpose of the problem

Studies in selectivity and choosing the right mechanism.

Suggested solution

In the first reaction, the nucleophile adds in the 'wrong' position (i.e. where the leaving group isn't) so a benzyne mechanism is likely. Notice that the nucleophile and the benzyne are formed with the same strong base, that the anion is recycled and that the nucleophile adds to the benzyne to put the negative charge next to OMe (p. 524 in the textbook).

The second reaction is a straightforward substitution by the addition-elimination mechanism activated by the nitro group. The amino group is a spectator.

PROBLEM 10

When we discussed reduction of cyclopentenone to cyclopentanol, we suggested that conjugate addition of borohydride must occur before direct addition of borohydride: in other words, the scheme below must be followed. What is the alternative scheme? Why is the scheme shown definitely correct?

Purpose of the problem

Serious thinking about mechanisms is an advantage when reactions get more complex.

Suggested solution

The alternative scheme would be to reduce the ketone first and the alkene second. This order must be wrong though, because simple alkenes are nucleophilic and are not reduced by NaBH₄. NaBH₄ is a nucleophilic reducing agent and attacks alkenes only if they are conjugated with an electron-withdrawing group. The conjugate addition must always occur first so as to keep the carbonyl group intact for the second step.

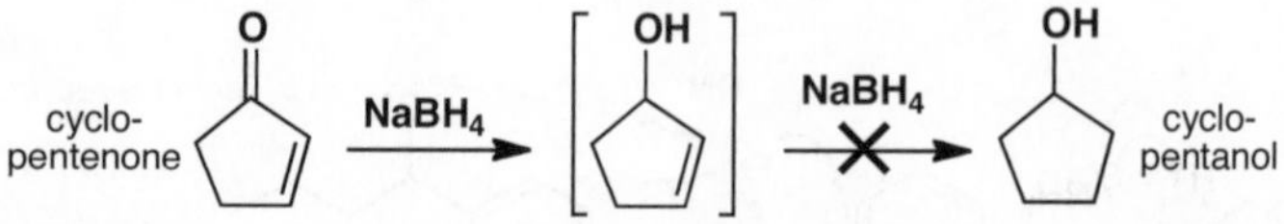

PROBLEM 11

Stirring thioacetic acid with acrolein (propenaldehyde) in acetone gives a compound with the NMR data shown below. What is the compound?

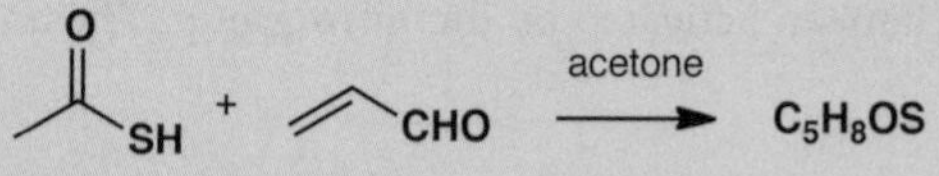

δ_H: 2.28 (3H, s), 3.58 (2H, d, *J* 8), 4.35 (1H, td, *J* 8, 6), 6.44 (1H, t, *J* 6, 7.67 (1H, d, *J* 6).

δ_C: 23.5, 31.0, 99.3, 144.2, 196.5.

Purpose of the problem

Using NMR to gain insight into a conjugate addition.

Suggested solution

The product formula is the sum of the reaction partners, and all 5 C and 8 H atoms are visible in the NMR spectra, so this looks like an addition reaction. The ^{13}C NMR tells us that there are two alkene carbons and one carbonyl, and the proton NMR clearly shows the aldehyde has gone. But it can't be direct addition to the C=O group, because the coupling pattern isn't right for a terminal alkene. The product is in fact the enol formed from conjugate addition of the sulfur, which is stable under these conditions. The low coupling constant across the alkene tells us it's formed unusually as the *Z*-isomer, probably because of an intramolecular proton transfer from the thioacid to the new OH group. The anhydrous conditions in dry acetone prevent the enol from tautomerizing back to the aldehyde.

■ This work is described by Lukas Hintermann in *J. Org. Chem.*, 2012, **77**, 11345.

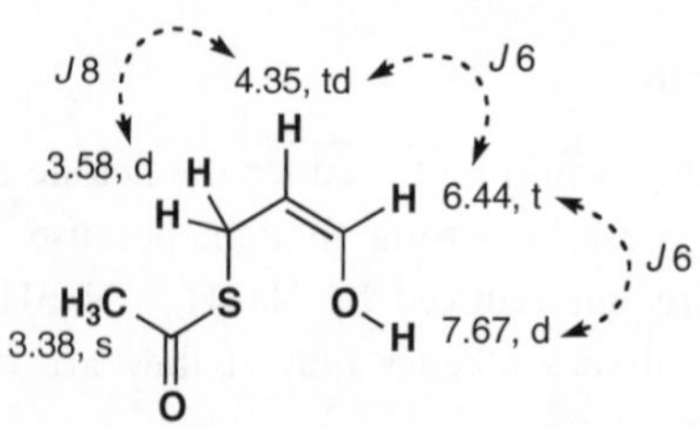

23

Suggested solutions for Chapter 23

Several problems in this chapter ask you to suggest ways to carry out conversions of one molecule into another. We always give one possible answer and sometimes comment on alternatives but you should realize that there are usually many possible 'right' answers to questions of this sort. Make sure you understand the principle behind the question and, if your answer is very different from ours, check with someone with experience of synthesis.

PROBLEM 1

How would you convert this bromo-aldehyde chemoselectively into the two products shown?

Purpose of the problem

A simple exercise in chemoselectivity and protection.

Suggested solution

You would like to add an organometallic reagent to go to the right and that's very simple as no protection is needed. A Grignard reagent will do the job.

The other product demands more care to avoid the reactions we have just done. The aldehyde needs to be protected, as an acetal, say, before we make the Grignard reagent from the aryl bromide. Then we can add to RCHO, and deprotect with acid, and we have our product.

PROBLEM 2

How would you convert this lactone selectively into either the hydroxyacid or the unfunctionalized acid?

Purpose of the problem

Exploration of chemoselectivity.

Suggested solution

The conversion into the hydroxy-acid is just hydrolysis and can be carried out in aqueous base. Conversion into the unfunctionalized acid demands selective reduction of the C–O at the secondary benzylic centre. Possibilities include catalytic hydrogenolysis (p. 539 in the textbook) or HBr followed by C–Br reduction.

■ You can read about an application of this chemistry in S. Torii *et al, Bull. Chem. Soc. Japan,* 1978, **51**, 3590.

PROBLEM 3

Predict the products of Birch reduction of these aromatic compounds.

Purpose of the problem

Exploring the principles of Birch reduction.

Suggested solution

In each case a unique product results if you draw a dianion intermediate placing the electron-withdrawing groups where they can stabilize the negative charges, and put the electron-donating groups on the alkenes, where they don't destabilize the negative charges.

■ See pp. 542–543 in the textbook for explanation of why this approach works.

PROBLEM 4

How would you carry out these reactions? In some cases more than one step may be required.

Purpose of the problem

Reduction, selectivity, and protection in the same sequence.

Suggested solution

Every step is straightforward except the final reduction where a less reactive ester must be reduced in the presence of a more reactive ketone. Protection is the answer and an acetal is suitable.

■ The final product was used to make an analogue of thromboxane, a human blood clotting agent, by M. Hayashi and group, *Tetrahedron Lett.*, 1979, 3661.

PROBLEM 5

How would you convert this nitro compound into the two products shown? Explain the order of events with special regard for reduction steps.

Purpose of the problem

Reduction, selectivity, and protection in two related sequences.

Suggested solution

The nitro group must be reduced to an amino group and cyclized onto the ketone or the carboxylic acid. Reductive amination (pp. 234–7 in the textbook) allows the amine to cyclize onto the more electrophilic ketone.

*these intermediates are not isolated—
the cyclization and reduction happen spontaneously*

Forming the six-membered ring requires more control. Protection of the ketone (say as the acetal) before reduction will give the six-membered cyclic amide. Now the amide carbonyl must be reduced with LiAlH₄ (p. 236 in the textbook) and the ketone deprotected. There are many good alternative answers to this problem.

PROBLEM 6

Why is this particular amine formed by reductive amination here?.

Purpose of the problem

Extending the concept of reductive amination by combining it with deprotection and cyclization.

Suggested solution

The two acetals will be hydrolysed at pH 5.5 to give the amine a choice between cyclization to one or other of the two aldehydes.

Cyclization to a five-membered ring is preferred to cyclization to a (strained) four-membered ring so reductive amination occurs to the right and not to the left (as drawn). Cyanoborohydride is stable under the weakly acidic conditions and does not reduce the remaining aldehyde.

■ This problem is based on work by G. W. Gribble and R. M. Soll, *J. Org. Chem.*, 1981, **46**, 2433.

PROBLEM 7

Account for the chemoselectivity of the first reaction and the stereoselectivity of the second. A conformational drawing of the intermediate is essential.

Purpose of the problem

Extending chemoselectivity into more subtle distinctions and conformational analysis.

Suggested solution

The two ketones are different only because one is conjugated. Since acetal formation is a thermodynamically controlled reversible reaction, the one that is formed retains the enone as the stabilizing effect of conjugation can be retained. A conformational diagram of the intermediate shows that there is inevitably one axial oxygen atom belonging to the acetal preventing the bottom face of the alkene from getting close to the catalyst. The axial methyl group is further away and more in the plane of the alkene. Hydrogen is delivered from the top face and the observed product results.

■ The conformation of decalin is discussed on pp. 378–9 of chapter 16 of the textbook. The 'flattening' effect of the alkene is dealt with in chapter 32.

PROBLEM 8

How would you convert this diamine to either of these two protected derivatives?

Purpose of the problem

Using protecting groups to reveal reactive groups selectively.

Suggested solution

The Boc group is a common acid-sensitive protecting group for amines, and making the first derivative is easy because the amine attached to the primary carbon is less hindered and more reactive. Treating the diamine with one equivalent of 'Boc anhydride' (Boc_2O) gives the correctly protected product.

The second is more of a challenge, because we have to protect the less reactive amino group. The solution is first to protect with a different protecting group: Cbz will do, using CbzCl (benzyl chloroformate) and base. Now the other amino group can be protected with Boc, and finally the Cbz protecting group removed by hydrogenation. Other protecting groups might be all right too, but they have to be removable without using acid, which would remove the Boc group.

■ This chemistry was used by chemists in Bordeaux and Manchester to build some new polymeric structures out of the two different amine products.

24

Suggested solutions for Chapter 24

PROBLEM 1

Two routes are proposed for the preparation of this amino-alcohol. Which do you think is more likely to succeed and why?

Purpose of the problem

Practical application of the choice of reagent to ensure the correct regioselectivity in a conjugate addition.

Suggested solution

Either route might give the product but enals are more likely to undergo direct addition to the carbonyl group rather than conjugate addition while conjugated esters are better at conjugate addition. So the ester is probably better.

■ See pages 505–510 and 581–582 in the textbook.

PROBLEM 2

Predict the products of these reactions.

Purpose of the problem

Practice at predicting the regioselectivity of a direct or conjugate addition.

Suggested solution

Both reactions involve addition of organometallic compounds to unsaturated carbonyl compounds. The key difference is the metal. With Cu(I) as catalyst, the Grignard reagent will give conjugate addition in the first case. MeLi will give direct addition in the second.

A =   B =

PROBLEM 3

Explain the different regioselectivity in these two brominations of 1,2-dimethylbenzene.

Purpose of the problem

Regioselectivity in electrophilic aromatic substitution and in radical substitution.

Suggested solution

AlCl$_3$ is a commonly used Lewis acid in electrophilic aromatic substitution reactions. Here it activates the bromine to form the electrophile 'Br$^+$', which attacks the aromatic ring. Methyl groups are *ortho,para* directors, so any of the four unsubstituted positions could be attacked, but steric hindrance directs the first bromine to go to one of the positions that does not lead to a 1,2,3-trisubstituted ring.

■ Make sure you understand *why* methyl groups direct *ortho, para*: if you need reminding, see p. 484 of the textbook.

Now we have three *ortho,para* directors, and bromine (with its lone pairs) is the strongest, so the next bromine will go *ortho* to the bromine in the less sterically hindered of the two possibilities.

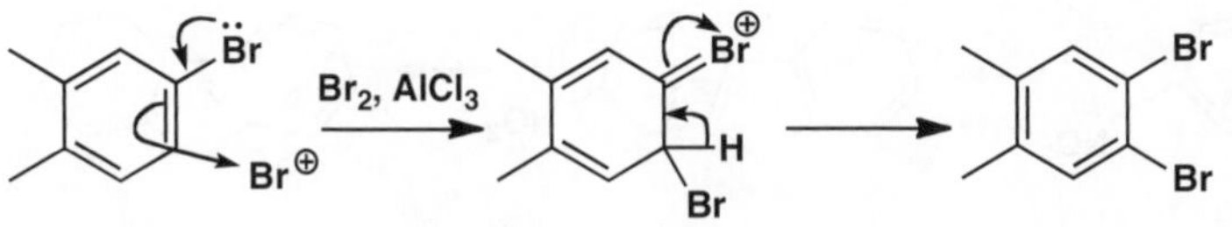

In the presence of light, bromine's weak Br–Br bond undergoes homolysis, and Br• radicals are formed. One of these can abstract a hydrogen atom, breaking the weakest C–H bond. The methyl groups' C–H bonds are weaker than those of the phenyl ring because the benzyl radical that forms is delocalized into the aromatic ring. The benzyl radical attacks another molecule of bromine, and the cycle continues.

■ A similar argument was used on p. 573 of the textbook to explain why Br abstracts H from an allylic, rather than a vinylic, position.

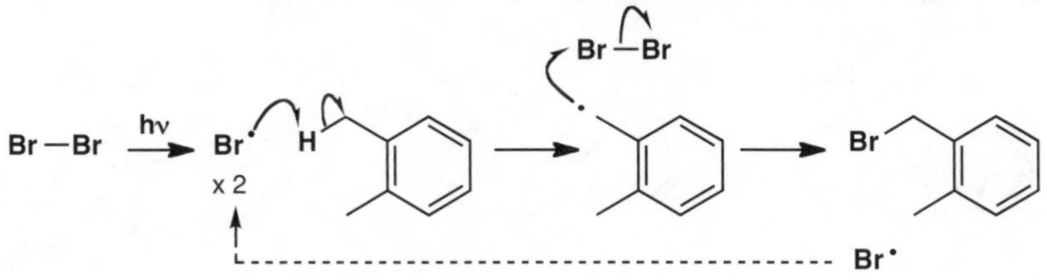

The mechanism is shown here for the first bromination; the same thing can happen on the other methyl group.

PROBLEM 4

The nitro compound below was needed for the synthesis of an anti-emetic drug. It was proposed to make it by nitration of the hydrocarbon shown. How successful do you think this would be?

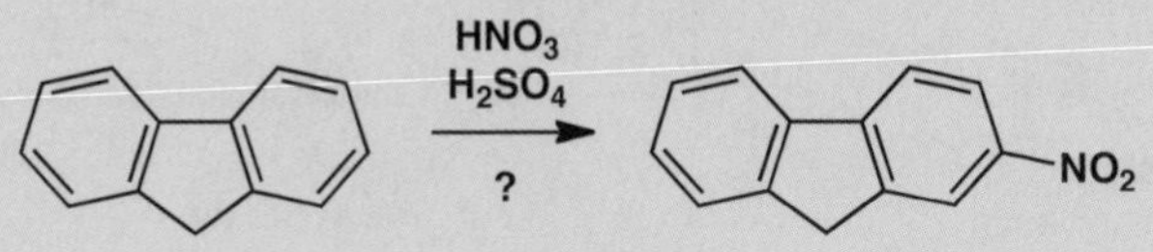

Purpose of the problem

Predicting regioselectivity in electrophilic aromatic substitution where directing effects are more subtle.

Suggested solution

The standard conditions for nitration generate the electrophile NO_2^+, and to get the product shown here, this species has to attack the ring as shown below. The intermediate cation looks quite all right, since the positive charge can be delocalized even into the other ring.

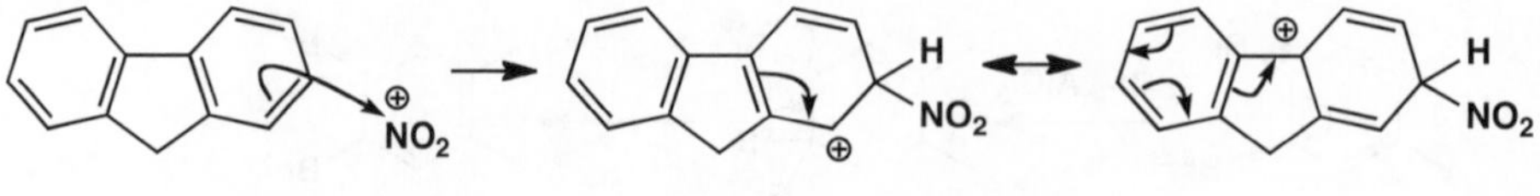

What about the alternatives? A similar cation is formed if the electrophile attacks the position labelled '1', but the nitro group is in a more hindered position here, so we don't expect this to contribute much to the product mixture. Position '2' gives the cation shown below, which although perfectly feasible as an intermediate, does not benefit from the same degree of stabilization as the one in the reaction we want (it can't be delocalized into the other ring). Position 4 is similar but more hindered. Overall we can reasonably expect the reaction to give the product we want.

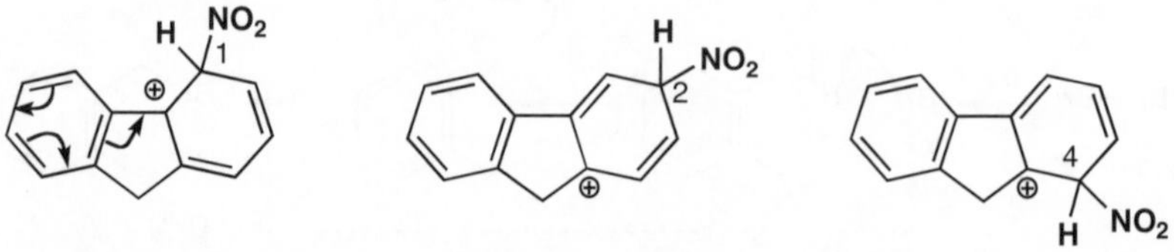

PROBLEM 5

Comment on the regioselectivity and chemoselectivity of the reactions shown below.

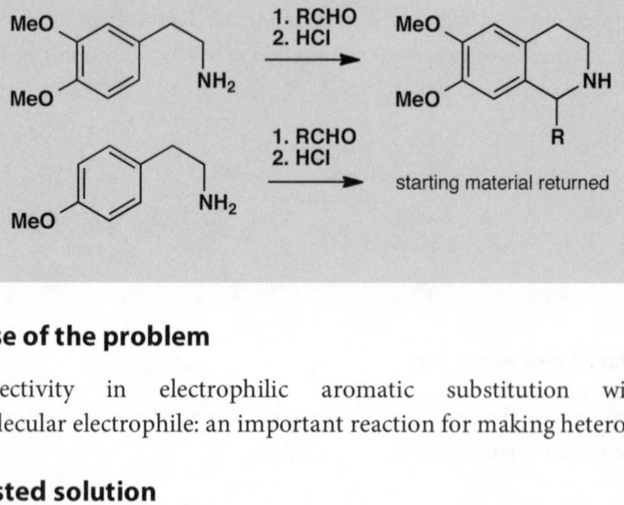

Purpose of the problem

Regioselectivity in electrophilic aromatic substitution with an intramolecular electrophile: an important reaction for making heterocycles.

Suggested solution

The reaction of an aldehyde with an amine gives an imine, and in acid (HCl), protonation gives an iminium ion, the electrophile that attacks the aromatic ring. The iminium ion is tethered to the ring, so it has only two choices of reaction site, since it can't reach any further than the positions *ortho* to the tether. The one it chooses is the less hindered. It is also *para* to an electron-donating methoxy group, so the reaction works well.

In the second case, there is only one methoxy group, and both the positions *ortho* to the tether are *meta* to it, where it can't activate substitution. The positions *ortho* to itself, where it can activate, are too far away for the iminium to reach, so no substitution takes place. Presumably the iminium ion forms, but it is just hydrolysed back to the aldehyde.

■ This reaction is a useful way of making some important alkaloid natural products (and indeed it mimics the way nature makes them). It is sometimes known as the 'Pictet-Spengler reaction'.

iminium ion can't reach   R

PROBLEM 6

Identify **A** and **B** and account for the selectivity displayed in this sequence of reactions.

Purpose of the problem

Analysing selectivity in a useful ring-forming sequence.

Suggested solution

The Friedel-Crafts acylation in the first step is controlled by the bromo substituent, which is an *ortho,para* director: here we get *para* selectivity as usual for steric reasons. Work through the mechanism and you find a ketoacid as the product **A**.

■ You have seen this sort of thing in the textbook on pp. 49–4

The next step is the reduction we introduced on p. 540 of the textbook, the 'Wolff-Kishner' reduction. The mechanism is there so we need not repeat it here; the product is the acid **B** (or, rather, its potassium salt). Now adding acid forms a ring in another Friedel-Crafts acylation. The electrophile must be the acylium ion: usually Friedel-Crafts acylations need more than just strong acid, but this one is fast because it is intramolecular. What about regioselectivity? Well, the only positions the electrophile can reach are *ortho* to the carbon chain, so it must react there (they are both the same) even though that means it has to attack *meta* to the Br group. It's still *ortho* to the alkyl chain though, which is *ortho,para* directing.

PROBLEM 7

The sequence of reactions below shows the preparation of a compound needed for the synthesis of a powerful anti-cancer compound. Explain the regioselectivity of the reactions. Why do you think two equivalents of BuLi are needed in the second step?

Purpose of the problem

Explaining regioselectivity in ortholithiation reactions.

Suggested solution

Both reactions involve ortholithiation—deprotonation of the aromatic ring to form an intermediate aryllithium. As we explain on pp. 563–4 of the textbook, the deprotonation occurs where the BuLi can be 'guided in' by coordinating oxygen atoms. The methoxymethyl acetal, with its two oxygen atoms, is very good at doing this, so we expect deprotonation at one of the two positions *ortho* to this group. The other acetal is also a complexing group, so the deprotonation happens in between the two oxygen atoms.

In the second step, deprotonation can again take place next to the methoxymethyl group. Two equivalents of BuLi are needed because the most acidic proton is in fact one of the protons of the methyl group: a benzyllithium forms first, and then a more reactive aryllithium. When the electrophile (DMF) is added, it reacts only with the last formed, more basic anion.

■ Phenyllithiums are more basic than benzyllithiums, because in benzyllithiums the 'anion' is conjugated with the ring; in phenyllithiums the 'anion' is perpendicular to the π system (like the lone pair in pyridine).

■ Selectivity in the reactions of dianions is described on p. 547 of the textbook.

■ This chemistry is from Corey's synthesis of ecteinascidin: E. J. Corey, D. Y. Gin and R. S. Kania, *J. Am. Chem. Soc.* 1996, **118**, 9202; E. J. Martinez and E. J. Corey, *Org. Lett.* 2000, **2**, 993.

PROBLEM 8

Comment on the regioselectivity and chemoselectivity of the reactions in the sequence below.

Purpose of the problem

Practice using simple principles of reactivity to explain why nucleophiles and electrophiles choose to react at particular sites, and using your knowledge of reactivity to deduce mechanisms for some unfamiliar reactions.

Suggested solution

Benzyl bromide is a good electrophile and it reacts well with alkoxides to make ethers. With neutral alcohols however the substitution is very slow, so only the more nucleophilic (and more basic) pyridine nitrogen is attacked, to make a pyridinium salt.

The pyridinium salt is a bit like an iminium ion, so sodium borohydride attacks it at the C=N$^+$ bond to make a neutral enamine. Looking at the product, you can see that another reduction must take place as well, but enamines are nucleophilic, so the borohydride can't atttack directly. What must happen instead is that the enamine is protonated to make another iminium, which can then be reduced. The final double bond is safe from attack, since it is an isolated, electron-rich alkene.

In the final step, another electrophile is added: it's an acid chloride, though a slightly unusual one, usually called 'methyl chloroformate'. As in the first step, the most nucleophilic atom is the pyridine N, so we use its lone pair to attack the carbonyl group and displace chloride. Now we have to lose the benzyl group, but the only reagent we have to help us is the chloride we just lost. Chloride is a nucleophile—a weak one, but powerful enough to attack the cationic species we have just generated. Which is the site most susceptible to nucleophilic substitution? The benzylic carbon, quite reasonably, because of the accelerating effect of the adjacent π system. Nucleophilic substitution here gives the final product.

■ See p. 431 of the textbook for a reminder about the reactivity of benzylic electrophiles.

PROBLEM 9

Explain the regioselectivity displayed in this synthesis of the drug tanomastat.

Purpose of the problem

Explaining why moderately complex molecules choose to react selectively.

Suggested solution

The first reaction is a Friedel-Crafts acylation. There are two rings and two carbonyl groups, so we must explain first of all the choice between each of these pairs. One ring is chlorinated: chlorine has a deactivating effect on electrophilic aromatic substitution, so the non-chlorinated ring reacts. The two carbonyls differ in that the top one is (a) less hindered and (b) not conjugated, both of which contribute to its greater reactivity. There is also the question of regioselectivity in the way that the acylation occurs at the *para* position of the non-chlorinated ring. Aryl substituents are *ortho,para* directing, because they can delocalize the positive charge formed from attack in this way; steric factors favour the *para* over the *ortho* positions.

■ The effect of halogens on the reactivity of aromatic rings is described on pp. 489–490 of the textbook.

■ The synthesis of tanomastat, a protease inhibitor, is from US patent 5,789,434.

In the second step, thiophenol gives the conjugate addition, rather than the direct addition product to either carbonyl group. Sulfur nucleophiles are soft, and this is typical behaviour for thiols.

PROBLEM 10

This compound is needed as a synthetic precursor to the drug etalocib. Suggest a synthesis. *Hint*: consider using nucleophilic aromatic substitution.

Purpose of the problem

Thinking about regioselectivity and reactivity in the synthesis of a moderately complex aromatic compound.

Suggested solution

There are lots of *ortho* relationships in this compound! And somehow we have to join the two aromatic rings together to make an ether. This can only really be done by nucleophilic aromatic substitution, so we need to look for an electron-withdrawing group to help us. The nitrile is in the right place, provided we have a leaving group (such as fluoride) *ortho* to it. So our last step can be as shown here:

■ See pp. 514–520 of the textbook for a reminder of nucleophilic aromatic substitution.

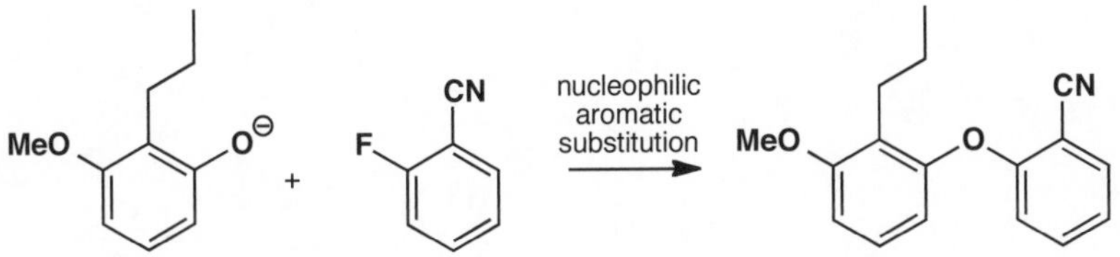

To make the left hand ring we have to consider what methods are available to introduce the three substituents we have. It's always easier to add C-substituents than O-substituents, so we might consider how to alkylate the phenol below. Both the OH and OMe groups are *ortho,para* directing, so we could consider a Friedel-Crafts reaction, but there are problems with this approach. One is the usual problem with primary alkyl groups—we would have to do an acylation and then reduce. Another is more serious: the less hindered positions the other side of the OMe or OH groups are also activated, so we will have a regioselectivity problem. The solution used by the chemists making this compound for the first time was to use ortholithiation, making the dianion with two equivalents of BuLi and making use of the fact that two O substituents guide the BuLi in to deprotonate the position between them.

■ The work is described by Sawyer *et al., J. Med. Chem.* 1993, **38**, 4411.

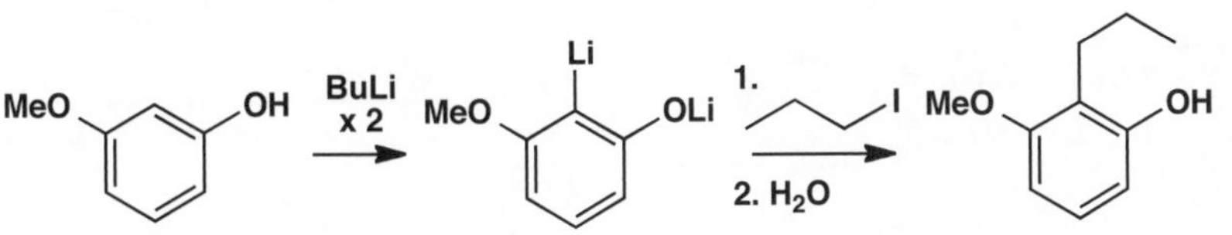

Suggested solutions for Chapter 25

25

PROBLEM 1

Suggest how these compounds might be made by alkylation of an enol or enolate.

Purpose of the problem

An exercise in choosing good routes to simple compounds.

Suggested solution

As you can see from the carbonyl groups in these compounds, it is pretty obvious which is the new bond to be made. In both cases, the electrophile will need to be an allylic halide. These are good electrophiles for S_N2 reactions so they will work well here. We need to use the electrophile twice in the first case and the enolate is that of diethyl malonate. The second case will require an enol or enolate equivalent to prevent self-condensation: a silyl enol ether (p. 595 in the textbook) or an enamine (p. 591 in the textbook) is ideal. If you use a silyl enol ether, don't forget the Lewis acid!

■ The reason why allylic halides make good electrophiles is discussed in the textbook on p. 341.

PROBLEM 2

How might these compounds be made using alkylation of an enol or enolate as one step in the synthesis?

Purpose of the problem

An exercise in using enolate chemistry to make carbonyl compounds disguised as acetals.

Suggested solution

■ See p. 227 of the textbook for a reminder of how to make cyclic acetals.

The only functional group in either compound is an acetal. Cyclic acetals are made from diols and carbonyl compounds so we need to have a look at the deprotected molecules before taking any further decisions.

If we are going to use enolate chemistry, we have to make the diols by reduction of carbonyl compounds. As both diols have a 1,3-relationship between the OH groups, the carbonyl precursors will be the very enolizable 1,3-dicarbonyl compounds, which can be alkylated and reduced. We have chosen arbitrarily to use ethyl esters here, so we should use ethoxide as the base in the alkylation step.

PROBLEM 3

How might these amines be prepared using enolate-style alkylation as part of the synthesis?

Purpose of the problem

An exercise in using enolates and related compounds in the synthesis of amines.

Suggested solution

The first amine could be made by reduction of a nitrile, and that could be made by alkylation of the 'enolate' from $PhCH_2CN$.

The second amine could be made by reductive amination of a ketone so we need to make the ketone by alkylation of an enolate. You could have chosen various specific enol equivalents for this job—we have chosen an enamine.

PROBLEM 4

This attempted enolate alkylation does not give the required product. What has gone wrong? What products would actually be formed? How would you make the required product?

Purpose of the problem

An exercise in trouble-shooting—it is important for you to recognize what might go wrong and how to get round the problem.

Suggested solution

The intention was obviously to make the lithium enolate of the aldehyde and to alkylate it with *i*-PrCl, but BuLi will attack the aldehyde carbonyl group rather than remove a proton. Even if it did make some of the enolate, the enolate would react with the aldehyde and self-condense (p. 590 in the textbook).

There is also a problem with *i*-PrCl: it is a secondary halide and chloride is the worst leaving group among the halogens Cl, Br, I—it is prone to elimination rather than substitution reactions. To make the required product, an aza-enolate (p. 593 in the textbook) or a silyl enol ether (p. 595 in the textbook) would be a better bet.

PROBLEM 5

Draw mechanisms for the formation of this enamine, its reaction with the alkyl halide, and the hydrolysis of the product.

Purpose of the problem

Exploration of the details of enamine formation and reaction. These are often misunderstood.

Suggested solution

The first step of the mechanism for enamine formation is not acid-catalysed—amines need no help in attacking carbonyl compounds. But the dehydration step is acid-catalysed as HO^- is not a good leaving group. The selectivity for elimination into the unbranched chain is because the enamine is planar and there would be a bad steric clash between the methyl group and the nitrogen substituents (all of which are in the same plane) if elimination occurred the other way.

■ The mechanism of enamine formation is given on p. 233 of the textbook.

The reaction of the enamine with the alkyl halide goes as expected—these very good S$_N$2 electrophiles work particularly well with enamines and the first product under the reaction conditions is another enamine.

Finally the enamine is hydrolysed by reprotonation to the same iminium salt and addition of water. These steps are the exact reverse of what happens in enamine formation.

PROBLEM 6

How would you produce specific enols or enolates at the points marked with the arrows (not necessarily starting with the ketones themselves)?

Purpose of the problem

First steps in making enol(ate)s with regiochemical control.

Suggested solution

The last two ketones have two different α-positions so there is a good chance of controlling enol formation from the parent ketone. But the first ketone has two primary α-positions and the difference appears only in the two β-positions. The obvious solution is conjugate addition and trapping (described in the textbook on p. 603). The thermodynamic enol is needed from the second ketone and direct silylation is a good bet. The third requires kinetic enolate formation and LDA is a good way to do that.

PROBLEM 7

How would the enol(ate) equivalents we have just made react with (a) bromine and (b) a primary alkyl halide RCH$_2$Br?

Purpose of the problem

Moving on from the formation of enol(ate)s to their reactions.

Suggested solution

The two silyl enol ethers will react well with bromine and won't need Lewis acid catalysis as bromine is such a powerful electrophile—so powerful that it might be dangerous to react the lithium enolate directly with bromine and making the silyl enol ether first might be advisable.

In the reaction with the primary alkyl halide, the boot is on the other foot as there will be a good reaction with the lithium enolate but no reaction with the more stable silyl enol ethers. Lewis acid won't help here either as primary cations are unstable. Preliminary conversion into a lithium enolate or a 'naked' enolate (using fluoride ion) would be better.

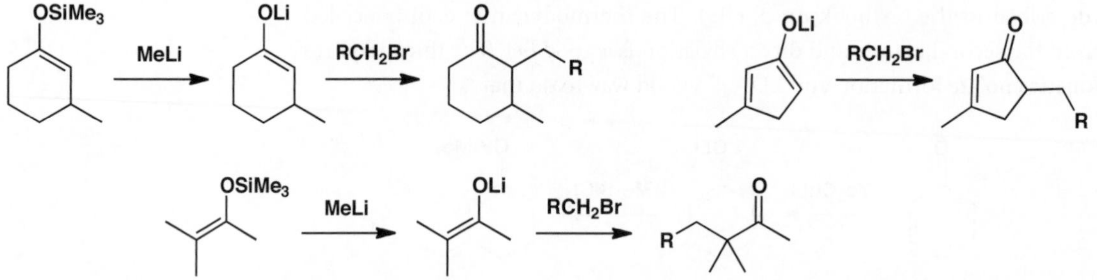

PROBLEM 8

Draw a mechanism for the formation of this imine:

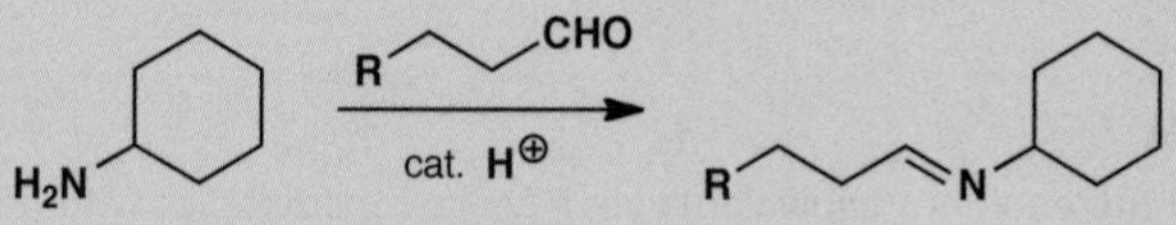

Purpose of the problem

Revision of the often forgotten mechanism for imine formation.

Suggested solution

■ The mechanism of imine formation is given on p. 230 of the textbook.

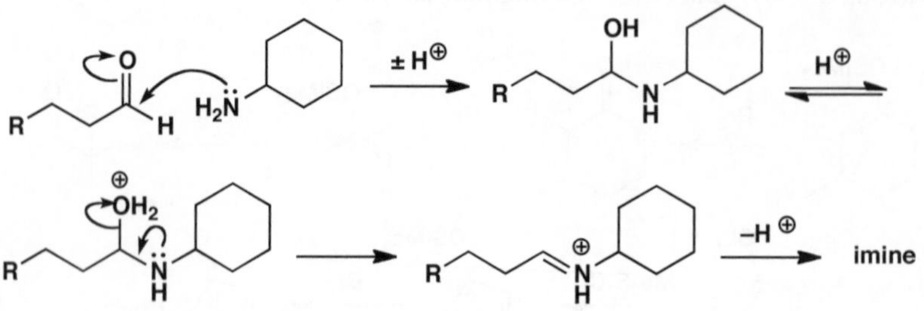

The main points in the mechanism are addition of the amine to the carbonyl group *without* catalysis and dehydration of the intermediate *with* acid catalysis.

PROBLEM 9

How would the imine from problem 8 react with the reagents below? Draw mechanisms for each step: the reaction with LDA, the addition of BuBr, and the work-up.

Purpose of the problem

Checking you know how to make and use an aza-enolate.

Suggested solution

LDA removes the most acidic proton of the imine so that the Li atom is transferred to the nitrogen atom to give the aza-enolate. Electrophiles, even alkyl halides, then add to the 'enolate' position and the work-up is hydrolysis of the imine with aqueous acid.

- Aza-enolate chemistry is described on p. 593 of the textbook.

- Imine hydrolysis is the reverse of imine formation and is discussed on p. 231 of the textbook.

PROBLEM 10

What would happen if you tried this short cut for the reactions in problems 8 and 9?

Purpose of the problem

Reminder of the problems with lithium enolates of aldehydes.

Suggested solution

Some aldehydes can be converted directly into lithium enolates but this is not usually very successful because the rate of reaction of the lithium enolate with the very electrophilic aldehyde is too great and at least some aldol reaction will occur.

■ The problem of aldol reactions competing with alkylations is mentioned in the green box on p. 584 and on p. 590 of the textbook. The aldol reaction itself is the subject of chapter 26.

PROBLEM 11

Suggest mechanisms for these reactions.

Purpose of the problem

Learning to unravel complicated looking sequences that are quite easy when you get into them.

Suggested solution

Double alkylation of the malonate enolate gives the four-membered ring and hydrolysis and decarboxylation gives the carboxylic acid product.

■ The hydrolysis-decarboxylation sequence is explained on p. 597 of the textbook.

PROBLEM 12

How does this synthesis of a cyclopropyl ketone work?

Purpose of the problem

Enols and enolates are involved in an unlikely looking sequence that you can work out if you persist.

Suggested solution

Alkylation of the enolate with the epoxide gives an alkoxide that cyclizes to give the lactone.

Now S_N2 opening of the protonated lactone with the soft nucleophile (bromide ion) gives the γ-bromoketone that cyclizes through its enolate. The formation of three-membered rings is favoured kinetically.

PROBLEM 13

Give the structures of the intermediates in the following reaction sequence and mechanisms for the reactions.

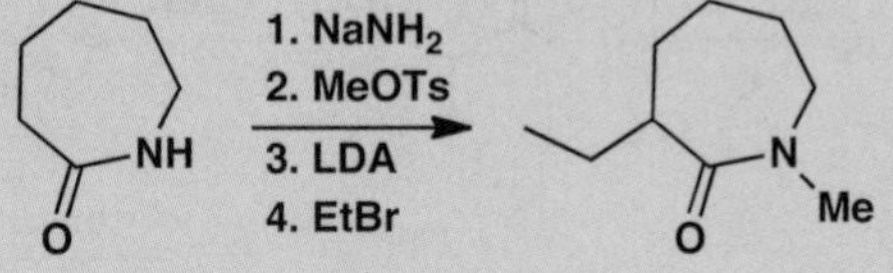

1. NaNH₂
2. MeOTs
3. LDA
4. EtBr

Purpose of the problem

A reminder that enolate-like intermediates can be formed at nitrogen as well as carbon providing that an oxygen atom can carry the negative charge.

Suggested solution

The first base removes the proton from nitrogen to make an enolate-like intermediate that reacts at nitrogen. Now that the NH is blocked, the second base makes the amide enolate that is alkylated on carbon.

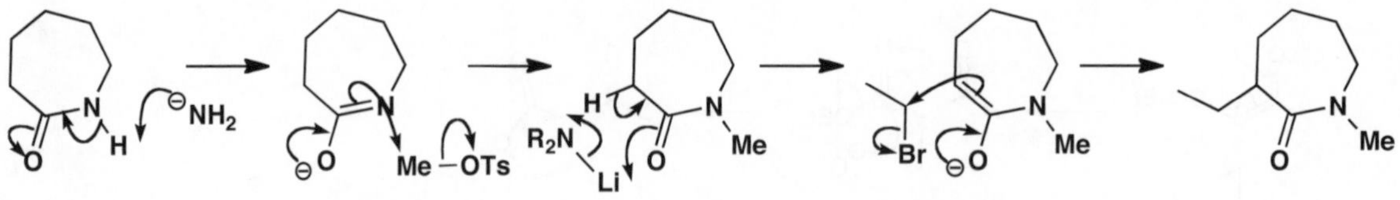

Suggested solutions for Chapter 26

26

PROBLEM 1

The aldehyde and the ketone below are self-condensed with aqueous NaOH so that an unsaturated carbonyl compound is the product in both cases. Give a structure for each product and explain why you think this product is formed.

Purpose of the problem

Drawing mechanisms for the simplest of aldols: self-condensation of aldehydes and ketones.

Suggested solution

In both cases only one compound can form an enolate and only one compound—the same one—can be the electrophile. This is very obvious with the aldehyde

With the ketone, there is a question of regioselectivity in enolate formation, but the aldol product can lose water only if the enolate from the methyl group is the nucleophile. If we draw both enolates and combine them with the ketone in an aldol reaction, it is clear that one can dehydrate as it has two enolizable H atoms but the other cannot dehydrate as it has no H atoms on the vital carbon atom (in grey). The mechanism is the same as the one with the aldehyde and the elimination in both cases is by the E1cB mechanism.

■ See p. 399 and p. 616 in the textbook for the E1cB mechanism.

can dehydrate

cannot dehydrate

PROBLEM 2

Propose mechanisms for the 'aldol' and dehydration steps in the termite defence compound presented on p. 623 in the textbook.

Purpose of the problem

Revision of elimination reactions and the mechanism for 'an aldol that can't go wrong.'

Suggested solution

■ See p. 587 in the textbook: a nitro group acidifies ajacent C–H bonds as much as two carbonyl groups.

The nitro group is twice as electron-withdrawing as a carbonyl group so it will readily form an 'enolate.' It cannot self-condense as nucleophilic attack rarely occurs on nitro groups so it attacks the aldehyde instead. Notice that the alkoxide product is basic enough to deprotonate another molecule of nitromethane so the reaction is catalytic in base.

■ E1cB elmination is on pp. 399 and 616 in the textbook.

The elimination step involves acylation of the hydroxyl group and an E1cB elimination again driven by the 'enolate' of the nitro group. Note that pyridine, a weak base, is strong enough.

PROBLEM 3

How would you synthesize the following compounds?

Purpose of the problem

Application of the aldol reaction to make unsaturated carbonyl compounds.

Suggested solution

Just find the conjugated alkene and so find the hidden carbonyl group. In the first case, cyclohexanone provides two enols to react with benzaldehyde. The phenyl rings in the product lie *trans* to the carbonyl group so that they can be planar.

In the second case, more options are available. Our solution suggests using a Wittig reaction for the first as we need the enolate of acetaldehyde (p. 628 in the textbook), and malonic acid for the second (p. 630 in the textbook). There are many alternatives such as using an aldol reaction for the first step, but with an excess of acetaldehyde, to compensate for self-condensation.

PROBLEM 4

How would you use a silyl enol ether to make this aldol product? Why is it necessary to use this particular intermediate? What would be the products be if the two carbonyl compounds were mixed and treated with base?

Purpose of the problem

Exploring control, and the lack of it, in different styles of aldol reaction.

Suggested solution

This is about the most difficult type of aldol reaction: two slightly different aldehydes, both enolizable, both capable of self-condensation. The only solution is to couple the silyl enol ether of one aldehyde with the other aldehyde using a Lewis acid as catalyst. This gives the aldol itself that can be dehydrated to the enal.

Without this control, each aldehyde would self-condense and would condense with the other aldehyde giving four products in unpredictable amounts. One of the cross-condensation products is, of course, the enal we are trying to make.

self-condensation   self-condensation   cross-condensation

+ the enal
required in
the problem

cross-condensation

PROBLEM 5

In what way does this reaction resemble an aldol reaction? Comment on the choice of base. How can the same product be made without using phosphorus chemistry?

Purpose of the problem

Showing that there are reactions closely related to the aldol reaction that give similar products.

Suggested solution

The formation of an alkene and the loss of phosphorus are typical of a Wittig reaction but the formation of an unsaturated carbonyl compound using an enolate is very like an aldol reaction. The phosphonate ester reagent is also like a 1,3-dicarbonyl compound, with P replacing C. The very weak base used shows how stable the enolate must be. The enolate attacks the aldehyde, perhaps to form an intermediate oxyanion.

■ This type of Wittig reaction was introduced on p. 628 of the textbook.

There is no doubt that the next intermediate is formed. It is a stable four-membered ring (phosphorus likes 90° bond angles). Finally phosphorus captures oxygen (the P–O bond is very strong) eliminating the alkene in its preferred *trans* stereochemistry.

The final product could also be made by the aldol condensation of a silyl enol ether and the same aldehyde. The silyl enol ether is the less substituted possibility so it will have to be made via the lithium enolate. The product will be the aldol itself and this can be dehydrated to the enone with TsOH.

PROBLEM 6

Suggest a mechanism for this attempted aldol reaction. How could the aldol product be made?

expected aldol product

Purpose of the problem

A demonstration of one way that aldol reactions with formaldehyde may fail.

Suggested solution

The aldol reaction appears to have taken place and then the ketone has been reduced. The only possible reducing agent is more formaldehyde and the reduction takes place by the Cannizarro reaction (p. 620 in the textbook). The aldol can be successful if a weaker base such as Na_2CO_3 is used as the Cannizarro requires a dianion intermediate.

PROBLEM 7

The synthesis of six-membered ketones by intramolecular Claisen condensation was described in the chapter where we pointed out that it doesn't matter which way round the cyclization happens as the product is the same.

Strangely enough, five-membered heterocyclic ketones can be made by a similar sequence. The starting material is not symmetrical and two cyclized products are possible. Draw structures for these products and explain why it is unimportant which is formed.

Purpose of the problem

To make sure you understand how extra ester groups can solve apparently complex acylation problems.

Suggested solution

The cyclization can occur in two different ways to give two different products as either ester can form an enolate that attacks the other in an intramolecular acylation. We should draw the two products.

Though these compounds are different, each gives the same ketone after hydrolysis and decarboxylation as the ketone carbonyl group is on the same position in the ring in both compounds.

PROBLEM 8

Attempted acylation at carbon often fails. What would be the actual products of these attempted acylations and how would you successfully make the target molecules?

Purpose of the problem

Revision of simple enolate reactions (chapter 20) and encouragement to clear thinking about what happens when you put carbonyl compounds in basic solutions.

Suggested solution

In the first case we want the aldehyde to form an enolate and then attack the ester. The first part is all right: the aldehyde will form an enolate more readily than the ester. But under these equilibrating conditions, the small amount of enolate that is formed will react faster with the aldehyde than with the less electrophilic ester. The aldehyde will self-condense in an aldol reaction.

To make the required compound we shall need to convert the aldehyde into a specific enol equivalent. There are various alternatives of which the best are an enamine or a silyl enol ether. Esters fail to acylate either and an acid chloride should be used instead. Don't forget the Lewis acid if you use the silyl enol ether.

The enolate formation in the second example is a separate step and will work well because the two carbonyl groups cooperate in forming a stable enolate and NaOMe is quite strong enough to convert the diketone entirely into the enolate. The problem is the acylation step. With a sodium enolate and a reactive acylating agent such as PhCOCl, a charge-controlled (hard/hard) interaction will occur at the oxygen atom to give an enol ester.

■ Reactions of enolate at oxygen and the role of hard and soft reagents are discussed on p. 467 of the textbook. Acylation at O appears on p. 648.

The escape route from this problem suggested in the chapter (p. 648) was to use a lithium or magnesium enolate. Magnesium is chelated by the two oxygen atoms of the stable enolate and blocks attack there so that C-acylation occurs even with acid chlorides.

PROBLEM 9

Acylation of the phenolic ketone gives compound **A**, which is converted into an isomeric compound **B** in base. Cyclization of **B** gives the product shown. Suggest mechanisms for the reactions and structures for **A** and **B**.

Purpose of the problem

Predicting products of acylation reactions. This is always more difficult than just drawing mechanisms but here you might work backwards from the final product as well as forwards.

Suggested solution

The starting material is $C_8H_8O_2$ so **A** has an extra C_7H_4O. This looks like the addition of PhCOCl with the loss of HCl. The most obvious reaction is acylation of the phenolic oxygen rather than enolate formation as OH is much more acidic than CH and pyridine is a weak base. This phenol is unusually acidic as the carbonyl group helps to stabilize the anion. Compound **A** is simply the benzoate ester of the phenol. Treatment with KOH isomerizes **A** to **B** and this is the heart of the problem. An intramolecular acylation of the only possible enolate can be catalysed by KOH even though it produces only a little enol as cyclization to form a six-membered ring is so easy.

The final step is acid-catalysed and clearly involves the attack of the phenolic OH group on one of the ketones. This intramolecular reaction much prefers to form a six-membered ring rather than a strained four-membered ring, and dehydration gives an aromatic ring—two electrons each from the double bonds and two from a lone pair on oxygen making six in all. Drawing the delocalization my help you to see this.

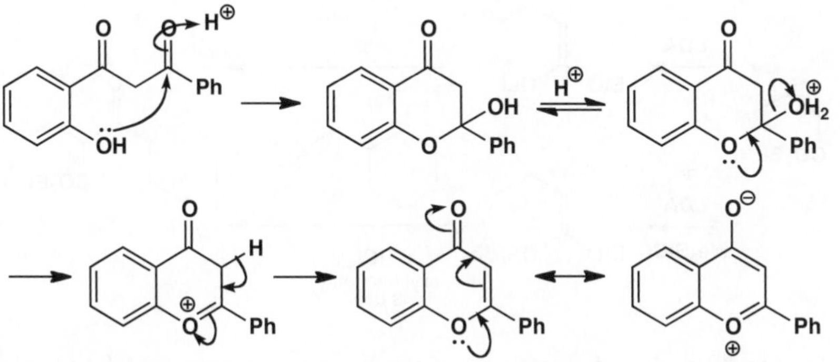

PROBLEM 10

How could these compounds be made using the acylation of an enol or enolate as a key step?

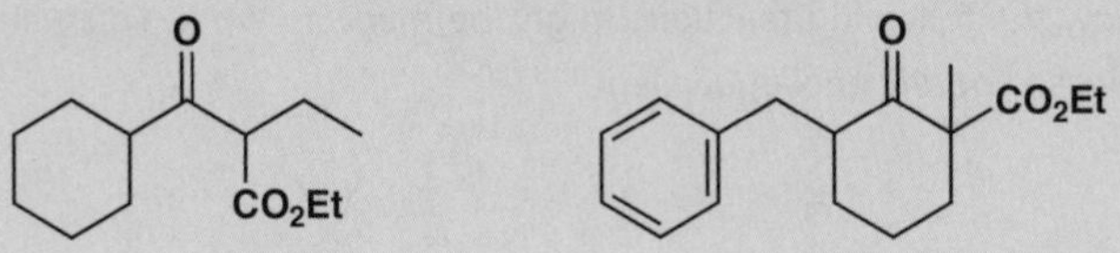

Purpose of the problem

Practice in using acylation at carbon to make compounds.

Suggested solution

The first problem has two possible solutions by direct acylation, labelled A and B in the diagram. A would have to be controlled as the straight chain ester could self-condense. B needs no control as only the ketone can enolize. Diethyl carbonate $(EtO)_2CO$ is more electrophilic than a ketone and only the wanted product can enolize again and form a stable enolate under the reaction conditions. However, route B adds only one carbon atom.

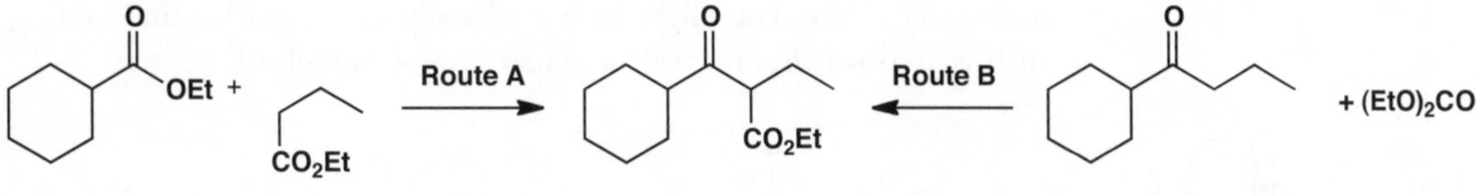

Route A can be realized with either a lithium enolate or a silyl enol ether, as explained on p. 649 of the textbook, using an acid chloride as the electrophile.

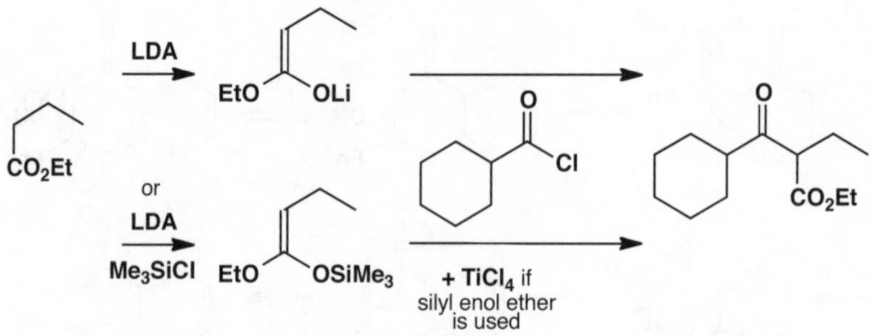

Route B requires the synthesis of the ketone starting material and this could be done by Grignard methods (chapter 9) or by acylation of an organo-copper compound with an acid chloride. Acylation with diethyl carbonate requires no special control.

PROBLEM 11

Suggest how the following reactions might be made to work. You will probably have to select a specific enol equivalent.

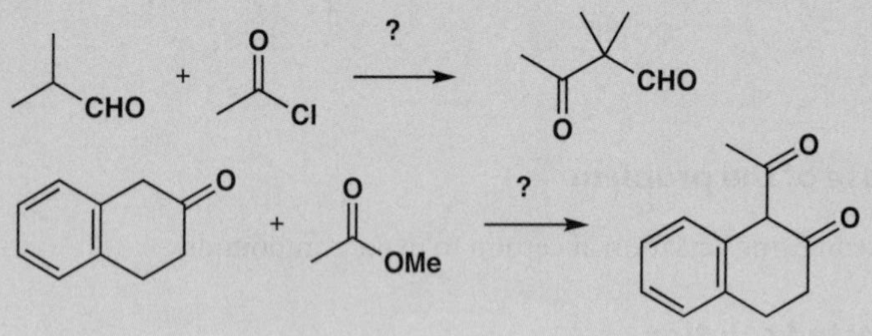

Purpose of the problem

Making reactions work is an important part of organic chemistry.

Suggested solution

The first reaction is a standard acylation of an aldehyde creating a quaternary centre. You might have used a silyl enol ether but an enamine, such as one made from a cyclic secondary amine, is probably better.

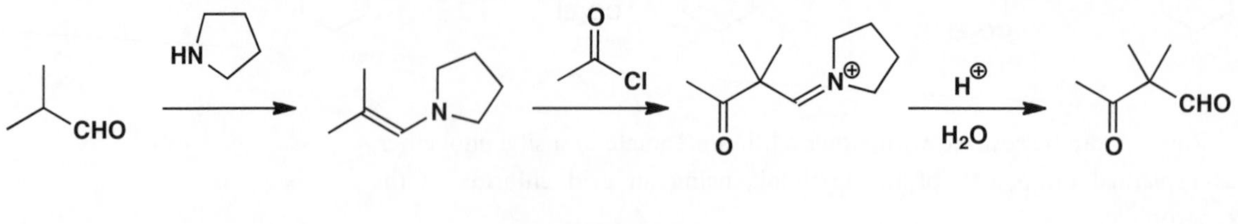

The second example might just go with simple base (MeO⁻) catalysis as the conjugated ketone enolate is much more stable than the enolate of the ester. However, it's probably safer to use a lithium enolate (or a silyl enol ether—though you'd then have to use an acid chloride as the electrophile).

PROBLEM 12

Base-catalysed reaction between these two esters allows the isolation of a product **A** in 82% yield.

The NMR spectrum of this product shows that two species are present. Both show two 3H triplets at about $\delta_H = 1$ and two 2H quartets at about $\delta_H = 3$ ppm. One has a very low field proton and an ABX system at 2.1–2.9 with J_{AB} 16 Hz, J_{AX} 8 Hz, and J_{BX} 4 Hz. The other has a 2H singlet at 2.28 and two protons at 5.44 and 8.86 coupled with J 13 Hz. One of these protons exchanges with D_2O. Any attempt to separate the mixture (for example by distillation or chromatography) gives the same mixture. Both compounds, or the mixture, on treatment with ethanol in acid solution give the same product B.

Compound B has IR 1740 cm⁻¹, δ_H 1.15–1.25 (four t, each 3H), 2.52 (2H, ABX system J_{AB} 16 Hz), 3.04 (1H, X of ABX split into a further doublet by J 5 Hz), and 4.6 (1H, d, J 5 Hz). What are the structures of **A** and **B**?

Purpose of the problem

Revision of enol structure by NMR and a further exploration of what happens to acylation products.

■ The couplings between H_A and H_X and between H_B and H_X are not quoted in the paper, but this should not prevent you identifying **B**. AB and ABX systems are discussed on pp. 296–8 of the textbook

Suggested solution

Only the diester can form an enolate and ethyl formate (HCO_2Et—it is half an ester and half an aldehyde) is much more electrophilic than the diester. We should expect the diester to be acylated by ethyl formate.

The compound **A1** fits the formula for **A** and the ^1H NMR spectrum of the compound with the low field signal (assigned to the CHO proton). This structure would also show an ABX system in its ^1H NMR spectrum. But what is the other compound (**A2**)? It is obviously in equilibrium with **A1** and it lacks both the aldehyde proton and the ABX system and it sounds like an enol. Compound **A1** is chiral so the CH_2 group appears as an ABX system but **A2** is not chiral so the CH_2 group is a singlet. Here are the structures with their NMR assignments. In both cases the 3H triplets and 2H quartets are ethyl groups.

Treatment with acidic ethanol simply makes the acetal from the aldehyde group of **A1**. Since **A1** and **A2** are in equilibrium, all **A2** is eventually converted into **A1** and then into **B**. Compound **B** is again chiral so the ABX system reappears with further coupling of X with the acetal proton. There are now four triplets and four quartets from the four ethyl groups.

Purpose of the problem

The opportunity to explore the consequences of the intramolecular version of an important reaction.

■ The original work can be found in R. S. Matthews and T. E. Meteyer, *J. Chem. Soc., Chem. Commun*, 1971, 1536.

Suggested solution

The ylid forms in the usual way but can't reach across the ring to attack the carbonyl group directly so it has to do conjugate addition instead. It also has to attack from the top face as it is tethered there. Completion of the cyclopropane forming reaction leaves the sulfur still attached to the angular methyl group. Raney nickel reduces the C–S bond (this reagent is commonly used for this purpose). This reaction shows that simple sulfonium ylids can do conjugate addition—they just prefer to add to carbonyl groups if that possibility is available.

PROBLEM 2

Explain the regiochemistry and stereochemistry of this reaction.

Purpose of the problem

Exploration of sulfur ylid chemistry.

Suggested solution

The ylid is stabilized by conjugation with the ester group—you can think of it also as an enolate. We can expect reversible addition to the carbonyl group and hence conjugate addition under thermodynamic control. The stereochemistry of the ring junction is inevitable: only a *cis* ring can be made (a *trans*-fused ring would be too strained). The interesting centre is that of the ester on the three-membered ring. It too is in a more stable configuration: on the outside of a folded molecule. The intermediate is probably a mixture of diastereoisomers, but as the conjugate addition is reversible the cyclopropane may be formed by cyclization of only the diastereoisomer that can give the more stable product.

■ Stereochemistry and stereoselectivity in fused rings is discussed in detail in chapter 32 of the textbook.

PROBLEM 3

Give mechanisms for these reactions, explaining the role of sulfur.

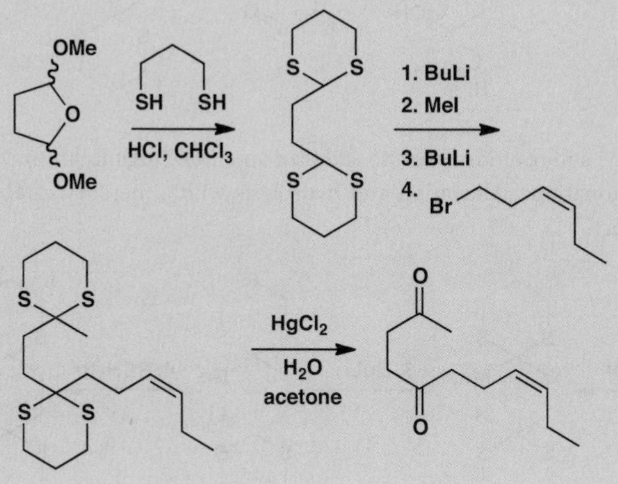

Purpose of the problem

Sulfur acetals as good nucleophiles: in the terminology of chapter 28, 'acyl anion equivalents' or 'd¹ reagents'.

Suggested solution

The first reaction is an acetal exchange controlled by entropy: three molecules go in and four come out (the product, two molecules of methanol and one of water). We show just part of the mechanism.

■ This chemistry was used to make the perfume *cis*-jasmone by R. A. Ellison and W. D. Woessner, *J. Chem. Soc., Chem. Commun.*, 1972, 529.

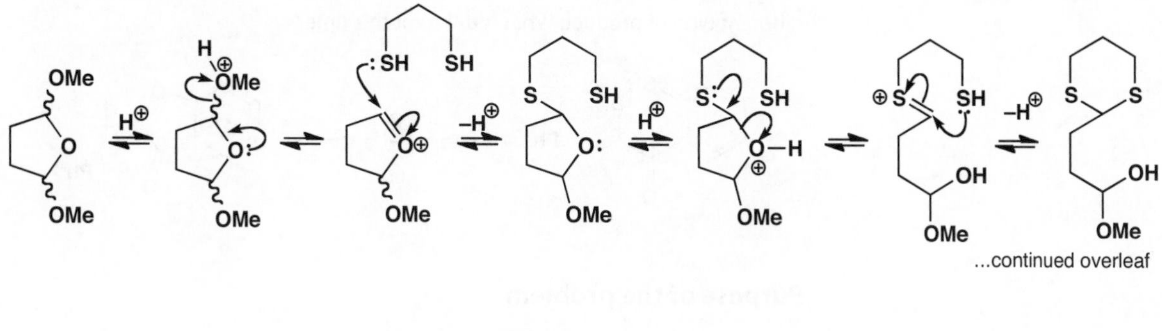

...continued overleaf

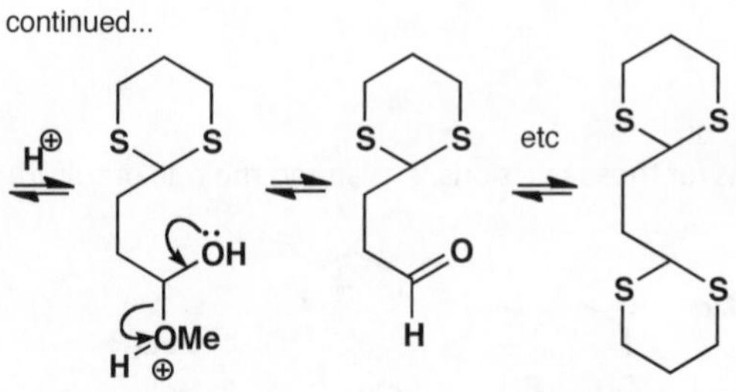

continued...

Now the sulfur atoms work to stabilize an anion (organolithium) formed by deprotonation. Alkylation and hydrolysis with a mercury catalyst gives the product.

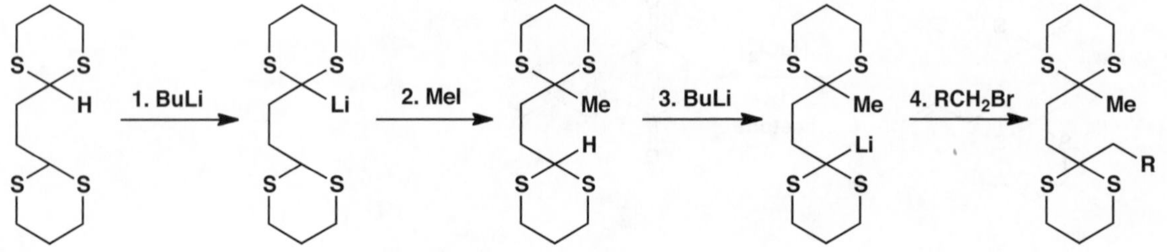

PROBLEM 4

Suggest a mechanism by which this cyclopropane might be formed.

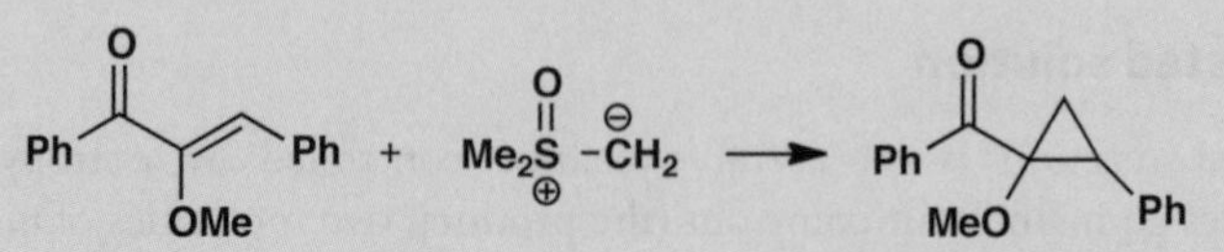

Attempts to repeat this synthesis on the related compound below led to a different type of product. What is different this time?

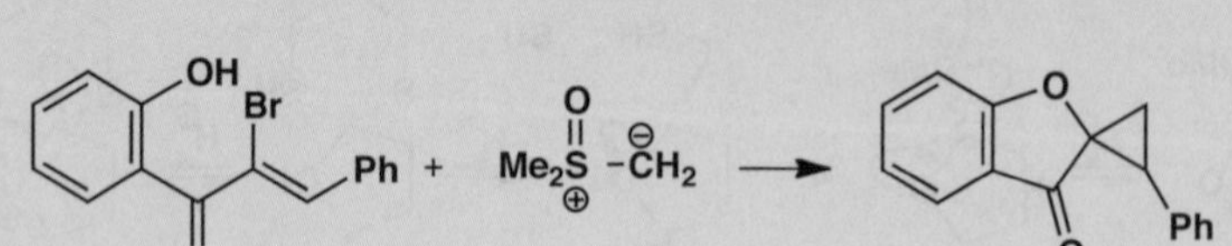

Purpose of the problem

Can you disentangle a curious variation on a simple mechanism?

Suggested solution

The first reaction is a straightforward cyclopropane formation with a sulfoxonium ylid and a conjugated ketone. The only unusual feature, the MeO group, makes no difference.

In the second example, the bromine atom and the phenolic OH evidently *do* make a difference. No doubt the reaction starts in the same way and a cyclopropane is formed. Under the reaction conditions, the phenol will exist as an anion and this displaces the bromine. This unusual S_N2 reaction at a tertiary centre is possible because of activation by the carbonyl group.

■ The accelerating effect of the carbonyl group on S_N2 reactions is demonstrated on p. 342 of the textbook.

PROBLEM 5

Deduce the structure of the product of this reaction from the NMR spectra and explain the stereochemistry. Compound **A** has δ_H 0.95 (6H, d, *J* 7 Hz), 1.60 (3H, d, *J* 5), 2.65 (1H, double septuplet, *J* 4 and 7), 5.10 (1H, dd, *J* 10 and 4), and 5.35 (1H, dq, *J*, 10 and 5).

Purpose of the problem

A simple way to make Z-alkenes, with a bit of NMR revision.

Suggested solution

This is obviously a Wittig reaction and we should expect a Z-alkene as the ylid is not stabilized by further conjugation. The evidence is plain: the signals at 5.10 and 5.35 are the alkene hydrogens and the coupling constant between them is 10 Hz. This is definitely a Z-alkene.

PROBLEM 6

A single geometrical isomer of an insect pheromone was prepared in the following way. Which isomer is formed and why?

Purpose of the problem

Testing your knowledge of the stereochemistry of the Wittig reaction.

Suggested solution

The first Wittig with a stabilized ylid gives the *E*-enal **A**. The second, with an unstabilized ylid, gives a *Z*-alkene so the final product is an *E,Z*-diene.

PROBLEM 7

How would you prepare samples of both geometrical isomers of this compound?

Purpose of the problem

A simple stereocontrolled alkene synthesis but both isomers are needed.

Suggested solution

There are many methods that can be used to tackle this question. The only snags are protecting the OH group if necessary and care in isolating the Z-compound as it may isomerize easily to the E-compound by reversible conjugate addition. One way to the Z-alkene uses reduction of an alkyne to control the stereochemistry. The OH group is protected as a benzyl ether removed by hydrogenation, perhaps under the same conditions as the reduction of the alkyne.

The E-alkene might be produced by reduction of the alkyne with an alkali metal in liquid ammonia but a Wittig reaction is probably easier. Either a phosphonium ylid or a phosphonate ester could be used. Protection of the alcohol as an ester allows hydrolysis of both esters in one step.

PROBLEM 8

Which alkene would be formed in each of these elimination reactions? Explain your answer mechanistically.

Purpose of the problem

Revision of the three main methods for stereoselective (or stereospecific) alkene bond formation.

Suggested solution

The first is sort of a Wittig reaction (the starting material is made by opening an epoxide with Ph$_3$P), the second a Julia reaction and the third and the fourth are Peterson reactions under different conditions. Each is described in detail in chapter 27 of the textbook. The Wittig reaction is under kinetic control and is a stereospecific *cis* elimination. In this case the product is a Z-alkene.

The Julia reaction is under thermodynamic control as equilibration occurs under the reaction conditions. The stereoselective product is the *E*-alkene.

The Peterson reaction is a *syn*-elimination under basic conditions, giving the Z-alkene from this starting material, but an E2 *anti*-elimination under acidic conditions, giving the *E*-alkene from this starting material.

PROBLEM 9

Give mechanisms for these reactions, explaining the role of silicon.

Purpose of the problem

Reminder of the anion-stabilizing role of sulfones and the excellence of the mesylate leaving group plus the special role of fluoride as a nucleophile for silicon.

Suggested solution

Sodium hydride removes a proton from the sulfone to give an anion that can act as a nucelophile. Displacement of mesylate gives an allyl silane, which is converted into an allylic anion by fluoride. Addition to the ketone gives a 5/5 fused system with the more stable *cis* ring junction.

PROBLEM 10

Give mechanisms for these reactions, explaining the role of silicon. Why is this type of lactone difficult to make by ordinary acid- or base-catalysed reactions?

Purpose of the problem

Basic organosilicon chemistry: the Peterson reaction and allyl silanes as nucleophiles.

Suggested solution

Acylation of the Grignard reagent is followed by a second attack on the ketone as expected but the tertiary alcohol is a Peterson intermediate and eliminates to give the alkene.

Now a Lewis acid catalysed reaction of the allyl silane via a β-silyl cation gives the lactone. The double bond in these 'exo-methylene' lactones easily moves into the ring in acid or base so mild conditions are ideal for these reactions.

PROBLEM 11

How would you carry out the first step in this sequence? Give a mechanism for the second step and suggest an explanation for the stereochemistry. You may find that a Newman projection helps.

Purpose of the problem

An important way to make an allyl silane and an important reaction of the product.

Suggested solution

The best route to the allyl silane is the Wittig reaction (p. 675 of the textbook). The ylid is not stabilized by extra conjugation so the Z-isomer is favoured.

The reaction with EtAlCl$_2$ is a Lewis acid-catalysed conjugate addition of the allyl silane to the enone. Conjugate addition is preferred because the nucleophile (the allyl silane) is tethered to the electrophile (enone) and the five-membered ring is preferred to a seven-membered ring.

The stereochemistry comes from the way the molecule prefers to fold and the Newman projection below should make that clear. The hydrogen atom

on the allyl silane tucks underneath the six-membered ring while the double
bond of the allyl silane projects out into space to give the stereochemistry
found in the product. The ratio between this diastereoisomer and the other
varies from 2:1 to 7.5:1 depending on conditions so the preference is really
quite weak.

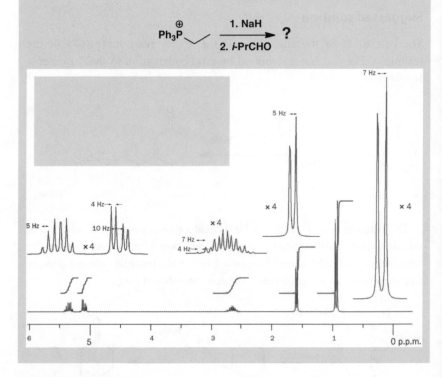

PROBLEM 12

The following reaction between a phosphonium salt, base, and an aldehyde gives
a hydrocarbon C_6H_{12} with the 200 MHz ^1H NMR spectrum shown. Give a structure
for the product and comment on its stereochemistry.

Purpose of the problem

Confirming the stereochemistry of the product from a Wittig reaction.

Suggested solution

We'll approach this as a spectroscopic problem, rather than predicting the outcome and then making the data fit. First analyse the data, measuring chemical shifts, integrals and J values.

δ_H ppm	integral	multiplicity	J values Hz	comments
0.97	6H	d	7	CHMe$_2$
1.60	3H	d	5	MeCHX
2.70	1H	double septuplet	7,4	Me$_2$CH-CH
5.15	1H	dd	10, 4	alkene
5.35	1H	1:3:4:4:3:1?	5?	alkene

From this alone we can see an alkene with two vicinal Hs, a methyl group, and an isopropyl group. That adds up to C$_6$H$_{12}$ so we have found everything and we can join it up in two ways as the alkene could be *cis* or *trans*. There is also a puzzle over the J values—there are too many 5 Hz couplings and the second 10 Hz coupling is missing, but we'll unravel that later.

The isopropyl group contains a 7 Hz coupling between the two methyl groups and the H at 2.70 ppm which is coupled to one of the alkene protons with $J = 4$ Hz. The remaining coupling of the alkene proton at 5.15 ppm (10 Hz) must be to the other alkene proton and that fits with a *cis* double bond. Working from the other end, the methyl group at 1.60 ppm is coupled to the alkene proton at 5.35 Hz with $J = 5$ Hz. Now we can solve the coupling constant mystery. We know that the alkene proton at 5.35 is actually a double quartet but it just so happens that the double coupling is exactly twice the quartet coupling. So we have two 1:3:3:1 quartets overlapping so that the inner lines coincide and give six lines in the ratio 1:3:4:4:3:1. The compound is *cis* (Z) 4-methylpent-2-ene. This is a Wittig reaction with an unstabilized ylid, so you should expect to find a *cis* double bond.

Suggested solution

1. ...

28

Suggested solutions for Chapter 28

PROBLEM 1

How would you make these four compounds? Give your disconnections, explain why you chose them and then give reagents for the synthesis.

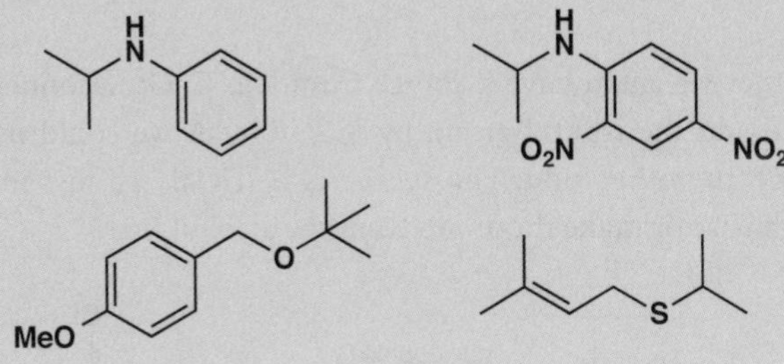

Purpose of the problem

Exercises in basic one-group C–X disconnections.

Suggested solution

We wish to disconnect one of the C–N bonds and prefer the one not to the benzene ring as we aim to use reductive amination (p. 234 of the textbook) as the best way to make amines.

analysis

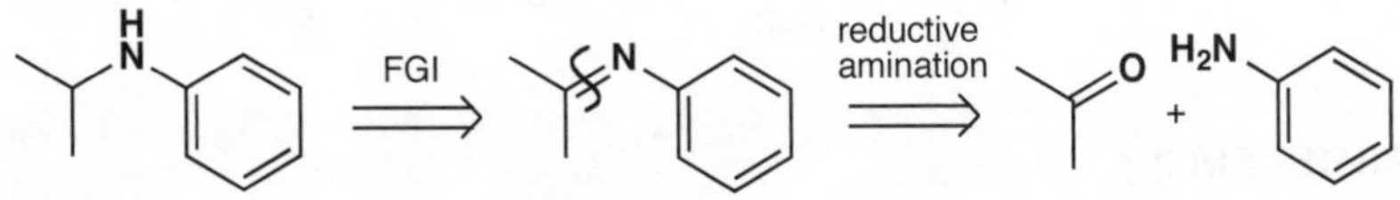

synthesis

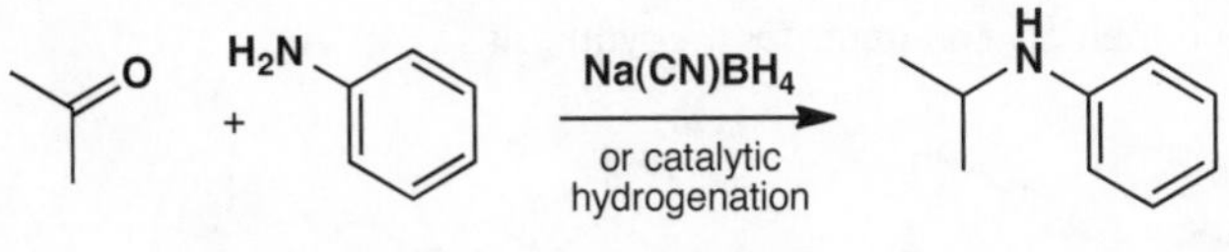

However the second aromatic amine can be made a different way. The two nitro groups promote nucleophilic aromatic substitution (p. 514 of the textbook) and the compound can be made by the addition-elimination mechanism from the dinitro chloro compound that can be made by direct nitration.

analysis

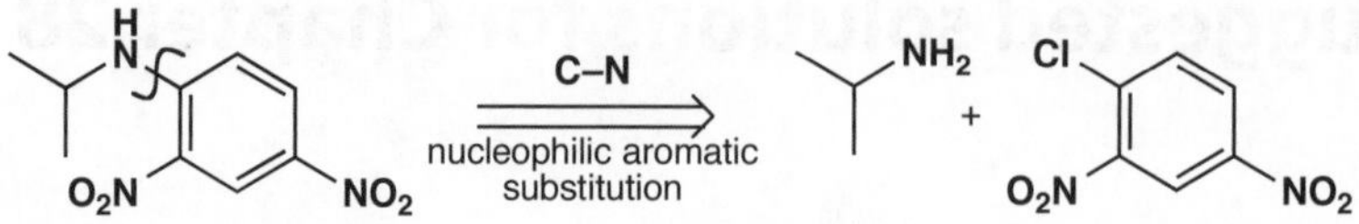

synthesis

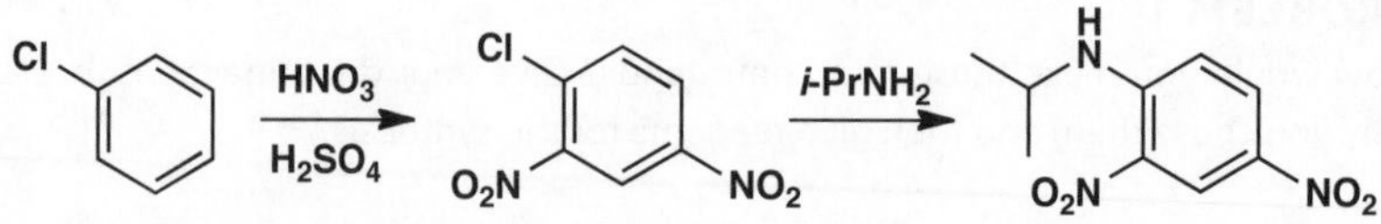

For the ether we again have a choice from two C–O disconnections. We prefer not to add the *t*-butyl group by S_N2 (though we could by S_N1) and disconnect on the other side. The synthesis is trivial: we just mix the two reagents with base or make the anion from the alcohol first.

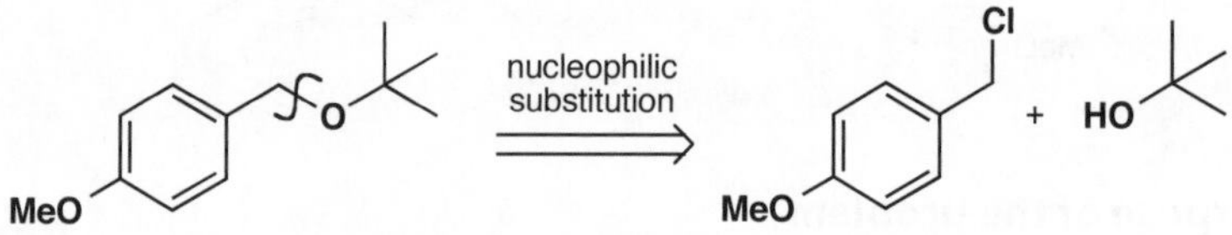

For the sulfide we shall want to use an S_N2 reaction and there is a slight preference for the disconnection we show as the allylic halide is very reactive. You would not be wrong if you had chosen the alternative C–S bond. This time only a weak base will be needed as the SH group is much more acidic than the OH group.

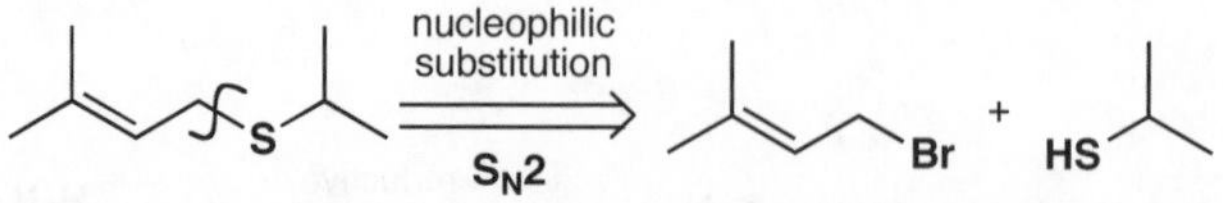

PROBLEM 2

How would you make these compounds? Give your disconnections, explain why you chose them and then give reagents for the synthesis.

Purpose of the problem

Exercises in basic one-group C–C disconnections.

Suggested solution

There are obviously more choices when you use C–C disconnections, but choose wisely! We suggest a solution, but you may have thought of others. The first compound is an alkyne and disconnection next to the alkyne (but not on the side of the benzene ring) makes a simple synthesis.

The alcohol has some symmetry: you will want to use Grignard or organolithium chemistry (chapter 9) and you could disconnect one or two of the identical groups using a ketone or an ester as the electrophile. The double disconnection leads to a shorter synthesis.

PROBLEM 3

Suggest ways to make these two compounds. Show your disconnections and make sure you number the functional group relationships.

Purpose of the problem

First steps in using two functional groups to design a synthesis.

Suggested solution

Both compounds have two oxygens singly bonded to the same carbon atom: they are acetals so they come from a carbonyl compound. Disconnecting the acetals helps us see what we are really trying to make.

The diol has a 1,3-relationship between the two alcohols so we need aldol or Claisen ester chemistry (chapter 26). One alcohol will have to be changed into a carbonyl group, perhaps an aldehyde or ester. Since we shall reduce all carbonyl groups to alcohols, it doesn't really matter whether we have aldehydes, ketones, or esters.

We prefer to make the disconnection between C2 and C3 to cut the molecule more or less in half and simplify the problem. There are various ways to do this—either the lithium or the zinc enolate would do, and below we show the use of zinc in a Reformatsky reaction.

If the keto-ester is used as a starting material it can be made by the same strategy (disconnection A) or alternatively (disconnection B) by first removing just one methyl group to reveal a symmetrical keto-ester made by a Claisen ester condensation.

Disconnection A

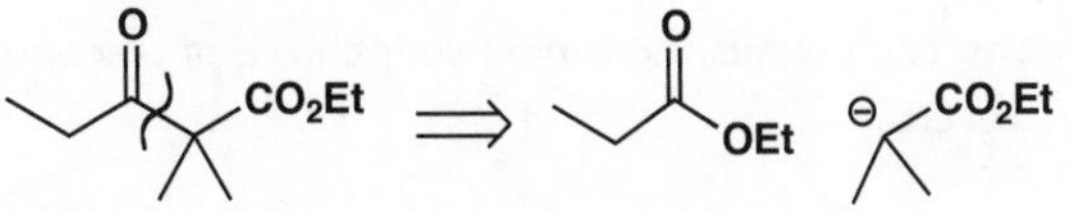

Disconnection B

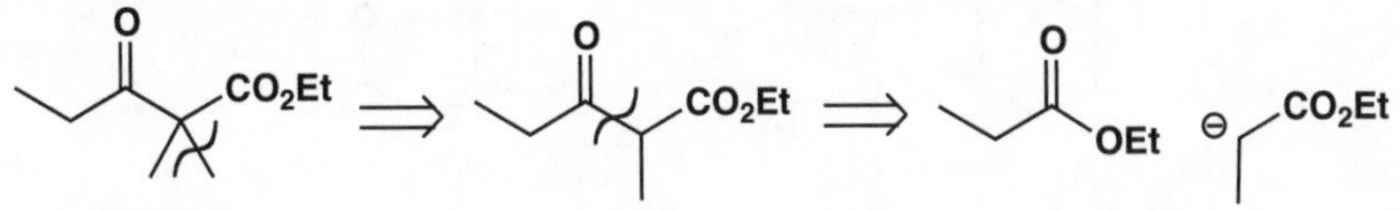

The advantage of disconnection B is that the synthesis involves a simple self-condensation of ethyl propionate. Methylation of the resulting keto-ester followed by reduction to the diol and acetal formation gives the target molecule.

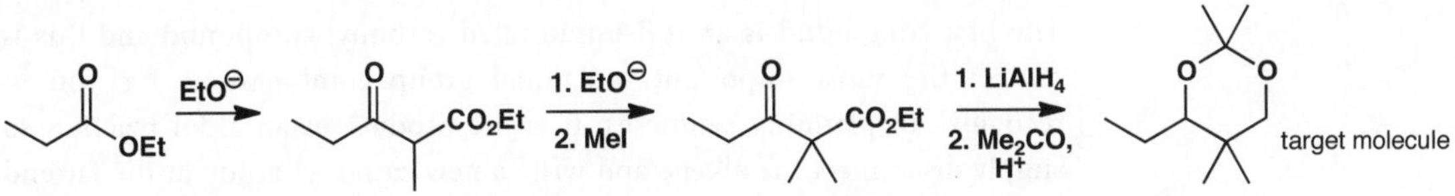

The other compound has a 1,5-relationship between the two functional groups and will need some sort of conjugate addition of an enolate (chapter 25). This time we want to reduce only one of the two carbonyl groups so we must make sure they are different. We already have an aldehyde so we choose an ester for the other one.

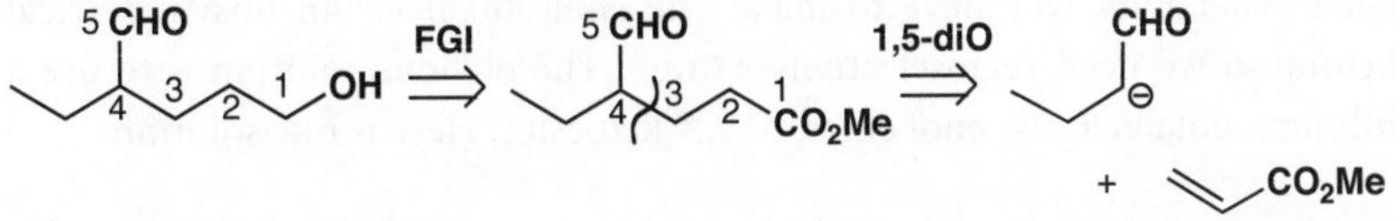

We must use a specific enol equivalent for the aldehyde enolate to avoid self-condensation: an enamine or a silyl enol ether would be fine. Since we must reduce the ester in the presence of the aldehyde, it makes sense to put the acetal in before we do this.

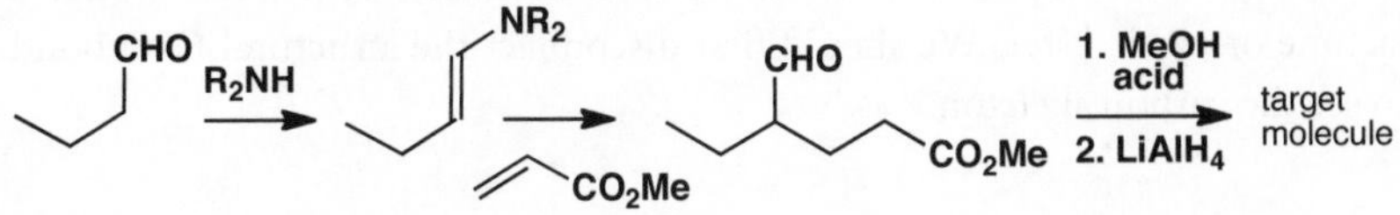

PROBLEM 4

Propose syntheses of these two compounds, explaining your choice of reagents and how any selectivity is achieved.

Purpose of the problem

First steps in designing syntheses in which selectivity is required.

Suggested solution

The first compound is an α,β-unsaturated carbonyl compound and this is one of the most important functional group combinations for you to recognize in planning syntheses. It is the product of an aldol reaction so simply disconnect the alkene and write a new carbonyl group at the far end of the old one. Don't lose any carbon atoms!

We need a crossed aldol reaction between two ketones so we also need chemoselectivity. We have to make one enol(ate) from an unsymmetrical ketone so we need regioselectivity as well. The obvious solution is to use a lithium enolate, a silyl enol ether, or a β-ketoester. Here is one solution.

The second compound contains another common functional group: a lactone or cyclic ester. We should first disconnect the structural C–O bond to see the carbon skeleton.

We discover that we have a 1,5-relationship between the functional groups and so we shall need conjugate addition. We must change the alcohol into a ketone, and the acid group to an ester. Notice that there are two reasonable disconnections and that we have added an ester group to each potential enolate as the way of making a specific enolate.

One possibility is to add malonate to the unsaturated ketone, which is an aldol dimer of acetone and readily available. We can reduce the ketone, expecting cyclization to be spontaneous, and decarboxylate to give our target molecule.

■ The strategy of using decarboxylation of a β-dicarbonyl compound is described on p. 597 of the textbook.

PROBLEM 5

The reactions to be discussed in this problem were planned to give syntheses of these three molecules.

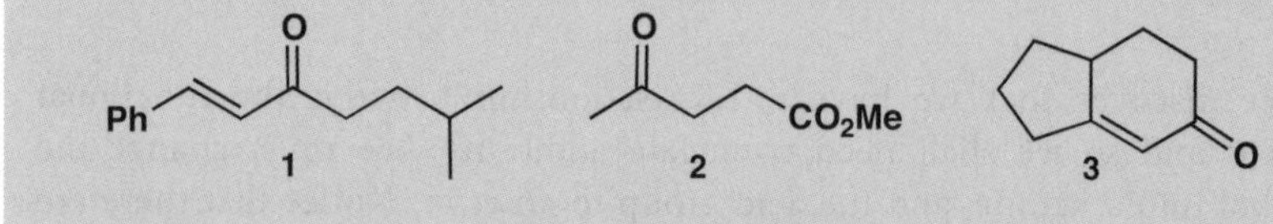

In the event each reaction gave a different product from what was expected, as shown below. What went wrong? Suggest syntheses that would give the target molecules above.

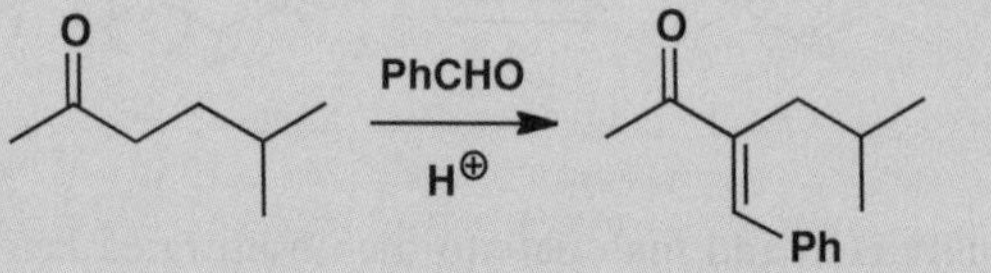

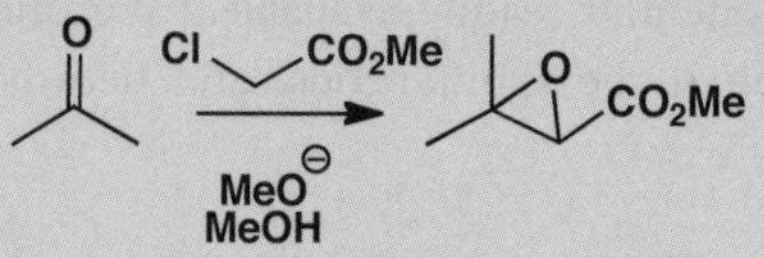

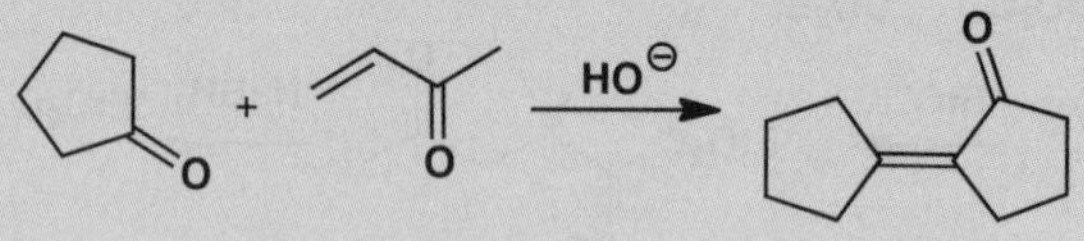

Purpose of the problem

Finding out what might go wrong is an important part of planning a synthesis.

Suggested solution

The aldol reaction planned for target molecule **1** looks all right but enol formation has occurred on the wrong side. This is not surprising in acid solution, so use base instead.

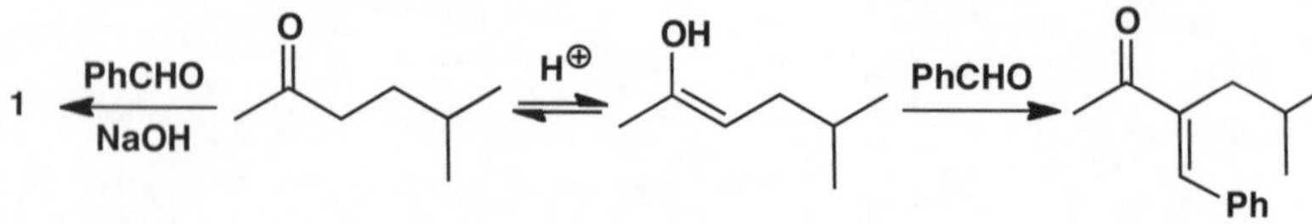

In the second case, alkylation of the enolate of the ketone was planned but evidently it is easier to form the enolate of the chloro-ester. The reaction that occurred is the Darzens condensation. To avoid this problem use a specific enolate of the ketone such as an enamine or a β-ketoester.

■ The Darzens condensation is on p. 639 of the textbook.

In the third case, the cyclopentanone has self-condensed and ignored the enone. The answer again is to use a specific enolate, such as the easily made β-keto-ester below. The six-membered ring is then easily formed by intramolecular aldol reaction. These two reactions together make a Robinson annelation. Finally the CO₂Me group must be removed by hydrolysis and decarboxylation.

■ The Robinson annelation: p. 638 of the textbook.

PROBLEM 6

The natural product nuciferal was synthesized by the route summarized here.

BrMg

?

OH

?

CHO

?

CHO

(a) Suggest a synthesis of the starting material.

(b) Suggest reagents for each step.

(c) Draw the retrosynthetic analysis giving the disconnections that you consider the planners may have used and label them suitably.

(d) Which synthon does the starting material represent?

Purpose of the problem

Practice at an important skill—learning from published syntheses—as well as a popular style of exam question.

Suggested solution

(a) Grignard reagents are made from the corresponding halide and the rest of the analysis used simple C–X disconnections.

BrMg

FGI

Br

1,1-diX

acetal

Br

HO   CHO

1,3-diX   HBr +

HO   HO

CHO

It turns out that the addition of HBr to the unsaturated aldehyde (trivially known as acrolein) and the protection as an acetal can be carried out in a single step as both are acid-catalysed.

CHO
acrolein

HO   OH

HBr

Br

Mg

Et$_2$O

BrMg

(b) The Grignard has obviously been added to a ketone to give the tertiary alcohol, but how do we replace OH by H? One way is direct catalytic hydrogenation but an easier way is to eliminate the tertiary (and benzylic) alcohol and hydrogenate the alkene. The acid used for dehydration will also remove the acetal.

The last step is an aldol reaction between two aldehydes. The easiest way to do this is by a Wittig reaction but a specific enol of propanal would also be fine.

(c) and (d) The retrosynthetic analysis is straightforward except for the last step. It is not obvious what reagent to use for the synthon in brackets. But you already know what was used: a Grignard reagent with a protected aldehyde, i.e. a d^3 reagent. This is needed because the 1,4 relationship between OH and CHO requires *umpolung* (p. 720 of the textbook).

d^3 synthon

PROBLEM 7

Show how the relationship between the alkene and the carboxylic acid influences your suggestions for a synthesis of these three compounds.

Purpose of the problem

An exploration of the importance of functional group relationships.

Suggested solution

The first is an α,β-unsaturated carbonyl compound and can best be made by an aldol reaction using some sort of specific enol equivalent for the acid part. A Wittig reagent, a malonate, or a silyl enol ether look the best.

analysis

synthesis

The second synthesis is difficult because the alkene can easily slip into conjugation with the carbonyl group. Perhaps the easiest strategy is to use cyanide ion as synthetic equivalent of $^-CO_2H$ since then the electrophile is an allylic halide. Other alternative routes could include alkyne reduction.

analysis

synthesis

The third is best approached by alkylation of a malonate with allyl bromide itself followed by hydrolysis and decarboxylation.

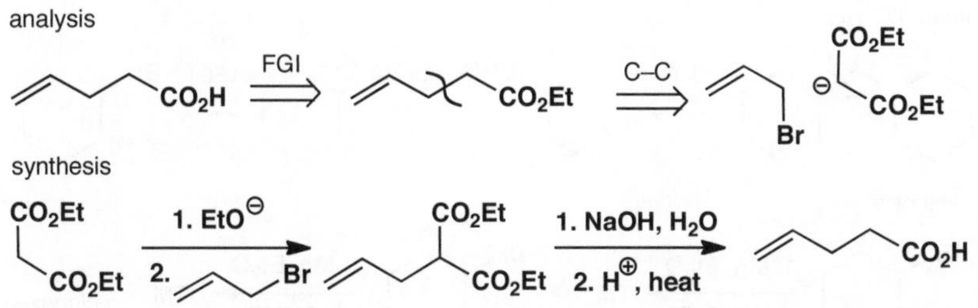

analysis

synthesis

PROBLEM 8

How would you make these compounds?

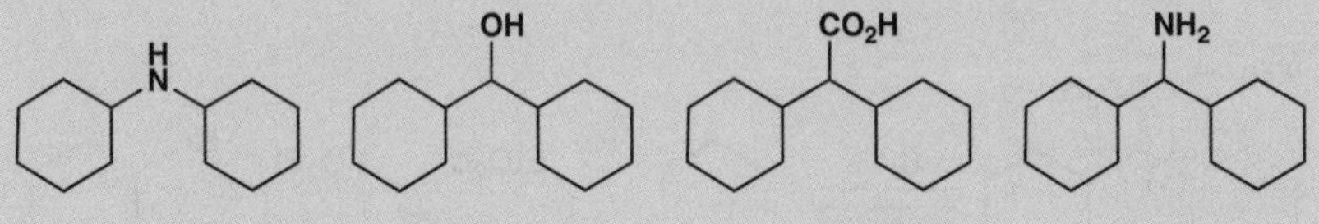

Purpose of the problem

A reminder of reductive amination and that simple syntheses of apparently related compounds may require very different chemistry.

Suggested solution

The secondary amine is best made by reductive amination via the imine (not usually isolated).

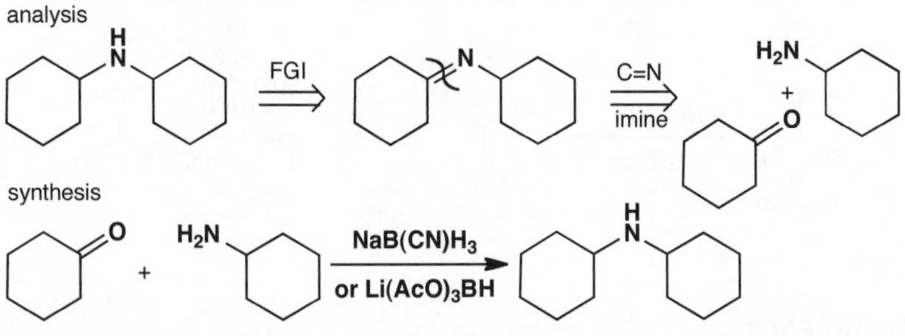

analysis

synthesis

The secondary alcohol can be made by some sort of Grignard chemistry. Cyclohexyl Grignard could be added twice to ethyl formate or once to the cyclohexane aldehyde.

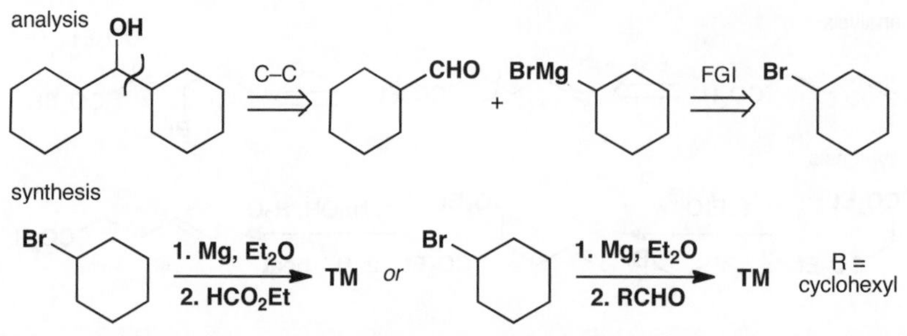
analysis
OH
C–C
CHO
BrMg
FGI
Br
synthesis
Br
1. Mg, Et_2O
2. HCO_2Et
TM
or
Br
1. Mg, Et_2O
2. RCHO
TM
R = cyclohexyl

The carboxylic acid could be made by double alkylation of malonate or some other specific enol equivalent.

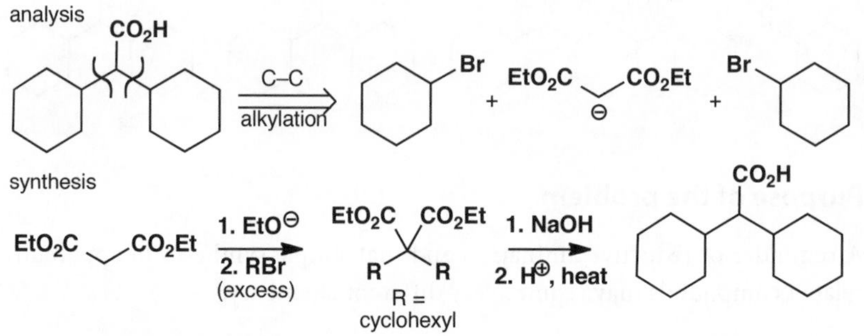
analysis
CO_2H
C–C
alkylation
Br
EtO_2C
CO_2Et
Br
synthesis
CO_2H
EtO_2C
CO_2Et
1. EtO⊖
2. RBr
(excess)
EtO_2C
CO_2Et
R
R
R = cyclohexyl
1. NaOH
2. H⊕, heat

Finally the primary amine could be made by reductive amination of a ketone that could in turn be made by oxidation of the secondary alcohol we have already made. Among many alternatives is the displacement of the tosylate of the same alcohol with azide ion and reduction of the azide.

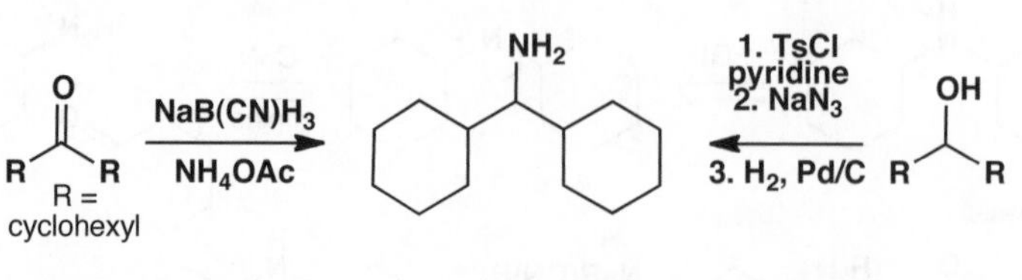
O
R
R
R = cyclohexyl
$NaB(CN)H_3$
NH_4OAc
NH_2
1. TsCl
pyridine
2. NaN_3
3. H_2, Pd/C
OH
R
R

PROBLEM 9

Show how the relationship between the two carbonyl groups influences your choice of disconnection when you design a synthesis for each of these ketones.

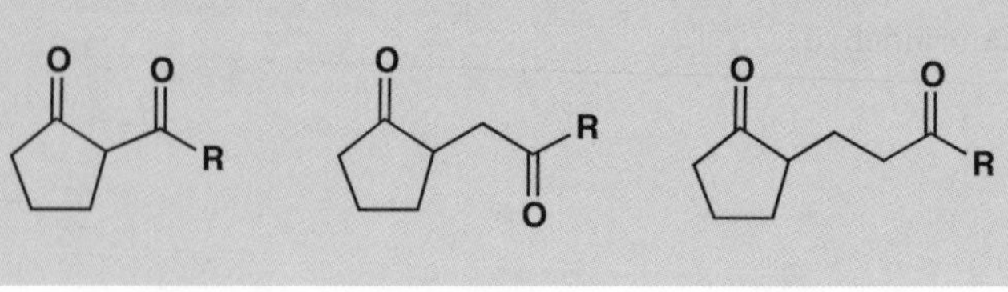
O
O
R
O
R
O
O
O
R

Purpose of the problem

An exercise in counting to reinforce the way that odd and even relationships affect the choice of a synthetic route.

Suggested solution

The three diketones have 1,3-, 1,4-, and 1,5-dicarbonyl relationships. In each case the obvious disconnection is of the bond joining the ring to the chain. But the chemistry is very different in each case. The 1,3-diketone can be made by acylation of a specific enolate. An enamine or a silyl enol ether is a good choice.

analysis

1,3-diCO

specific enol(ate) equivalent needed

synthesis

Me₃SiCl / Et₃N

RCOCl / TiCl₄

The same disconnection on the 1,4-diketone leads to different chemistry (alkylation of an enolate) and requires an enamine as the specific enol.

analysis

1,4-diCO

non-basic specific enol(ate) equivalent needed

synthesis

R₂NH / Et₃N

The 1,5-diketone requires conjugate addition of the same enolate and we suggest a different specific enolate equivalent though others would be just as good. This time the specific enol equivalent is needed to stop self-condensation of the cyclopentanone.

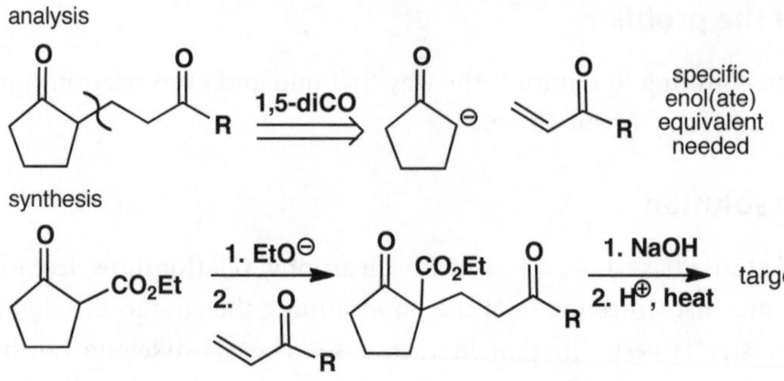

analysis

synthesis

PROBLEM 10

A synthesis of this enantiomerically pure ant pheromone was required for the purposes of pest control. Given a supply of the enantiomerically pure alkyl bromide as a starting material, suggest a synthesis of the pheromone.

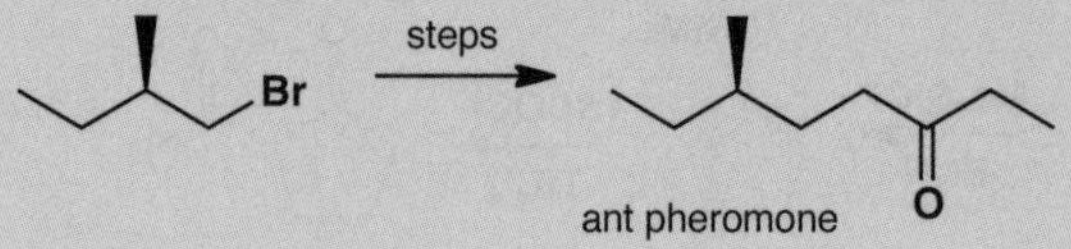

ant pheromone

Purpose of the problem

Planning the strategy of a synthesis from a given starting material.

Suggested solution

We know what the disconnection must be, since we have been given one starting material. This looks like an enolate alkylation, and we need to use a specific enolate to stop the ketone self-condensing. The best enolate equivalent will be one that is not too basic, to avoid competing elimination. The simplest solution is probably to use a keto-ester, easily made by Claisen condensation with diuethyl carbonate. After alkylation, the ester group is removed by decarboxylation.

analysis

synthesis

29

Suggested solutions for Chapter 29

PROBLEM 1

For each of the following reactions (a) state what kind of substitution is suggested and (b) suggest what product might be formed if monosubstitution occured.

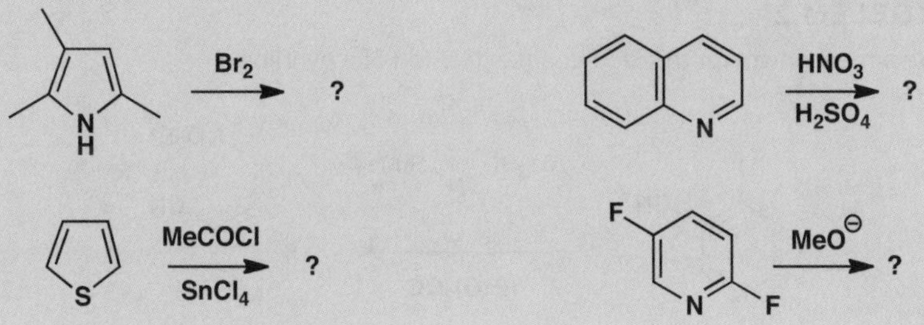

Purpose of the problem

A simple exercise in aromatic substitution on heterocycles.

Suggested solution

The first three reactions are all electrophilic substitutions: a bromination of a pyrrole, the nitration of quinoline, and a Friedel-Crafts reaction of thiophene. Bromination of the pyrrole occurs at the only remaining site. Nitration of quinoline occurs on the benzene rather than the pyridine ring (actually giving a mixture of 5- and 8-nitroquinolines) and the acylation occurs next to sulfur.

■ You needn't be concerned with the mixture of 5- and 8-nitroquinoline here, but p. 749 of the textbook has more detail.

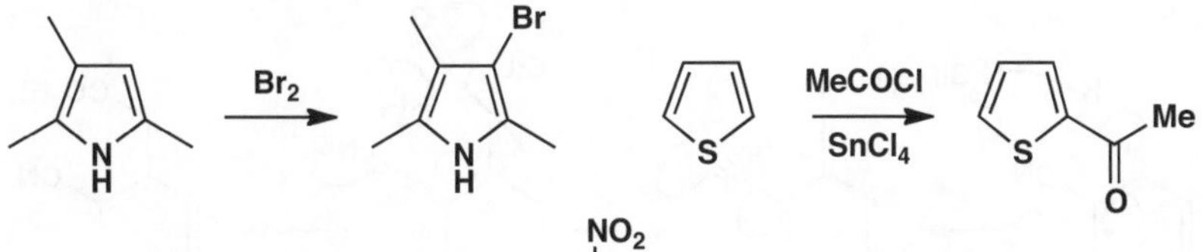

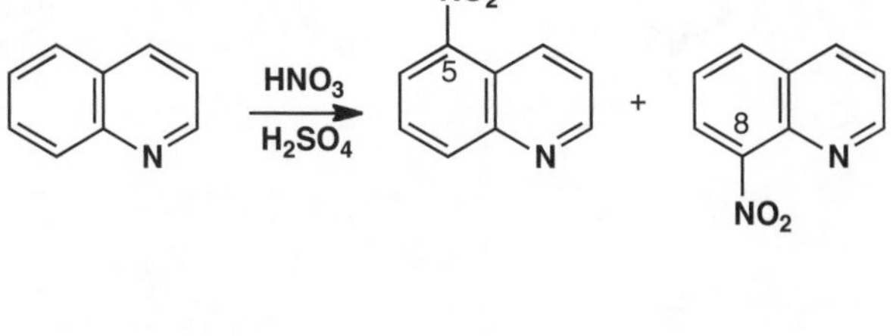

The last reaction is a nucleophilic aromatic substitution on a pyridine. It occurs only at the site where the negative change in the intermediate can be delocalized onto the nitrogen.

PROBLEM 2

Give a mechanism for this side-chain extension of a pyridine.

Purpose of the problem

An exercise in thinking about the reactivity of alkylated pyridines.

Suggested solution

The strong base (LHMDS, lithium hexamethyldisilazide) removes a proton from the methyl group so that the anion is stabilized both by the nitrile and the pyridine nitrogen atom. Acylation occurs outside the ring to preserve the aromaticity. If you drew the lithium atom covalently bound to nitrogen, your answer is better than ours.

■ This sort of chemistry was introduced by D. J. Sheffield and K. R. H. Wooldridge, *J.Chem. Soc., Perlin 1*, 1972, 2506 and by A. S. Kende and T. P. Demuth, *Tetrahedron Lett.*, 1980, **21**, 715 in a synthesis of the antileukaemic sesbanine.

PROBLEM 3

Give a mechanism for this reaction, commenting on the position in the furan ring that reacts.

Purpose of the problem

An unusual electrophilic substitution on furan with interesting selectivity.

Suggested solution

Furans normally prefer substitution at the α-positions (2 or 5) but one α-position is already blocked and the other is too far away to reach the allyl cation. Attack at the other end of the allylic system would give an eight-membered ring with a *trans* alkene in it. This would theoretically be possible but closure of a six-membered ring is much faster. In other words, the electrophile and nucleophile are *tethered*.

PROBLEM 4

Suggest which product might be formed in these reactions and justify your choice.

Purpose of the problem

Regioselectivity test with contrasted electrophilic aromatic substitution.

Suggested solution

In each case we have a choice between reaction on a benzene ring or an aromatic heterocycle. The pyrrole is more reactive than the benzene and the pyridine less so. The pyrrole does a Vilsmeier reaction (p. 734 of the textbook) in the remaining free position while nitration occurs on the benzene. Pyridine acts as an electron-withdrawing and deactivating substituent, and therefore directs *meta*.

PROBLEM 5

Explain the formation of the product in this Friedel-Crafts alkylation of an indole.

Purpose of the problem

Checking up on your understanding of indole chemistry.

Suggested solution

The Lewis acid combines with allyl bromide to give either the allyl cation or the complex we show here. In either case, electrophilic attack occurs at the 3-position of the indole. The benzyl group migrates to the 2-position where there is a proton that can be lost to restore aromaticity.

■ The textbook, pp. 745–6, explains why it is the 3- and not the the 2-position that is attacked.

PROBLEM 6

Suggest what the products of these nucleophilic substitutions might be.

Purpose of the problem

Checking your understanding of nucleophilic aromatic substitution involving decisions on chemoselectivity and regioselectivity.

Suggested solution

Each compound has potential nucleophilic and electrophilic sites. In the first case the benzene ring is not activated towards nucleophilic substitution but the pyridine is, both by the pyridine nitrogen atom and by the ester group. The NH_2 on the benzene ring is much more nucleophilic than the pyridine nitrogen atom.

In the second case, the chlorine on the heterocyclic ring is much more reactive towards nucleophilic substitution as the intermediate is stabilized by two nitrogen atoms and the benzene ring is not disturbed. The saturated heterocycle (piperazine) can be made to react once only as the product under the reaction conditions is strictly the hydrochloride of the unreacted amino group. This is much more basic than the one that has reacted as that lone pair is conjugated with the heterocyclic ring.

PROBLEM 7

Suggest how 2-pyridone might be converted into the amine shown. This amine undergoes nitration to give compound **A** with the NMR spectrum given. What is the structure of **A**? Why is this isomer formed?

NMR of **A**: δ_H 1.0 (3H, t, *J* 7 Hz), 1.7 (2H, sextet, *J* 7 Hz), 3.3 (2H, t, *J* 7 Hz), 5.9 (1H, broad s), 6.4 (1H, d, *J* 8 Hz), 8.1 (1H, dd, *J* 8 and 2 Hz), and 8.9 (1H, d, *J* 2 Hz). Compound **A** was needed for conversion into the enzyme inhibitor below. How might this be achieved?

Purpose of the problem

Revision of proof of structure together with electrophilic and nucleophilic substitution on pyridines and a bit of synthesis.

Suggested solution

The first step requires nucleophilic substitution so we could convert the pyridine into 2-chloropyridine and displace the chlorine with the amine.

The nitration occurs only because this pyridine is activated by the extra amino group so you could start by predicting which compound might be made. Alternatively you could work out the structure from the NMR. The key points are (i) **A** has only three aromatic protons so nitration has occurred on the ring, (ii) there is only one coupling large enough to be between *ortho* hydrogens (8 Hz), and (iii) there is a proton that has only *meta* coupling (2 Hz) a long way downfield (at large chemical shift). The pyridine nitrogen causes large downfield shifts at positions 2, 4, and 6, the nitro group causes large downfield *ortho* shifts, and the amino group causes upward *ortho* shifts (to smaller δ). All this fits the structure and mechanism shown. The amino group directs *ortho, para* and *para* is preferred sterically.

To get the enzyme inhibitor we need to reduce the nitro group to an amine and add the new chain to the other amine. This conjugate addition is best done first while there is only one nucleophilic amine. The ester is probably the best derivative to use, but you may have chosen something else.

PROBLEM 8

The reactions outlined in the chart below were the early stages in a synthesis of an antiviral drug by the Parke-Davis company. Consider how the reactivity of imidazoles is illustrated in these reactions, which involve not only the skeleton of the molecule but also the reagent **D**. You will need to draw mechanisms for the reactions and explain how they are influenced by the heterocycles.

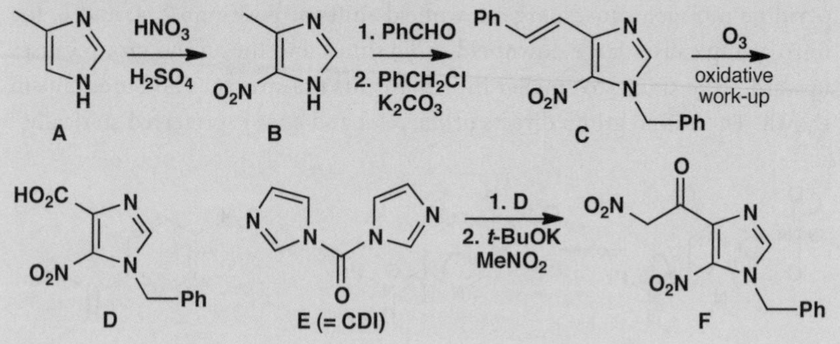

Purpose of the problem

An exploration of the chemistry of imidazole beyond that considered in chapter 29.

Suggested solution

The first reaction is the nitration of an imidazole in one of only two free positions. The position next to one nitrogen is more nucleophilic than the one between the two nitrogens. Imidazole has one pyridine-like and one pyrrole-like nitrogen so it is more nucleophilic than pyridine but less so than pyrrole.

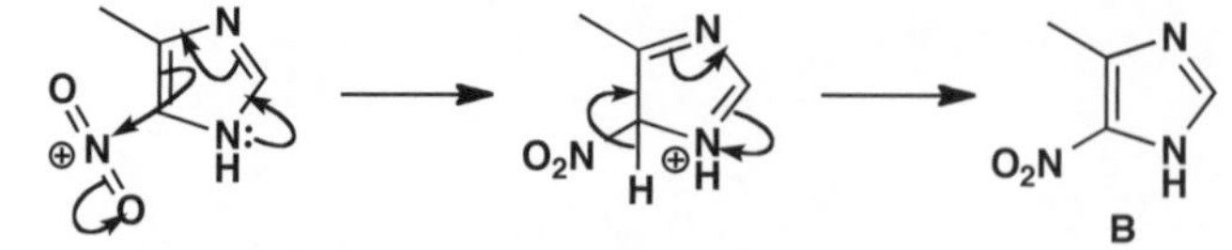

The second reaction is like an aldol condensation between the methyl group on the ring and the benzaldehyde as the electrophile. The nitro group provides some stabilization for the 'enolate' but that would not be enough without the imidazole—*ortho*-nitro toluene would not do this reaction. The elimination is E1cB-like, going through a similar 'enol' intermediate.

Next, alkylation occurs on one of the nitrogen atoms in the imidazole ring. We need the anion of the imidazole which could be alkylated on either nitrogen. Alkylation on the lower N is preferred because the product has the longer conjugated system—we've put in the curly arrows to show it.

Ozonolysis of the alkene of **C** frees the carboxylic acid of **D** which reacts with carbonyl diimidazole **E** (CDI) in a nucleophilic substitution at the carbonyl group, with the relatively stable imidazole anion as the leaving group. The product is an 'activated ester', like an anhydride, from which the anion of nitromethane displaces the second molecule of imidazole to give the product **F**.

■ See pp. 622–3 of the textbook for some chemistry of the anion of nitromethane.

E = carbonyl diimidazole (CDI)

PROBLEM 9

What aromatic system might be based on this ring system? What sort of reactivity might it display?

Purpose of the problem

A chance for you to think creatively about aromatic heterocycles.

Suggested solution

The aromatic system has the poetic name 'pyrrocoline' and you will have found it by trial and error. One ring looks like a pyridine and one like a pyrrole but counting the electrons should have made you realize that you need the lone pair on nitrogen to give a ten electron system. The nitrogen is therefore pyrrole-like and so if you predicted that this compound would react well in electrophilic substitutions on the five-membered ring you would be right: that is exactly what it does. The easiest pyrrocolines to make have alkyl groups at position 3 and these compounds are nitrated to give the 4-nitro compounds. Friedel-Crafts reactions happen at the same atom.

pyrrocoline

PROBLEM 10

Explain the order of events and the choice of bases in this sequence.

Purpose of the problem

The use of selective lithiation in furan chemistry.

Suggested solution

The allylic group evidently goes into the 2-position so deprotonation of the starting material by LDA must occur there, directed by both the oxygen and bromine atoms. The second electrophile (MeI) talkes the place of the Br atom, so BuLi must lead to bromine-lithium exchange rather than deprotonation. The alternative order of events would require selective lithiation adjacent to the methyl group—not something you would expect to work reliably.

■ Ortholithiation is introduced on pp. 563–4 of the textbook, and the lithiation of furan is on p. 737. Halogen-metal exchange is on p. 188.

■ The product is related to a constituent of the perfume of roses and was made by N. D. Ly and M. Schlosser, *Helv. Chim. Acta* 1977, **60**, 2085.

PROBLEM 1

Suggest a mechanism for this synthesis of a tricyclic aromatic heterocycle.

Purpose of the problem

A simple exercise in the synthesis of a pyridine fused to a pyrrole (or an indole with an extra nitrogen atom).

Suggested solution

The first step must be the formation of an enamine between the primary amine and the ketone. Now, because we have a pyridine and not a benzene ring, nucleophilic aromatic substitution can occur. These 'aza-indoles' are more easily formed than indoles.

PROBLEM 2

Is the heterocyclic ring created in this reaction aromatic? How does the reaction proceed? Comment on the regioselectivity of this cyclization.

Purpose of the problem

Exploring the synthesis and aromaticity of an unfamiliar heterocycle.

Suggested solution

The left-hand ring is obviously aromatic as it is a benzene ring. The right-hand ring has four electrons from the double bonds and can have two from a lone pair on oxygen, making six in all. This is more obvious in a delocalized form. Alternatively the whole system can be considered as a 10-electron molecule. Strangely enough, this is easier to see in the other Kekulé form.

a ten-electron π system

■ This is a very old reaction discovered by H. von Pechmann and C. Duisberg, *Ber.*, 1883, 2119.

■ See p. 483 of the textbook for more on the selectivity between the *ortho* and *para* positions.

The first step in the reaction is a transesterification and cyclization then occurs in the *ortho* position, *para* to the other hydroxyl group. Cyclization might have happened to the position in between the two substituents, as the other OH is *ortho, para*-directing, but the position chosen is more reactive for both steric and electronic reasons.

PROBLEM 3

Suggest mechanisms for this unusual indole synthesis. How does the second mechanism relate to electrophilic substitution on indoles (p. 746) ?

Purpose of the problem

A combination of a Fischer indole synthesis with revision of a bit of indole chemistry from the last chapter.

Suggested solution

The first step starts off as a normal Fischer indole synthesis (we have omitted the first step); you just have to draw the molecules carefully to show the *spiro* ring system, and you have to stop before an indole is formed as the quaternary centre prevents aromatization.

Treatment with a Lewis acid initiates a rearrangement very like those occurring when 3-substituted indoles are attacked by electrophiles (p. 746 of the textbook). The aromatic ring is a better migrating group than the primary alkyl alternative and an indole can finally be formed.

■ The new seven-membered heterocycle (an azepine) is found in some tranquilizers: see T. S. T. Wang, *Tetrahedron Lett.*, 1975, 1637.

PROBLEM 4

Explain the reactions in this partial synthesis of methoxatin, the coenzyme of bacteria living on methanol.

Purpose of the problem

A combination of Fischer indole synthesis with revision of indole chemistry from chapter 29.

Suggested solution

■ Diazotization: see p. 521 of the textbook.

There is clearly a Fischer indole synthesis in the second step but the first step makes the usual hydrazone in a most unusual way. The first reaction is a diazotization so we have to combine the diazonium salt with the enolate of the keto-ester. That creates a quaternary centre and the KOH deacylates it to give the aryl hydrazone needed for the next step.

Now that we have the hydrazone, the Fischer indole step is straightforward and gives the indole-2-carboxylic acid derivative. There is only one site for an enamine and the indole is formed on the side of the benzene ring away from the other substituents.

The next stage must involve the primary amine as nucleophile and the conjugated keto-diester as electrophile. You may have expected direct addition of the amine to the ketone as that gives the product by a reasonable mechanism. In fact, conjugate addition must occur first as the tertiary alcohol **A** can be isolated. The dehydration is obviously acid-catalysed and the oxidation by air [or Ce(IV)] is also acid-catalysed.

PROBLEM 5

Explain why these two quinoline syntheses from the same starting materials give (mainly) different products.

Purpose of the problem

An exercise in regioselectivity in a heterocyclic synthesis controlled by pH.

Suggested solution

■ This selective route to quinolines by the Friedländer synthesis was discovered by E. A. Fehnel, *J. Org. Chem.*, 1966, **31**, 2899.

You have a choice here: either you first form an enol(ate) from butanone and do an aldol reaction with the aromatic ketone or you first make an imine and then form enamines from that. In either case, you would expect enol or enamine formation on the more substituted side in acid but the less substituted side in base.

PROBLEM 6

Give mechanisms for these reactions used to prepare a fused pyridine. Why is it necessary to use a protecting group?

Purpose of the problem

Saturated and aromatic heterocycles combined with stereochemistry make an interesting synthesis for you to explore.

Suggested solution

The first starting material is a stable cyclic enamine and conjugate addition is what we should expect with an enone. Of course, if the aldehyde were unprotected, direct addition might occur there as well as carbonyl condensations. The product is in equilibrium with both its enols, one of which can cyclize to form the new six-membered ring.

The enol must attack the five-membered ring in a *cis* fashion as the tether is too short to reach the other side. There is no control over one stereogenic centre (represented with a wiggly line) but that is unimportant as it is soon to disappear.

Now the reaction with hydroxylamine in acid solution. Formation of the oxime of the ketone produces one molecule of water—just enough to hydrolyse the acetal—and the pyridine synthesis is completed by cyclization and a double dehydration (p. 765 of the textbook).

PROBLEM 7

Identify the intermediates and give mechanisms for the steps in this synthesis of a triazole.

Purpose of the problem

Revision of aromatic nucleophilic substitution and a chance to unravel an interesting mechanism.

Suggested solution

The first reaction forms **A**, just the enamine from the ketone and the secondary amine (morpholine). Below we have diazotization of an aromatic amine and replacement by azide to give **B**. This nucleophilic substitution could occur by the addition-elimination mechanism activated by the nitro group or by the S_N1 mechanism (chapter 22).

Now comes the interesting bit. The two reagents **A** and **B** combine without losing anything—it is evident that the enamine must be the nucleophile and so the azide must be the electrophile. We can see from the final product that the enamine attacks one end or the other of the azide. Trial and error takes over! Here is one possible solution with some side chains in the intermediate abbreviated for clarity. This product **C** can be isolated but its stereochemistry is not known.

■ An alternative is a 1,3-dipolar cycloaddition, see chapter 34.

Finally, the new aromatic system (a triazole) is formed by elimination of the aminal. Protonation of the most basic nitrogen is followed by expulsion of morpholine and aromatization by deprotonation.

■ This synthesis was discovered in Milan during a mechanistic study of the reactions between enamines and azides: R. Fusco *et al.*, *Gazz. Chim. Ital.*, 1961, **91**, 849.

PROBLEM 8

Give detailed mechanisms for this pyridine synthesis.

Purpose of the problem

Revision of aldol and conjugate addition reactions of enol(ate)s and a synthesis involving two furans and one pyridine.

Suggested solution

The first reactions are an aldol condensation and a conjugate addition. We have shown just the first steps, but make sure that you can draw full mechanisms for both. The last step is a standard pyridine synthesis.

PROBLEM 9

Suggest a synthesis for this compound.

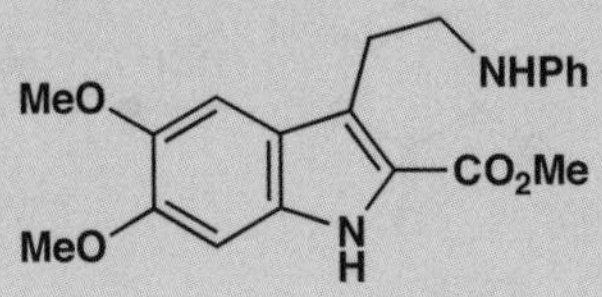

Purpose of the problem

The synthesis of an indole with a slight twist.

Suggested solution

This looks very much like a perfect subject for the Fischer indole synthesis. Let's see.

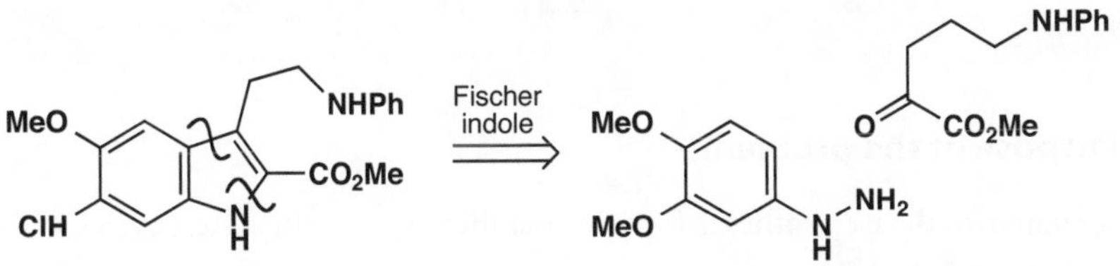

This looks fine, though we may wonder how we are going to have an amino group in that position on the keto ester. Surely it will cyclize onto the ester to form a lactam? One solution would be to protect it with something like a Boc group, but the solution found by the Sterling drug company was partly motivated by a desire to make a variety of compounds with different amine substituents. They chose hydroxyl as an easily replaceable group and accepted that the starting material would exist as a lactone. They made it like this:

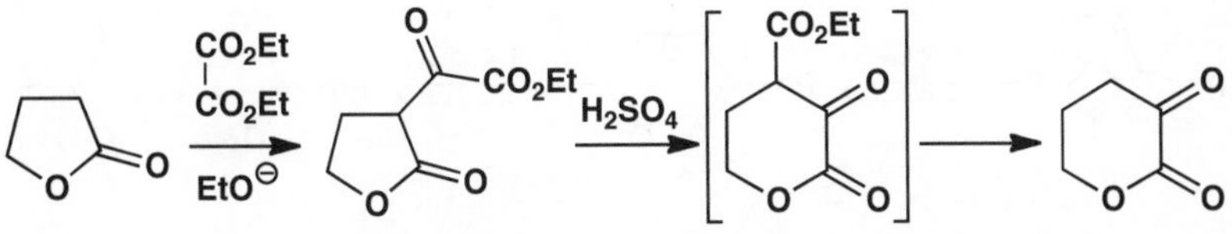

The first step is a typical Claisen ester condensation and the second is an acid-catalysed thermodynamically controlled transesterification (the lactone and ethyl ester exchange alcohol partners) to give the more stable six-membered lactone, followed by decarboxylation. Now the Fischer indole synthesis works well and work-up with dry HCl in methanol gave the alkyl

■ This chemistry is in the patent
literature but see S. Archer, *Chem.
Abstr.*, 1971, **78**, 29442..

chloride that could be displaced with amines to give a series of anti-
depressants.

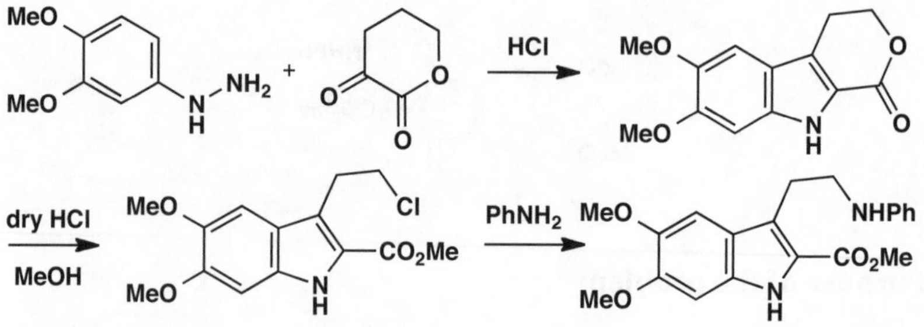

PROBLEM 10

How would you synthesize these aromatic heterocycles?

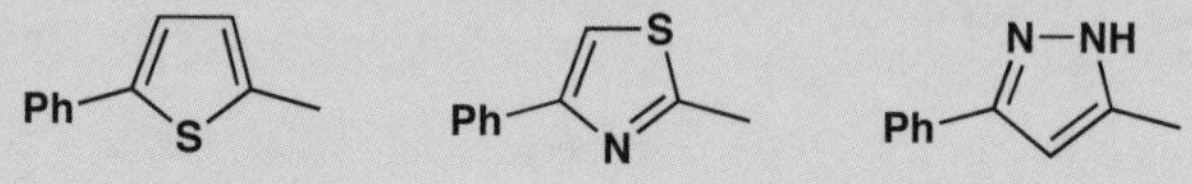

Purpose of the problem

A chance to devise syntheses for five-membered aromatic heterocycles with
one or two heteroatoms.

Suggested solution

These compounds all look much the same but the strategies needed for each
are rather different. Removing the heteroatom from the thiophene reveals a
1,4-diketone to be made by one of the methods in chapter 28. We have
chosen to propose an enamine and an α-bromoketone though there are
many other good choices.

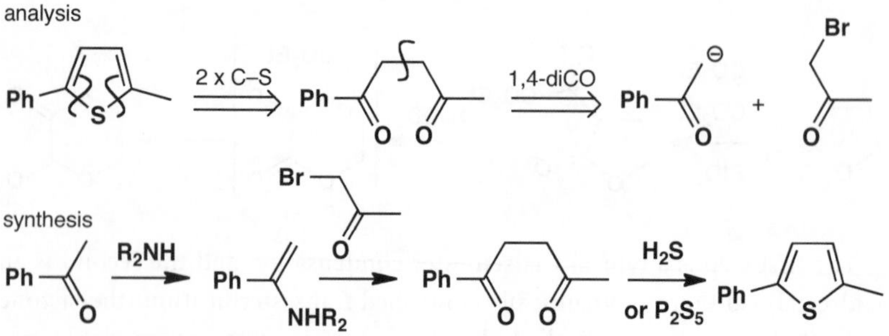

analysis

synthesis

The second compound is a thiazole and we want to use a thioamide to
make it (see p. 771 of the textbook). We should disconnect C–N and C–S

bonds to give the thioamide and another α-bromoketone remembering to let the nucleophiles exercise their natural preferences: sulfur attacking saturated carbon and nitrogen attacking the carbonyl group.

analysis

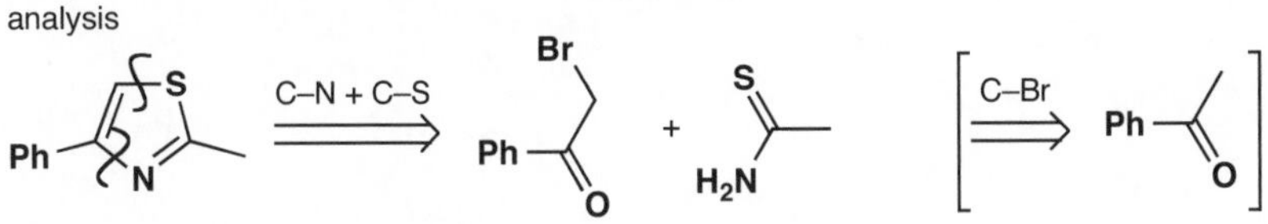

synthesis

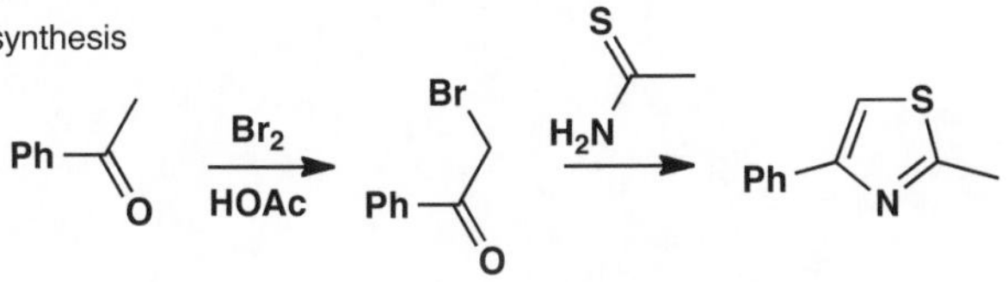

The third compound has the two heteroatoms joined together so we should keep them that way. We disconnect both C–N bonds revealing the hidden molecule of hydrazine (NH_2NH_2). We then need a 1,3-diketone so we need Claisen ester chemistry (chapter 26).

analysis

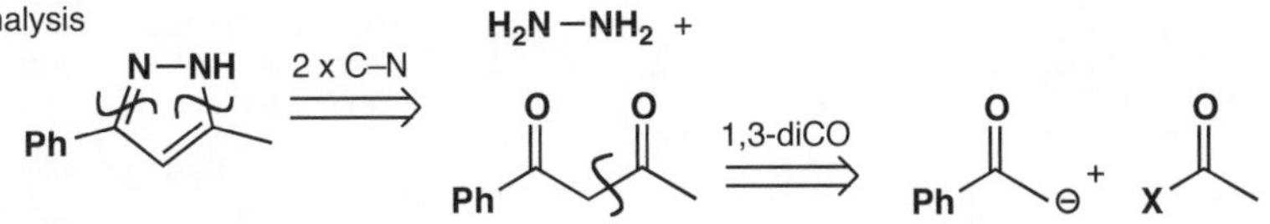

synthesis

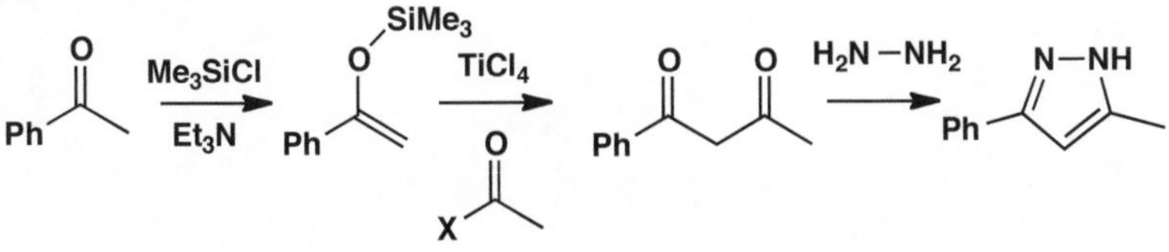

Suggested solutions for Chapter 31

PROBLEM 1

Predict the most favourable conformation for these insect pheromones.

Purpose of the problem

Practice at drawing the conformations of cyclic acetals.

Suggested solution

There are many good ways to draw these conformations and yet more not quite so good. The one thing you must do is place each acetal oxygen atom *axial* on the other ring to enjoy the full anomeric effect. We show three ways of drawing each compound. You get extra credit if you noticed that these compounds can each exist as two diastereoisomers and each diastereoisomer as two enantiomers.

■ The natural products are two compounds in the middle according to R. Baker *et al., J. Chem. Soc.,* 1982, 601

PROBLEM 2

The *Lolium* alkaloids have a striking saturated heterocyclic skeleton. One way to make this skeleton appears below. Suggest a mechanism and explain the stereochemistry.

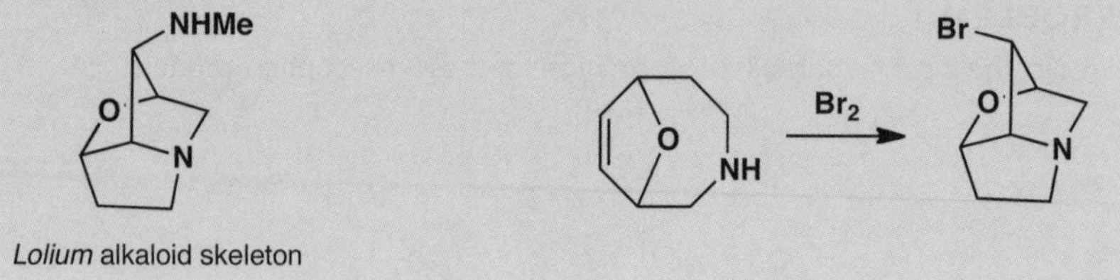

Lolium alkaloid skeleton

Purpose of the problem

Analysis of a reaction to make a bicyclic heterocycle stereospecifically.

Suggested solution

Bromine, of course, attacks the alkene to form a bromonium ion. If it has the right stereochemistry, it cyclizes but, if it doesn't, it reverts to starting materials. The reaction may remind you of halolactonization (p. 568 of the textbook).

■ This particular reaction was used by S. R. Wilson *et al., J. Org. Chem.,* 1981, **46**, 3887, to help establish the correct structure of the *Lolium* alkaloids.

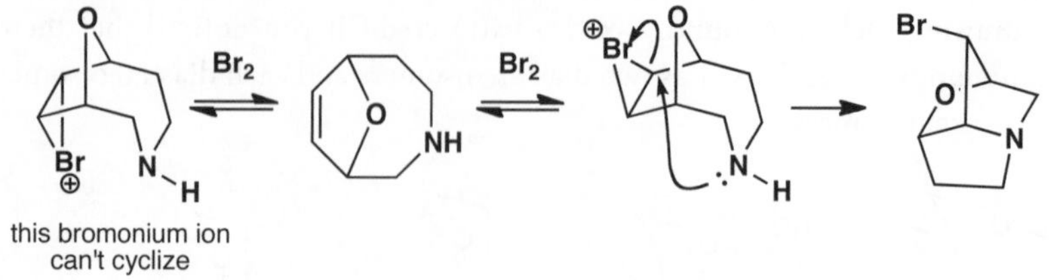

this bromonium ion
can't cyclize

PROBLEM 3

One of the sugar components of the antibiotic kijanimycin has the basic structure shown here and NMR spectrum given below. What is the stereochemistry? When you have deduced the structure, suggest which conformation the molecule will prefer.

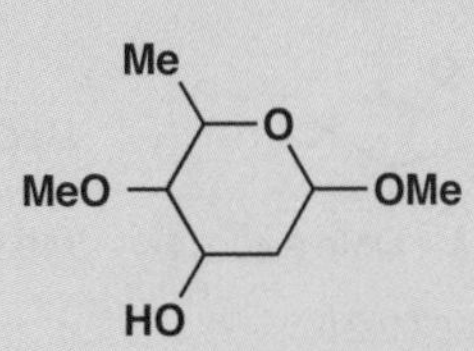

δ_H 1.33 (3H, d, J 6 Hz), 1.61* (1H, broad s), 1.87 (1H, ddd, J 14, 3, 3.5 Hz), 2.21 (1H, ddd, J 14, 3, 1.5 Hz), 2.87 (1H, dd, J 10, 3 Hz), 3.40 (3H, s), 3.99 (1H, dq, J 10, 3 Hz), 3.40 (3H, s), 1.33 (3H, d, J 6 Hz), 4.24 (1H, ddd, J 3, 3, 3.5 Hz) and 4.79 (1H, dd, J 3.5, 1.5 Hz). The signal marked * exchanges with D_2O.

Purpose of the problem

Using NMR to deduce stereochemistry and seeing how stereoelectronics decide the conformation of a cyclic acetal.

Suggested solution

You can make some preliminary assignments from a combination of shift and coupling:

Signal	Integral and splitting	Comments	Assignment
1.33	3H, d, J 6	3H, d must be CHMe	Me^7
1.61*	1H broad s	exchanges so must be OH	OH
1.87	1H, ddd, J 14, 3, 3.5	14 Hz looks like CH_2	H^2 or H^3
2.21	1H, ddd, J 14, 3, 1.5	2.21 and 1.87 are CH_2	H^2 or H^3
2.87	1H, dd, J 10, 3	must be axial H (10 Hz)	H^4 or H^5
3.40	3H, s	one OMe group	OMe
3.47	3H, s	the other OMe group	OMe
3.99	1H, dq, J 10, 6	q means H^6 (axial)	H^6
4.24	1H, ddd, J 3, 3, 3.5	small J must be equatorial	H^4 or H^5
4.79	1H, dd, J 3.5, 1.5	small J must be equatorial	H^1

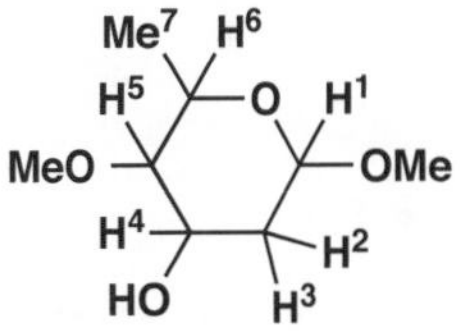

We don't mind which is H^2 or H^3 as they don't affect the stereochemistry, but we do mind which is H^4 or H^5. Since H^6 is a 10 Hz doublet coupled with H^5, we know that H^5 is at 2.87 and is axial. This gives the entire assignment

■ For the original analysis, see A. K. Mallams *et al.*, *J. Am. Chem. Soc.*, 1981, **103**, 3938.

and the stereochemistry: H^5 and H^6 are axial; H^1 and H^4 are equatorial. That is why there are no large vicinal (^3J) couplings to the diastereotopic CH_2 group (H^2 and H^3). All couplings not shown on the second diagram are <4 Hz.

PROBLEM 4

Explain why this cyclization gives a preponderance (3:1) of the oxetane, though the tetrahydrofuran is much more stable.

Purpose of the problem

A reminder that Baldwin's rules may apply to any cyclization.

Suggested solution

Clearly iodine attacks the alkene and the OH group adds to the intermediate iodonium ion. Let's draw this first without stereochemistry to see what happens. The starting material is in the middle, with the pathway to the tetrahydrofuran (THF) running to the left and the oxetane to the right.

Whether the oxetane or the tetrahydrofuran is formed depends on which end of the iodonium ion is attacked by the OH group. In terms of Baldwin's rules, oxetane formation is a simple 4-*exo-tet* reaction and is favoured. The THF formation is slightly more complicated. It is a 5-*exo-tet* as far as the S_N2 reaction is concerned, but in the transition state the nucleophile, the carbon atom under attack and the leaving group are also all in the same six-membered ring—there is disfavoured 6-*endo-tet* character. It is very difficult to get the two dotted lines in the transition state diagram at the required 180° to each other.

Now what about the stereochemistry? Did you notice that each product has an all-*trans* arrangement of substituents around the ring? And what about the second alkene? The two alkenes are in fact diastereotopic and which one is attacked by iodine as well as on which face determines the stereochemistry of the product. This is rather like iodolactonization. Iodine adds randomly and reversibly to both faces of both alkenes. Only when cyclization can gives the most stable all-*trans* product does the reaction continue.

PROBLEM 5

Draw a mechanism for the following multistep reaction. Do the cyclization steps follow Baldwin's rules?

Purpose of the problem

Baldwin's rules at work in the synthesis of a bicylic heterocycle with one nitrogen atom in both rings.

Suggested solution

Hydrolysis of the acetal releases an aldehyde and Mannich-style condensation leads to the product. The iminium ion forms by (favoured) 5-*exo-trig* attack on the aldehyde. The cyclization step in which the enol attacks the iminium ion is 6-*endo-trig* and is thus also favoured. By folding

■ This new synthesis of a bicyclic amine was reported by F. D. King, *Tetrahedron Lett.*, 1983, **24**, 3281.

the molecule into a chair a reasonable overlap between the required p orbitals is possible.

PROBLEM 6

Consider the question of Baldwin's rules for each of these reactions. Why do you think they are both successful?

Purpose of the problem

Developing judgement in using Baldwin's rules in the synthesis of heterocyclic compounds.

Suggested solution

The first ring system is the same as the one we have just been considering but the route to it is decidedly different and is more demanding of Baldwin's rules, though we should still describe it as 6-*endo-trig*. Manganese dioxide is a specific oxidant for allylic alcohols and conjugate addition of the amine to the enone gives the bicyclic amine. This works because *endo* reactions are just about all right when six-membered rings are formed and because conjugate addition is under thermodynamic control: as long as *some* of the reaction occurs, the product is the most stable compound in the mixture— any competing attack of the amine on the ketone gives a much less stable four-membered ring.

■ J. J. Tufariello and R. C. Gatrone, *Tetrahedron Lett.*, 1978, 2753.

The second example is again 6-*endo-trig* but it is acid-catalysed: protonation increases the reactivity of the enone and reduces its rigidity. Both these 6-*endo-trig* reactions occur through chair-like transition states rather like the example in the previous problem.

PROBLEM 7

A revision problem in spectroscopy. A Pacific sponge contains 2.8% dry weight of a sweet-smelling oil with the following spectroscopic details. What is its structure and stereochemistry?

Mass spectrum gives formula: $C_9H_{16}O$. IR 1680 and 1635 cm^{-1}.

δ_H 0.90 (6H, d, J 7), 1.00 (3H, t, J 7), 1.77 (1H, m), 2.09 (2H, t, J 7), 2.49 (2H, q, J 7), 5.99 (1H, d, J 16), and 6.71 (1H, dt, J 16, 7).

δ_C 8.15 (q), 22.5 (two qs), 28.3 (d), 33.1 (t), 42.0 (t), 131.8 (d), 144.9 (d), and 191.6 (s).

Purpose of the problem

A reminder of stereochemistry in two dimensions.

Suggested solution

The IR suggests a conjugated carbonyl compound, confirmed by the carbonyl and two alkene signals in the carbon NMR with the additional information that the carbonyl group is an aldehyde or ketone (δ_C about 200). The proton NMR shows it is a ketone (no CHO proton), that the alkene has two protons (5.99 and 6.71), and that they are *trans* (J = 16 Hz). We also see an ethyl group (2H q and 3H t) attached to something with no Hs (could it be the carbonyl group?). This suggests the unit in the margin which leaves only C_4H_8. We know we have Me$_2$CH- from the 6H d and that leaves only CH$_2$. We have a structure.:

■ The distinctive features of ^{13}C NMR spectra of C=O compounds are described on p. 408 of the textbook.

δ$_H$ 6.71

δ$_H$ 1.77 (m)

δ$_C$ 191.6   δ$_C$ 144.9

δ$_H$ 1.00 Me   O   H   Me   H

16 Hz   7 Hz   7 Hz

Me

7 Hz   H   H   H   H   H   δ$_H$ 0.90

δ$_H$ 2.49   δ$_H$ 2.09   7 Hz

δ$_H$ 5.99

Me   δ$_C$ 22.5

Me

δ$_C$ 8.15 Me

δ$_C$ 42.0   δ$_C$ 33.1   δ$_C$ 28.3

δ$_C$ 131.8

PROBLEM 8

Reaction between this aldehyde and ketone in base gives a compound **A** with the proton NMR spectrum: δ$_H$ 1.10 (9H, s), 1.17 (9H, s), 6.4 (1H, d, *J* 15), and 7.0 (1H, d, *J* 15). What is its structure? (Don't forget stereochemistry!). When this compound reacts with HBr it gives compound **B** with this NMR spectrum: δ$_H$ 1.08 (9H, s), 1.13 (9H, s), 2.71 (1H, dd, *J* 1.9, 17.7), 3.25 (dd, *J* 10.0, 17.7), and 4.38 (1H, dd, *J* 1.9, 10.0). Suggest a structure, assign the spectrum, and give a mechanism for the formation of **B**.

O   O

     +   H $\xrightarrow{\text{base}}$ **A** $C_{11}H_{20}O$ $\xrightarrow{\text{HBr}}$ **B** $C_{11}H_{21}BrO$

Purpose of the problem

Slightly more difficult determination of stereochemistry moving from two to three dimensions. Revision of the Karplus relationship and of conjugate addition.

Suggested solution

The structure of **A** is easy. It has a *trans* alkene with two H's (*J* 15) and two tertiary butyl groups. There isn't much else except a carbonyl group so it must be an aldol product between the enolizable ketone and the unenolizable aldehyde.

O $\xrightarrow{\text{base}}$ O   O   H $\xrightarrow{\text{etc.}}$ O   H 7.0

6.4 H

B is more difficult. The alkene has obviously gone (no signals beyond 4.48) and there is one extra H. It looks as though HBr has added. The 17.7 coupling cannot be a *trans* alkene as the chemical shifts are too small, so it must be geminal coupling. This means that the molecule must be chiral so that the two hydrogens on the same carbon are diastereotopic. In fact, the expected conjugate additon of HBr to the enone has occurred.

■ See E. R. Kennedy and R. S. Macomber, *J. Org. Chem.*, 1974, **38**, 1952.

The three hydrogens that we have drawn in form an ABX system: A and B are the diastereotopic CH_2 group (J_{AB} = 17.7) and X is the CHBr proton (J_{AX} = 10 and J_{BX} = 1.9). It is not normally possible to say which proton is A and which B but here the large groups, along with the big difference between the two coupling constants, allow us to surmise there is one favoured conformation with dihedral angles of about 180° and 60°.

favoured conformation
has large groups antiperiplanar

^2J 17.7 Hz (geminal)

^3J 1.9 Hz (60° angle)

^3J 10 Hz (180° angle)

PROBLEM 9

Two diastereoisomers of this cyclic keto-lactam have been prepared. The NMR spectra have many overlapping signals but the marked proton can be seen clearly. In isomer **A** it is at δ_H 4.12 (1H, q, J 3.5) and in isomer **B** it is δ_H 3.30 (1H, dt, J 4, 11). Which isomer has which stereochemistry?

Purpose of the problem

Assignment of three-dimensional stereochemistry from NMR when only one signal can be clearly seen.

Suggested solution

The two isomers have *cis* and *trans* ring junctions so we should start by making conformational drawings of both. The *trans* compound is easy as it has a fixed *trans*-decalin shape. The *cis* compound can have two conformations as both conformers can flip.

The vital proton is clearly axial in isomer **B** as it has two large couplings (10 Hz) to other axial protons so this must be the *trans* isomer. Isomer **A** has three equal small couplings and this fits one conformation of the *cis* isomer.

PROBLEM 10

Given a sample of each of these two compounds, how would you determine the stereochemistry?

Purpose of the problem

An approach from the other end: how would you do the job? Also to remind you that we can determine the *relative* stereochemistry (i.e. which diastereoisomer do we have?) but not the absolute stereochemistry (i.e. which enantiomer do we have?) by NMR.

Suggested solution

By NMR of course. Both compounds are six-membered rings so we should first make conformational diagrams of all the possibilities. Both will have the *t*-butyl group equatorial. The first compound can have the methyl group *cis* or *trans* to the *t*-butyl group while the second compound can have both methyl groups on the same side, on the other side to the *t*-butyl group, or one on each side. Two of these are *meso* compounds though this doesn't affect the assignment.

■ It is better, if we can, to draw the carbonyl group at the 'end' of the ring, because then we can easily make it look planar. Compare the less pleasing structures in the solution to problem 9, where we were forced to put the carbonyl group at the back or front of the ring.

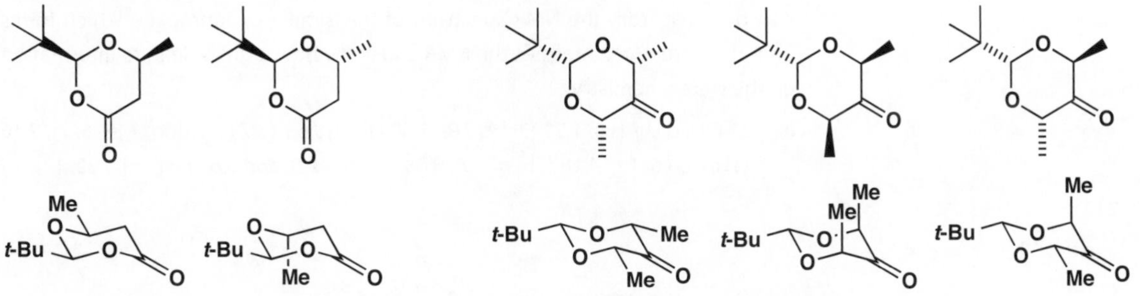

The key H atoms in the NMR spectrum are those shown below. In the first compound H^D tells us nothing as it has no neighbours and no coupling. H^B and H^C are useful as they tell us about H^A. H^A is easily identified by its quartet coupling to the methyl group. If it has a large axial-axial coupling (about 10 Hz) to H^B we have the *cis* compound, but if all its couplings are small (perhaps <4 Hz) then it is the *trans* compound.

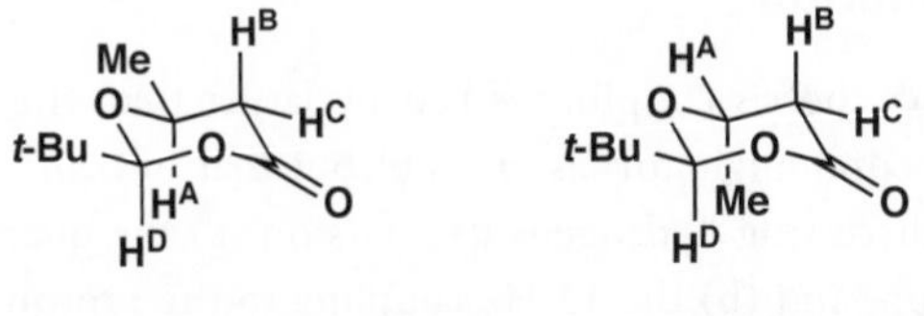

In the second compound a difficulty emerges: there is no coupling! We can tell by symmetry whether we have, on the one hand, the symmetrical *cis, cis-* or the *trans, trans-* compounds or, on the other hand, the non-symmetrical *cis, trans-* compound. The symmetrical compounds will of course show only one peak for the two methyl groups, for example. But how can we tell which of the symmetrical compounds we have? The chemical shifts will be different but we won't know which is which. However, if we irradiate the signal for the methyl groups, we should get a strong NOE (pp. 799–800 of the textbook) at H^A for the *trans* compound and not for the all-*cis* compound.

symmetrical compounds

unsymmetrical compound

NOE

Me *cis* to *t*-Bu

Me *trans* to *t*-Bu

unsymmetrical compound

PROBLEM 11

The structure and stereochemistry of the antifungal antibiotic ambruticin was in part deduced from the NMR spectrum of this simple cyclopropane which forms part of its structure. Interpret the NMR and show how it gives definite information on the stereochemistry.

δ_H 1.21 (3H, d, J 7 Hz), 1.29 (3H, t, J 9), 1.60 (1H, t, J 6), 1.77 (1H, ddq, J 13, 6, 7), 2.16 (1H, dt, J 6, 13), 4.18 (2H, q, J 9), 6.05 (1H, d, J 20), and 6.62 (1H, dd, J 20, 13).

Purpose of the problem

Assigning a more complex NMR and making decisions about stereochemistry in small rings.

Suggested solution

In cyclopropanes the *cis* coupling is usually larger than the *trans* coupling because the dihedral angle for *cis* Hs is 0° but that of *trans* Hs is not 180°. Assigning the three ring hydrogens depends on (a) the quartet coupling to the methyl group, and (b) the 13 Hz coupling to the proton on the alkene. This means that the third proton on the ring (t, J 7) must be next to the carbonyl group. The two *trans* couplings round the ring are the same (6 Hz) and smaller than the *cis* coupling (7 Hz). The double bond geometry is on more certain grounds as 20 Hz can be only a *trans* coupling.

■ This work is described in J. A. Donelly *et al.*, *J. Chem. Soc., Perkin Trans. 1*, 1979, 2629.

PROBLEM 12

A reaction produces two diastereoisomers of the product below: isomer **A** has δ_H 3.08 (1H, dt, J 4, 9, 9) and 4.32 (1H, d, J 9), while isomer **B** has δ_H 4.27 (1H, d, J 4). All other protons (except those of the Me groups) overlap in the NMR. Isomer **B** is converted into isomer **A** in base. What is the stereochemistry of **A** and **B**?

Purpose of the problem

Determining stereochemistry with the minimum of information.

Suggested solution

There are only two diastereoisomers and the difference in coupling constants is striking. The observed Hs must be those next to the functional groups. These compounds are not true cyclohexanes as they are flattened by the benzene ring and are best drawn as cyclohexenes. You should imagine the benzene ring coming towards you from the double bond.

trans diastereoisomer   cis diastereoisomer

The two protons we can see in isomer **A** must be H^1 and H^2 as they have the largest shifts. The proton with only one coupling must be H^1 as it has only one neighbour H^2. The coupling between these two is 9 Hz so they must both be axial. Isomer **A** is therefore the *trans* compound. H^2 is a double triplet because it has two axial neighbours and one equatorial neighbour (H^4). Isomer **B** shows H^1 alone and it is clearly equatorial (J 4) and so it must be the *cis* isomer.

A =   B =

■ A reminder: we are showing only relative configuration here: NMR tells us nothing about whether we have one or both enantiomers of each diastereoisomer.

Conversion to **A** occurs by enolate formation and the *trans* compound is more stable than the *cis*, as you might expect.

PROBLEM 13

Muscarine, the poisonous component of the death cap mushroom, has the structure below. We give the proton NMR spectrum. Can you see any definite evidence for the stereochemistry? Couplings are in Hz, m stands for multiplet, and * means that the proton exchanges with D_2O.

δ_H 1.16 (3H, d, J 6.5), 1.86 (1H, ddd, J 12.5, 9.5, 9.5), 2.02 (1H, ddd, J 12.5, 6.0, 2.0), 3.36 (9H, s), 3.54 (1H, dd, J 13, 9.0), 3.92 (1H, dq, J 2.5, 6.5), 4.03 91H, m), 4.30* (1H, d, J 3.5), and 4.68(1H, m).

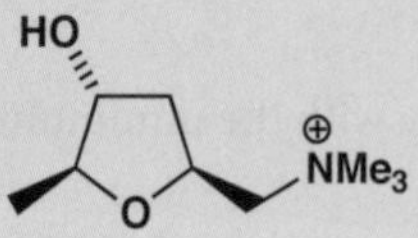

Purpose of the problem

Demonstrating that it can be difficult to determine stereochemistry even with all the information.

Suggested solution

Couplings round five-membered rings tend to be much the same whether they are ^2J (geminal), $^3J_{cis}$, or $^3J_{trans}$ (vicinal). Even so, the two diastereotopic CH_2 groups are easy to find with their large ^2J couplings of 13 and 12.5 Hz. The one with extra coupling must be in the side chain and the other in the ring. Here is the full analysis. You will see that it is very difficult to get conclusive evidence on stereochemistry from NMR alone without using NOE. You should see that, in general, *cis* couplings in five-membered rings tend to be larger than *trans*, though there are many, many exceptions!

■ The details are in J. Mulzer *et al.*, *Liebigs Ann. Chem.*, 1987, 7

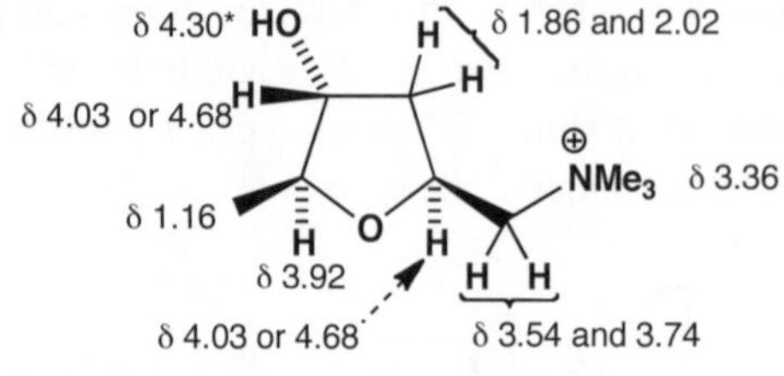

PROBLEM 14

Treatment with base of the two compounds shown here gives an unknown compound with the spectra given below. What is its structure?

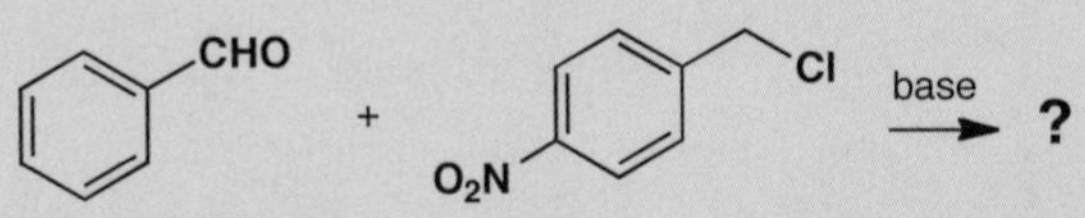

m/z: 241 (M^+, 60%), 90 (100%), 89 (62%)

δ_H (ppm in $CDCl_3$) 3.89 (1H, d, J 3 Hz), 4.01 (1H, d, J 3 Hz), 7.31 (5H, s), 7.54 (2H, d, J 10 Hz) and 8.29 (2H, d, J 10 Hz)

δ_C (ppm in $CDCl_3$) 62, 64, 122, 125, 126, 127, 130, 136, and 148 (the last three are weak).

Purpose of the problem

Further practice at structure determination, making use of the size of the coupling constant.

Suggested solution

The compound is an epoxide: the coupling constants around the three-membered ring are small (3 Hz: contrast 10 Hz on the benzene ring) because of ring size and the oxygen atom. All the Hs on the Ph ring happen to come at the same chemical shift. Those on the nitrated ring are at lower field and separated by the nitro group.

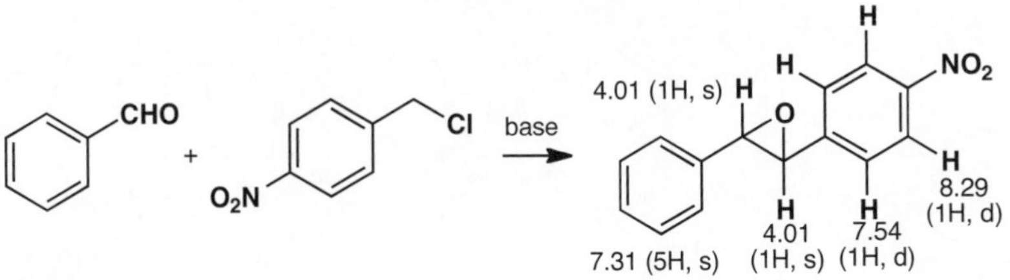

Purpose of the problem

An exercise to remind you how important conformational analysis is in any stereoselective reactions of saturated six-membered rings.

Suggested solution

The conformation of the cyclohexene will place the ester group in an equatorial position, almost in the plane of the alkene, so that it offers only slight steric hindrance. The opening of the epoxide is dominated by conformation: approach **a** would give a twist-boat product but approach **b** gives the chair cyclohexane observed.

Before cyclization, the compound must go into a boat form so that the amine and the ester can approach one another. This boat is fixed in the final bicyclic structure. The cyclization does not affect the stereochemistry.

PROBLEM 2

Explain the stereochemistry of this sequence of reactions, noting the second step in particular.

Purpose of the problem

To show how even 'non-reactions' can influence stereochemistry.

Suggested solution

The hydrogenation will add a molecule of hydrogen in *cis* fashion to give what appears at first to be the wrong product.

■ H. W. Pinnick and Y. Chang, *Tetrahedron Lett.*, 1979, 837.

The second step is important as it changes the stereochemistry. Ethoxide will form the enolate of the ester reversibly and allow it to move to the outside, convex face of the folded molecule. Though neutral nitrogen is not normally a chiral centre because it undergoes rapid pyramidal inversion, here it is fixed by by the need of the 5/5 fused system to have a *cis* ring junction. The last step is just reduction of the ester with no change in stereochemistry.

PROBLEM 3

Comment on the control over stereochemistry achieved in this sequence.

Purpose of the problem

An exploration of stereochemistry in rings.

Suggested solution

The reducing agent could attack either side of the ring in the first step but by reacting with the OH group it can deliver hydride intramolecularly from the bottom face. The mesylation does not affect the stereochemistry as no bonds are formed or broken at any of the stereogenic centres.

The reaction with ammonia probably starts with displacement of the primary mesylate and the second displacement is intramolecular. It is also stereospecific as S_N2 reactions must occur with inversion and, fortunately, the amine is on the bottom face.

■ As described by E. J. Corey and group in *Tetrahedron Lett.*, 1979, 671.

PROBLEM 4

What controls the stereochemistry of this product? You are advised to draw the mechanism first and then consider the stereochemistry.

Purpose of the problem

To show how ring-closing reactions, particularly those on the side of an already existing ring, can give excellent stereochemical control. And again, the importance of conformation in six-membered rings.

Suggested solution

We'd better draw the mechanism first, as the question suggests. Grignard reagents tend to do direct rather than conjugate addition to enones, and the product shows that the methyl group has done just that. But the OH group is in the wrong position to cyclize to the ester and there doesn't seem to be much scope for stereochemical control so we probably get a mixture of diastereoisomers.

The first product is a tertiary allylic alcohol so it will lose water under the acidic work-up conditions to form a carbocation. Readdition of water to the other end of the allylic cation gives an alcohol that could cyclize to the final product.

An alternative and probably better mechanism is that the ester, or the acid derived from it by hydrolysis, cyclizes onto the allylic cation. In our first mechanism, the OH group would have to be on the same side of the ring as the ester or acid to allow the lactone to form, but this cyclization gives the *cis* lactone directly from the allylic cation intermediate.

■ E. W. Colvin and R. A. Raphael, *J. Chem. Soc, Chem. Commun.*, 1971, 858.

PROBLEM 5

Why is one of these esters more reactive than the other?

Purpose of the problem

Reminder of the power of folded molecules with concave and convex sides.

Suggested solution

The molecule is folded along the ring junction with one of the esters inside the fold (on the concave side) and one out in the open (on the convex side). In the rate-determining step of ester hydrolysis, the attack of the hydroxide ion on the carbonyl group, the forming tetrahedral intermediate is larger than the starting ester. This means that the ester on the outside, which has more room to 'expand', reacts faster than the ester on the inside.

■ P. F. Hudrlik *et al.*, *J. Am. Chem. Soc.*, 1973, **95**, 6848.

PROBLEM 6

Explain the stereoselectivity in these reactions.

Purpose of the problem

Another cyclization reaction and an example of a controlled inversion of configuration not using the S_N2 reaction.

Suggested solution

The first stereoselective reaction is surprising as it may appear that the initial alkylation decides the stereochemistry. But that is not the case as you will see if you draw the mechanism. The ester enolate is very easily formed as it is stabilized by the pyridine ring and the nitrile as well as by the ester. Even a weakish base such as carbonate is good enough.

The first intermediate produced by alkylation with the primary alkyl bromide (or the epoxide) has two stereogenic centres and will no doubt be formed as a mixture of diastereoisomers. But this doesn't matter as the enolate has to be reformed for the next alkylation, and that destroys one of the chiral centres. We are back to a single compound again.

All now depends on the arrangement of the molecule for the cyclization step. The mechanism is straightforward enough but drawing the transition state is tricky. We offer two representations: an attempt at a conformational diagram and a Newman projection. The vital feature in both is that the enolate carbon and the C–O bond of the epoxide must be collinear. The molecule folds so that the five-membered ring bends upwards away from the large pyridine ring. This is not obvious even when you know the answer. You may have a better diagram.

The rest of the reactions are straightforward. Cyclization is spontaneous in base and the oxidation and reduction simply invert the OH group as the borohydride ion approaches the ketone from the side opposite the large pyridine ring.

■ This stereochemistry was not predicted by the chemists who did the work (A. S. Kende and T.P. Demuth, *Tetrahedron Lett.*, 1980, **21**, 715) who said 'The product was obtained as a single racemic isomer rather than the expected *cis/trans* mixture.'

PROBLEM 7

The synthesis of a starting material used in chapter 32 (p. 834 of the textbook) is a good example of how cyclic compounds can be used in a simple way to control stereo-chemistry. Draw mechanisms for each reaction and explain the stereochemistry.

Purpose of the problem

Reinforcement of material from chapter 32 and some important reactions.

Suggested solution

Tosylation of the primary alcohol is followed by ester exchange with methanol to release the anion of a secondary alcohol that promptly closes to an epoxide. There is no change at the stereogenic centre.

Now the vinyl cuprate attacks the epoxide at its less substituted end, releasing the same oxyanion, which promptly closes the lactone again. Once more there is no change at the stereogenic centre.

Finally, the double bond is introduced by selenium chemistry (p. 686 of the textbook). The steps are straightforward and the geometry of the alkene is dictated by the ring.

PROBLEM 8

Draw conformational drawings for these compounds. State in each case why the substituents have the positions you give. To what extent could you confirm your predictions experimentally?

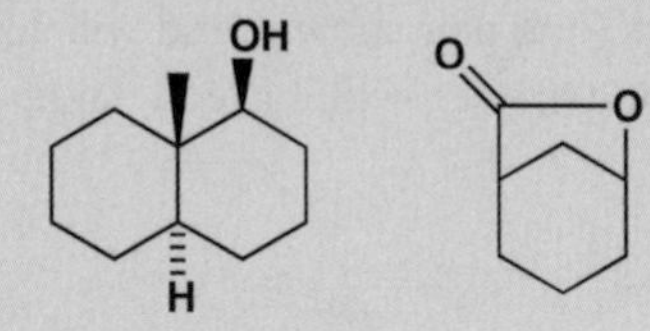

Purpose of the problem

Further exploration of conformation and establishing the link with NMR spectra.

Suggested solution

The first two compounds have no choice about their conformation but the third does. The two functional groups prefer to be equatorial rather than axial.

Confirming the conformations experimentally means measuring coupling constants in the proton NMR so we need to look at the vital protons (marked 'H' in the diagrams below) and consider whether they can be seen in the spectrum. Fortunately the interesting protons are all next to functional groups so they will all be easily recognizable. We may find it difficult to identify the axial proton at the ring junction in the first example but *trans* decalins must have axial groups at the ring junction so that it not important.

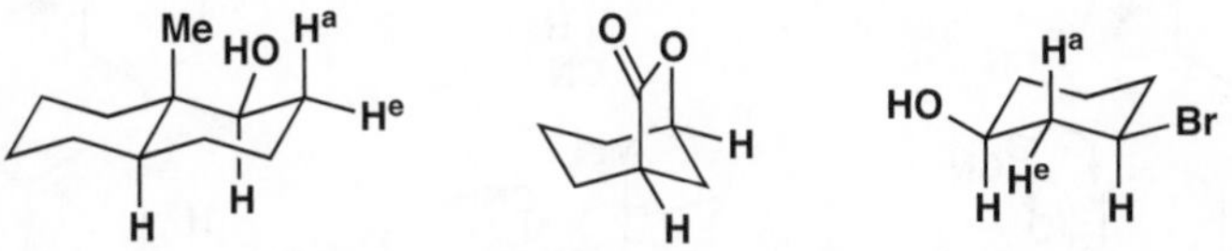

In the first molecule, proton H has two neighbours, one axial (H^a) and one equatorial (H^e) so it will appear as a double doublet with characteristic large axial/axial and small axial/equatorial couplings. By contrast the two marked equatorial Hs in the second compound have each got two axial and two equatorial neighbours and all the coupling constants will be about the same and small. They will both appear as narrow triplets of triplets but may be difficult to analyse. The important thing is that they have no large couplings. The two axial protons in the third example have each got two axial and two equatorial neighbours (one pair shown) and will again appear as triple triplets but this time one triplet will have a large axial/axial coupling.

PROBLEM 9

Predict which products would be formed on opening these epoxides with a nucleophile, say cyanide ion.

Purpose of the problem

Practice at choosing the correct regiochemistry of a stereoelectronically controlled epoxide opening.

Suggested solution

The opening of cyclohexene epoxides is controlled by the need to get the *trans* diaxial products. To get the right answer we need merely to draw the only possible *trans* diaxial (i.e with CN and O⁻ diaxial) product from each of these conformationally fixed *trans* decalins. Cyanide must, of course, open the epoxide with inversion so the OH group in the products is on the same side as the oxygen atom in the original epoxides.

PROBLEM 10

These two sugar analogues are part structures of two compounds used to treat poultry diseases. Which conformation will they prefer?

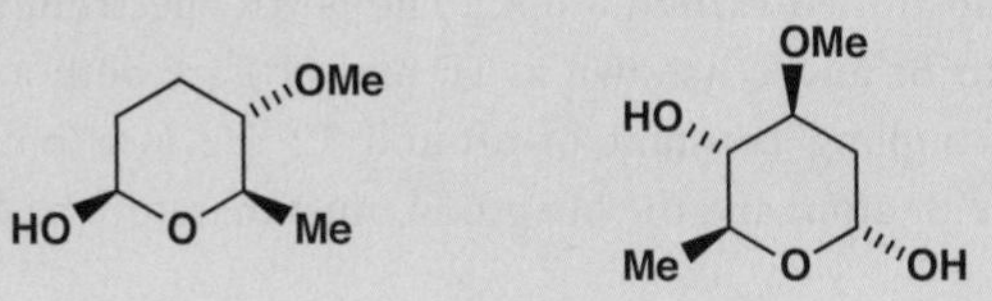

Purpose of the problem

Exploration of conformations in natural products with many substituents.

Suggested solution

Trial and error gives the conformation with the most equatorial substituents. The first compound can have all its substituents equatorial. The second can have three groups equatorial and one OH group axial, the preferred conformation due to the anomeric effect (pp. 801–3 of the textbook).

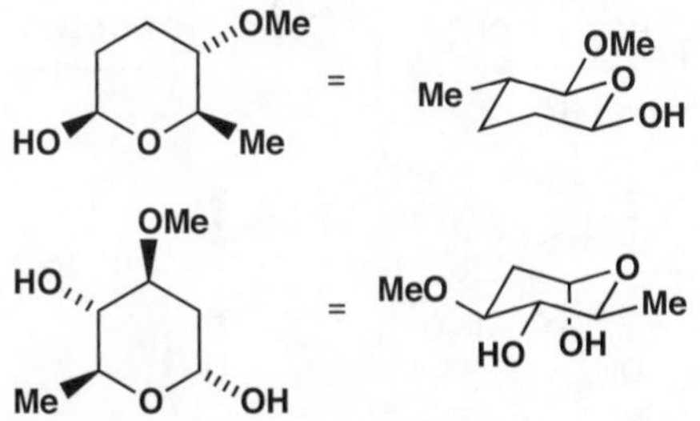

PROBLEM 11

Suggest a mechanism for the following reaction. The product has the following signals in its ^{1}H NMR spectrum: δ_H 3.9 (1H, ddq, J 12, 4, 7) and 4.3 (1H, dd, J 11, 3). What is the stereochemistry and conformation of the product?

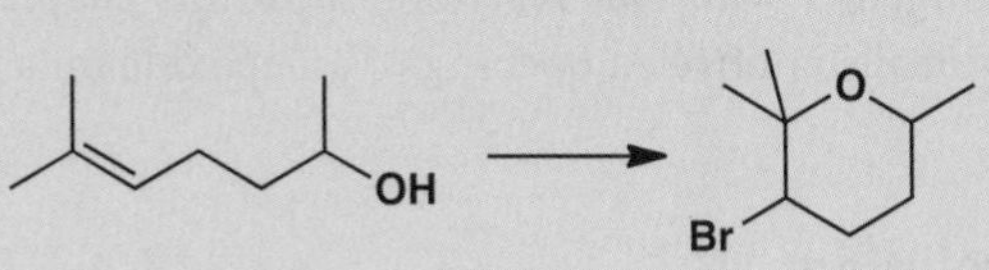

Purpose of the problem

A variation of a familiar mechanism, and illustration of how even limited spectroscopic data can help identify stereochemistry.

Suggested solution

■ This 'bromoetherification' is a variant of the more familiar mechanism of bromolactonization (pp. 568–9 of the textbook).

The mechanism is formation of a bromonium ion by attack on the electron-rich alkene, followed by intramolecular nucleophilic attack by the OH group at the more substituted carbon atom. The NMR spectrum shows that the protons next to Br and O (shown as H^a and H^b) are both axial, since they have a large coupling constant of around 12 Hz (see pp. 796–9 of the textbook). The Br atom and the Me group must therefore be equatorial.

PROBLEM 12

Revision problem. Give mechanisms for each step in this synthesis and explain any regio- or stereochemistry.

Purpose of the problem

Revision of chapters 11 (acetal formation and hydrolysis) and 17 (eliminations) together with two reactions from this chapter: epoxidation and conformationally controlled opening of an epoxide on a six-membered ring.

Suggested solution

■ Acetal formation and hydrolysis is on p. 226 of the textbook, conformational analysis is in chapter 16, and elimination reactions are in chapter 17.

The first step is the standard formation of an acetal from a ketone. The epoxidation occurs from the bottom face of the ring (as drawn) because the axial methyl group blocks the top face. Ring opening with HF gives the *trans* di-axial product. The elimination is by the E1cB mechanism and the hydroxyl group is easily lost as it is axial.

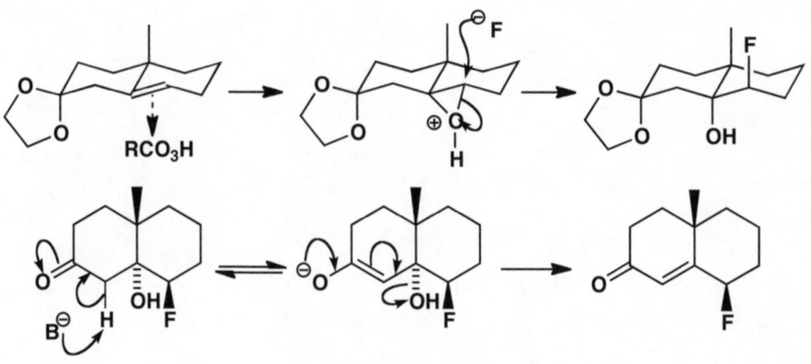
O
O
RCO$_3$H
F
O
O
O
H
O
O
F
OH
O
OH
F
H
B
O
OH
F
O
F

Suggested solutions for Chapter 33

PROBLEM 1

How would you make each diastereoisomer of this product from the same alkene?

Purpose of the problem

A gentle introduction to stereochemical control in open-chain compounds.

Suggested solution

The compounds are acetals and can be made from the corresponding diols with no change in stereochemistry. The question really is: how do you make *cis* and *trans* diols from the alkene?

The *cis* diol is best made by dihydroxylation with OsO_4 as the reagent and a co-oxidant to regenerate it. The *trans* diol comes from the epoxide by nucleophilic attack with water.

PROBLEM 2

Explain the stereochemistry shown in this sequence of reactions.

Purpose of the problem

Chelation-controlled reduction is an important method for stereochemical control in open-chain compounds.

Suggested solution

In both reductions the zinc atom is coordinated to the oxygen of the nearer functional group (CO_2Bn in the first and OH in the second) and the oxygen of the ketone being reduced. This fixes the conformation of the molecule and the borohydride ion attacks from the less hindered side. *Anti* stereochemistry results in both cases.

■ T. Nakata and group, *Tetrahedron Lett.*, 1983, **24**, 2657.

■ You might prefer to draw the zinc-chelated structures as Newman projections, as shown in the textbook on p. 863.

PROBLEM 3

How is the relative stereochemistry of this product controlled? Why was this method chosen?

Purpose of the problem

This may seem trivial but the principle is important.

Suggested solution

The relationship between the two chiral centres in the product is 1,5 and that is too remote for any realistic control. The only plan is to disconnect between the two centres and add a removable anion-stabilizing group to one side and a leaving group to the other. The starting materials must of course be single enantiomers—then only one diastereoisomer can be formed.

■ The synthesis of this compound, a precursor to the aggregation pheromone of flour beetles, is described by K. Mori *et al.* in *Tetrahedron*, 1983, **39**, 2439.

PROBLEM 4

When this hydroxy-ester is treated with a two-fold excess of LDA and then alkylated, one diastereoisomer of the product predominates. Why?

Purpose of the problem

Analysis of an apparently simple case where chelation has the last word.

Suggested solution

The first LDA molecule removes the OH proton and only the second gives the lithium enolate. The enolate is held in a ring by chelation to the first lithium atom so that the allyl group adds to the less hindered face—opposite the methyl group. We've rotated the right hand end of the product to compare the stereochemistry clearly with the structure in the problem: make sure you can see that there is no change at the ester-bearing centre when you do this.

PROBLEM 5

Explain the stereochemical control in this reaction, drawing all the intermediates.

Purpose of the problem

Aldols are versatile and important ways of controlling open-chain stereochemistry by way of a cyclic transitions state.

Suggested solution

The geometry of the enolate is all important (p. 868 of the textbook) and here the large *t*-butyl group will direct the formation of the *cis* lithium enolate. Then the aldol reaction goes through a six-membered cyclic transition state (Zimmerman-Traxler) with the R group of the aldehyde taking up an equatorial position. This gives the *syn* aldol product.

PROBLEM 6

Explain how the stereochemistry of this epoxide is controlled.

Purpose of the problem

An example of the important iodolactonization reaction.

Suggested solution

The bicarbonate ($NaHCO_3$) is a strong enough base to remove the proton from the carboxyic acid. Iodine attacks the alkene reversibly to give a mixture of diastereoisomers of the iodonium ion. If the I^+ and Me groups are on the same side of the chain, the carboxylate group can attack the iodonium ion from the back and set up a *trans* iodolactone. The iodolactone is cleaved by methoxide and the oxyanion displaces iodide to give the epoxide.

PROBLEM 7

Explain how these reactions give different isomers of the same product.

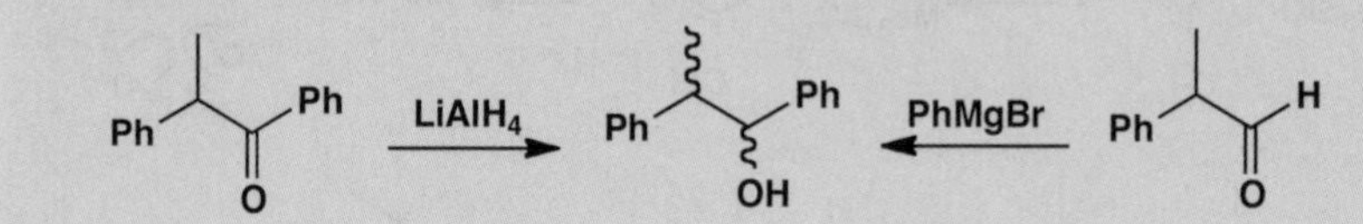

Purpose of the problem

Practice at analysing stereochemical control using the Felkin-Anh model.

Suggested solution

In each case we have nucleophilic attack on a carbonyl group with a neighbouring chiral centre. The Felkin-Anh analysis tells us first to put the largest group prependicular to the carbonyl group and then to bring the nucleophile in alongside the smaller substituent. This is best shown as a Newman projection. In the first case it is better to rotate the front atom in the product so that the two Ph groups are at 180° and we can then draw the structure in the same arrangement.

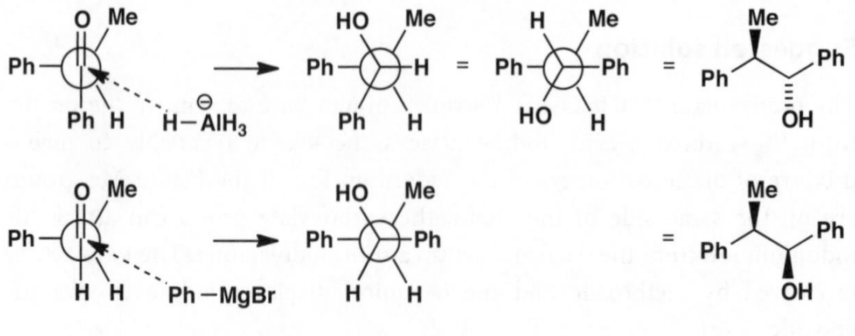

PROBLEM 8

Explain the stereoselectivity of this reaction. What isomer of the epoxide would be produced by treatment of the product with base?

Purpose of the problem

A stereoelectronically controlled Felkin-Anh analysis.

Suggested solution

In this case the chloro substituent dominates because it has an electronic interaction with the carbonyl group. The two alkyl chains come out opposite one another so it is easy to draw the product in a reasonable fashion by imagining yourself observing the Newman projection from the top right.

To draw the stereochemistry of the epoxide formation it is sensible to put the reacting groups in the plane of the paper and arranged so that the oxyanion can do an S_N2 displacement.

PROBLEM 9

How could this cyclic compound be used to produce the open-chain compound with correct relative stereochemistry?

Purpose of the problem

Practice at relating the stereochemistry of cyclic and open-chain compounds.

Suggested solution

We should first discover which atoms in the cyclic compound provide which atoms in the product. Numbering the atoms is the easiest way and it shows little change except that C9 has gone and C8 has become an aldehyde.

■ Oxidative cleavage of diols is on p. 443 of the textbook.

We need to hydrolyse the ester and the acetal and oxidize the 1,2-diol to cleave the C–C bond between the two OH groups. The stereochemistry at C3 and C7 is unchanged and neither is threatened by any of the reaction conditions.

PROBLEM 10

How would you transform this alkene stereoselectively into either of the diastereoisomers of the amino alcohol?

Purpose of the problem

A more difficult extension of problem 1.

Suggested solution

Opening the epoxide with a nitrogen nucleophile makes one isomer. At least the alkene is symmetrical so it doesn't matter which end of the epoxide is attacked by the nucleophile. We have chosen azide ion as the nucleophile. You were not asked to make both diastereoisomers so we can stop there.

PROBLEM 11

Explain the formation of essentially one stereoisomer in this reaction.

Purpose of the problem

A more difficult extension of problem 4 with added Felkin-Anh considerations.

Suggested solution

The *syn* selectivity of the aldol reactions comes from the chair conformation of the cyclic (Zimmerman-Traxler) transition state. Ignoring the stereochemistry of the aldehyde we have this simplified explanation. The transition state contains a chair in which the methyl group has no choice but to be axial while the aldehyde's R substituent chooses to be equatorial.

We have inevitably drawn the *syn* aldol product as one enantiomer but so far we have no explanation for the control of absolute stereochemistry. The aldehyde itself is a single enantiomer so the two faces of the carbonyl group are diastereotopic and we might expect one would be chosen by the normal Felkin-Anh argument.

■ This was also to the surprise of Satoru Masamune and his co-workers: see *Angew. Chem., Int. Ed. Engl.*, 1980, **19**, 557.

To our surprise this is not the preferred isomer. In fact the 'anti-Felkin' isomer predominates by about 3:1. The compound is entirely the *syn* aldol, as predicted, but attack has occurred on the aldehyde in the alternative conformation.

There is an important lesson to be learnt here. The principles we have been explaining are generally true but in any new individual case the result may not follow the principle. This is particularly true of Felkin-Anh control with aldehydes as the small size of the H atom allows other conformations to become relatively favourable.

PROBLEM 12

The following sequences show parts of the syntheses of two different HIV protease inhibitors. What reagents are required for steps 1–4? (For steps 1 and 3, consider carefully how the stereochemistry of the product might be controlled.)

Purpose of the problem

Choosing reagents to achieve desired stereo- and chemoselectivity.

Suggested solution

In step 1, we need to achieve diastereoselectivity during the addition of a nucleophile, CN^-, to a carbonyl group adjacent to a stereogenic centre. The question is: do we need chelation control, or just Felkin-Anh control? Drawn out below is the conformation in which the aldehyde would react with cyanide if simple Felkin-Anh control were operating: check for yourself that the product is the correct one (you may need to build a model if you cannot see this easily). No chelation is needed. In step 2, the nitrile is hydrolysed to the acid and the benzyl groups are hydrogenolysed. This has to be the order: hydrogenolysis first risks reducing the nitrile to an amine.

■ In fact , this step was carried out with KCN in the presence of a Lewis acid (Me_3Al) because the bulky benzyl groups prevent the nitrogen participating in chelation.

In the second sequence, the nucleophile in step 3 must be a vinyl anion equivalent, maybe vinylmagnesium bromide. Comparison of the relative configuration of this product with the one above it immediately suggests that Felkin-Anh control is not operative here, since the opposite diastereoisomer is formed. Drawn below is the expected reactive conformation for a reaction involving chelation control: note that the acidic NH proton must be removed by any basic nucleophile. The outcome is correct: we need to achieve chelation control, so a magnesium counterion is a good choice (Mg^{2+} readily takes part in chelated transition states). The final C=C bond cleavage in step 4 can be achieved by ozonolysis.

■ Methods for breaking a C=C bond to form a C=O bond are outlined on p. 443 of the textbook.

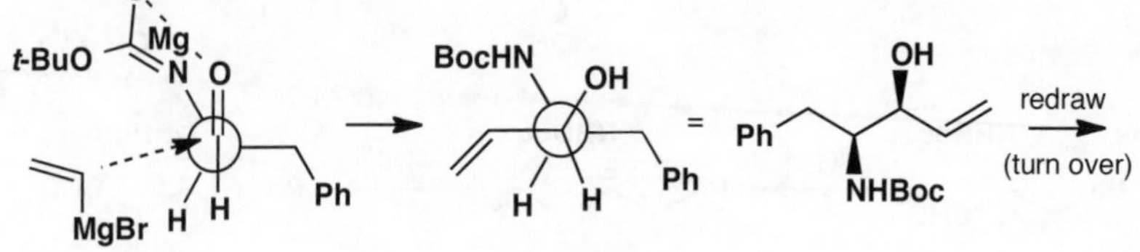

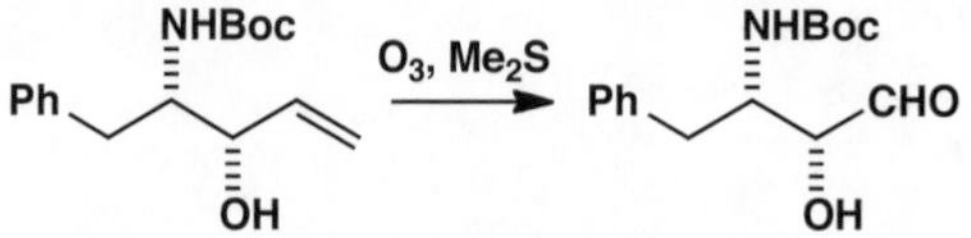

redraw (turn over)

Suggested solutions for Chapter 34

Purpose of the problem

Can you deal with a moderately complicated Diels-Alder?

Suggested solution

The diene is electron-rich and will use its HOMO in the cycloaddition. It will therefore prefer the alkene with the lowest LUMO and that must be the unsaturated ester. Both substituents on the diene direct reaction to the same end. We can predict this from electron donation from either of the oxygen atoms of the diene and in other ways.

both substituents put largest
coefficient of HOMO here

The stereochemistry of the alkene (H and CO$_2$Me *cis*) will be faithfully reproduced in the product. The stereochemistry at the OMe group comes from *endo* attack—we should tuck the ester group underneath (or above—makes no difference) the diene so that it can overlap with the orbitals of the middle two atoms of the diene. If you also said that this product would eliminate methanol on workup so that only the stereochemistry of the ring junction matters, you'd be right.

■ This chemistry is part of a synthesis of the antitumour agent vernolepin by S. Danishefsky and group, *J. Am. Chem. Soc.*, 1976, **98**, 3028.

PROBLEM 2

Comment on the difference in rate between these two reactions.

Purpose of the problem

More details of the intramolecular Diels-Alder reaction.

Suggested solution

The dienes are the same, the ring sizes are the same, and the only difference is the presence of a benzene ring in the faster reacting compound. We should draw a mechanism for one of the reactions to see what is happening.

■ This reaction is part of a synthesis of the taxane skeleton by K. J. Shea and P. D. Davis, *Angew. Chem. Int. Ed. Engl.*, 1983, **22**, 419.

We are making two new rings. The six-membered ring containing an alkene in the product presents no problem. The eight-membered ring with a ketone in it might present a problem, but the ten-membered ring containing a *trans* alkene is definitely a problem. It is much easier to make medium rings (8- to 14-membered) when there is a *cis* alkene in the ring and the benzene ring helps there. It also increases the population of conformers with the ends of their chains close together and probably lowers the LUMO energy by conjugation with the ketone.

PROBLEM 3

Justify the stereoselectivity in this intramolecular Diels-Alder reaction.

Purpose of the problem

Exploring the stereochemistry of an intramolecular Diels-Alder reaction.

Suggested solution

Intramolecular Diels-Alder reactions can give *endo-* or *exo-* products. We should first discover which this is. Drawing the transition state for the *endo* reaction, we find that it is correct—the *endo* product is formed. So electronic factors dominate, perhaps because the dienophile has such a low-energy LUMO and it has two carbonyl groups for secondary orbital overlap with the back of the diene.

■ J. D. White and B. G. Sheldon *J. Org. Chem.*, 1981, **46**, 2273.

PROBLEM 4

Explain the formation of single adducts in these reactions.

Purpose of the problem

Investigating the regio- and stereoselectivity of one inter- and one intra-molecular Diels-Alder reaction.

Suggested solution

The stereochemistry of the first reaction is straightforward: it gives the *endo* product.

The regiochemistry is not quite so simple. The diene has the larger HOMO coefficient at the top end as drawn, so we must deduce that the largest LUMO coefficient in the unsymmetrical quinone is at the top left as drawn. This would result from the electron-donating MeO group making the top carbonyl group and the right-hand alkene less electrophilic, while the bottom carbonyl activates the top end of the left-hand alkene. Or, if you use the mnemonic, this is an 'ortho' product.

■ These are early steps in Corey's synthesis of the plant hormone gibberellic acid. E. J. Corey *et al., J. Am. Chem. Soc.*, 1978, **100**, 8031.

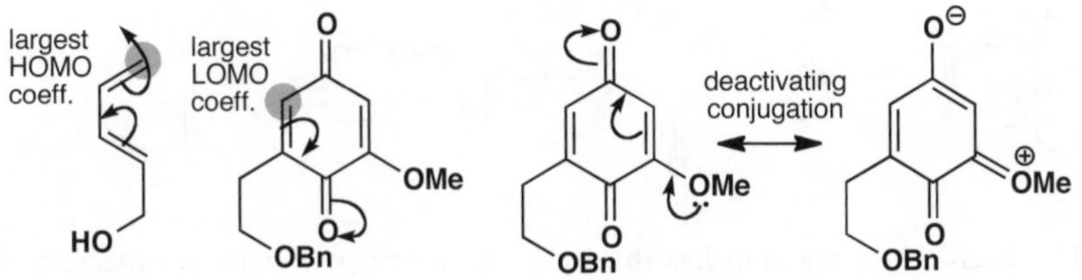

largest
HOMO
coeff.

largest
LOMO
coeff.

O

OMe

HO

OBn

deactivating
conjugation

O

OMe

OBn

O⊖

⊕
OMe

OBn

The second example is intramolecular so the regiochemistry is determined by that alone: the ester linkage between the diene and the dienophile is too short for any variation. This same link ('tether') also forces the dienophile to approach the diene from below. All that remains is the *endo/exo* question and the diagram shows that the product is *endo* with the carbonyl group tucked under the back of the diene.

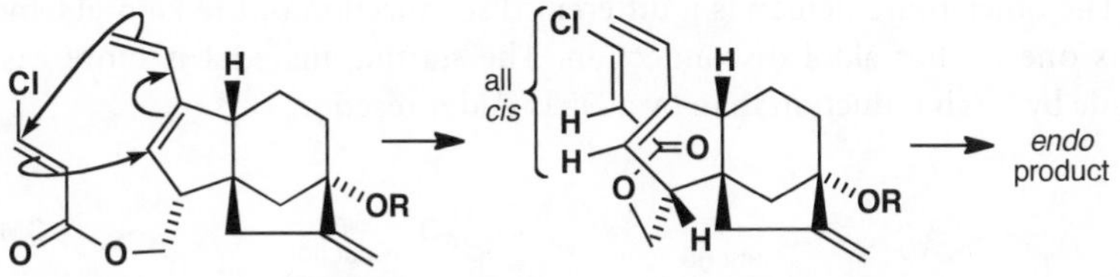

Cl

H

"'OR

all
cis

Cl

H

H

H

O

O

"'OR

H

endo
product

PROBLEM 5

Suggest two syntheses of this spirocyclic ketone from the starting materials shown. Neither starting material is available.

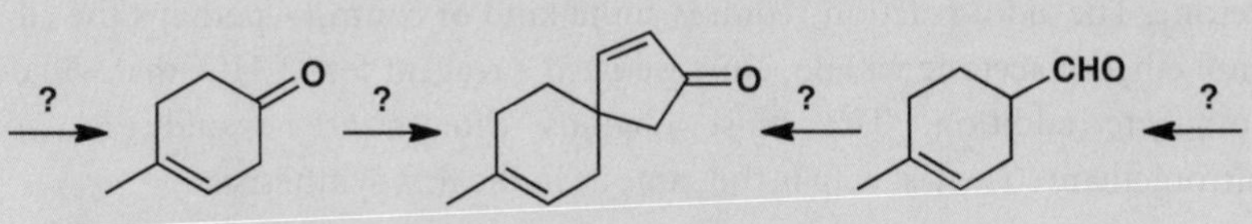

?

O

?

O

?

CHO

?

Purpose of the problem

Revision of synthesis (chapters 24 and 28) with some cycloaddition. Helping you to see that there are alternative ways of making six-membered rings.

Suggested solution

The most obvious disconnection is of the α,β-unsaturated ketone with an aldol reaction in mind. This reveals a 1,4-dicarbonyl compound. Direct disconnection to one of the starting materials is now possible and each can be made by a Diels-Alder reaction.

The Diels-Alder reaction has the right ('*para*') regioselectivity, especially if we use a Lewis acid catalyst such as $SnCl_4$, and we shall need a non-basic specific enol equivalent for the alkylation: an enamine will do fine.

The other route demands a different disconnection of the keto-aldehyde plus one further aldol disconnection. The starting material is more easily made by Birch reduction than by a Diels-Alder reaction.

■ The use of Birch reduction to make cyclohexenes is on p. 542 of the textbook.

The Birch reduction gives the enol ether of the ketone and demands careful hydrolysis to avoid the alkene moving into conjugation with the ketone. The aldol reaction requires some kind of control—perhaps the silyl enol ether of acetone will do. Now we need a reagent for '⁻CHO' that will do conjugate addition. The most obvious choices are cyanide ion or nitromethane. The last step is the same as in the first synthesis.

PROBLEM 6

Draw mechanisms for these reactions and explain the stereochemistry.

Purpose of the problem

Exploration of stereochemical control by 1,3-dipolar cycoaddition reactions. Revision of the importance of cyclic compounds in stereochemistry.

Suggested solution

The nitrile oxide adds in one step to the *cis* alkene to give a single diastereoisomer of the 1,3-dipolar cycloadduct. This is a [3+2] cycloaddition with the three-carbon dipole supplying four electrons. The two methyl groups on the alkene start *cis* and remain so in the adduct.

The first reduction must be of the imine as it is stereoselective, with hydride being transferred to the face of the five-membered ring opposite to the methyl groups. N–O reduction follows.

■ If reduction of the N–O bond occurred first, we should expect little control in the reduction of the open chain imine.

PROBLEM 7

Give mechanisms for these reactions and explain the regio- and stereochemical control (or lack of it!). Note that MnO₂ oxidizes allylic alcohols to enones.

mixture of diastereoisomers

Purpose of the problem

Selectivity and application of a 1,3-dipolar cycloaddition.

Suggested solution

■ The 1,3-dipolar cycloaddition was developed by J. J. Tufariello and R. G. Gatrone, *Tetrahedron Lett.*, 1978, 2753.

The first thing to do is to sort out the mechanism for the cycloaddition. The nitrone uses its LUMO (the π* of the C=N bond) to react with the HOMO of the diene whose largest coefficient is at the end away from the phenyl group (this is where an electrophile would react). There is no selectivity as there is no conjugation and no *exo/endo* selection.

Reduction with zinc cleaves the N–O bond and MnO₂ oxidizes the allylic alcohol to the enone. At this point there is only one chiral centre so the mixture of diastereoisomers has become one compound. Conjugate addition of the amine gives the new ring.

The stereochemistry is more difficult to explain. The product will choose a *trans* ring junction (the nitrogen can invert and *trans* 6,6-ring fusions are more stable), but that means the phenyl group has to be axial, which is presumably not the more stable arrangement. It seems likely that this is the kinetic product. It looks as though the ring closes with the best overlap between the nitrogen lone pair and the π^* orbital of the enone to give a *cis* ring junction that equilibrates by pyramidal inversion at nitrogen to the more stable *trans* ring junction. Axial phenyl is not so bad here as there is only one 1,3-diaxial interaction to the phenyl group, and even that is just with a hydrogen atom.

■ This is part of a synthesis of various alcohols by C. Kibayashi *et al.*, *J. Chem. Soc., Chem. Commun.*, 1983, 1143.

PROBLEM 8

Suggest a mechanism for this reaction and explain the stereo- and regiochemistry.

Purpose of the problem

Two non-routine Diels-Alder-type reactions.

Suggested solution

The reaction is clearly a cycloaddition but at first sight the selectivity is all wrong. The puzzle is solved when we realize that this is a reverse electron demand Diels-Alder. The diene is very electron-deficient with its two conjugated carbonyl groups so the dienophile needs to be electron-rich. It is not very electron rich as drawn, but its enol is. The first formed adduct loses carbon dioxide in a reverse cycloaddition.

■ This sequence was used by D. S. Watt and E. J. Corey in a synthesis of occidentalol (*Tetrehedron Lett.*, 1972, 4651).

PROBLEM 9

Photochemical cycloaddition of these two compounds is claimed to give the diastereoisomer shown. The chemists who did this work claimed that the stereochemistry of the adduct is simply proved by its conversion into a lactone on reduction. Comment on the validity of this deduction and explain the stereochemistry of the cycloaddition.

Purpose of the problem

Selectivity and application of photochemical [2 + 2] cycloadditions.

Suggested solution

Either of the two starting materials could absorb the light to provide the SOMO for the cycloaddition. This does not affect the stereochemistry of the reaction. There is no *endo* effect in [2 + 2] photocycloadditions so the molecules simply come together with the rings arranged in an *exo* fashion to give the least steric hindrance.

The stereochemistry is easy to explain as the molecule is folded in such a way that only the bottom face of the carbonyl group is open to nucleophilic attack. The oxyanion produced can immediately cyclize to form the lactone. Clearly this is possible only if the O⁻ group is up but also only if the CO_2Me groups are on the same side of the middle four-membered ring as the O⁻ group. The formation of the lactone does indeed prove the stereochemistry.

PROBLEM 10

Thioketones, with a C=S bond, are not usually stable. However, this thioketone is quite stable and undergoes reaction with maleic anhydride to give an addition product. Comment on the stability of the thioketone, the mechanism of the reaction, and the stereochemistry of the product.

Purpose of the problem

Exploration of a new structure, revision of aromaticity, and an encounter with [8 + 2] cycloadditions.

Suggested solution

This particular thioketone is stable because the C=S bond is very polarized by delocalization making the seven-membered ring an aromatic cation with six electrons in it. You can represent this in various ways.

The cycloaddition uses maleic anhydride as a two-electron component with a low LUMO. Although in principle this could undergo a Diels-Alder reaction with one of the dienes in the thioketone, it prefers to react by including the sulfur atom, using eight electrons in a component with a high HOMO coefficient. The tricyclic product is clearly folded back on itself so that the triene in the seven-membered ring and the carbonyl groups in the anhydride are close to each other. From the outcome, it seems there must be an *endo* effect in this [8 + 2] cycloaddition.

PROBLEM 11

This unsaturated alcohol is perfectly stable until it is oxidized with Cr(VI): it then cyclizes to the product shown. Explain.

Purpose of the problem

Discovery of a common effect in intramolecular cycloadditions.

Suggested solution

The starting material might undergo a Diels-Alder reaction but the diene and the dienophile are poorly matched. Both have high energy HOMOs and there isn't a low energy LUMO in sight. Once the enone is formed, the alkene becomes electron-deficient: now the energies match well and cycloaddition is fast. The stereochemistry comes from an *endo* arrangement.

PROBLEM 12

Give mechanisms for these reactions, explaining the stereochemistry.

Purpose of the problem

Looking at [2 + 2] cycloadditions of ketenes.

Suggested solution

Treatment of acid chlorides with tertiary amines produces ketenes. In this case an intramolecular [2 + 2] cycloaddition is possible. The stereochemistry is trivial: a *cis* ring junction is the only one possible.

If a more reactive alkene (in this case the electron-donating O makes the enol ether more reactive) is available, the ketene adds to that instead. Note that the alkene must be present as the ketene is generated. The mechanism and part of the stereochemistry are simple. Because the cyclic alkene has *cis* stereochemistry, the two hydrogens on the six-membered ring must be *cis* in the product. The regiochemistry arises because the alkene is an enol ether and the large coefficient in its HOMO interacts with the central atom of the ketene, the one with the larger LUMO coefficient.

The stereochemistry at the remaining centre comes from the way the two molecules approach one another. The two components are orthogonal and the dotted lines in the middle diagram below show how the new bonds are formed. The carbonyl group of the ketene will prefer to be in the midle of the ring and the side chain of the ketene will bend down away from the top ring. These [2 + 2] thermal cycloadditions normally give an all *cis* product.

■ There is rather more in this chemistry than we can discuss here: see R. H. Bisceglia and C. J. Cheer, *J. Chem. Soc., Chem. Commun.*, 1973, 165.

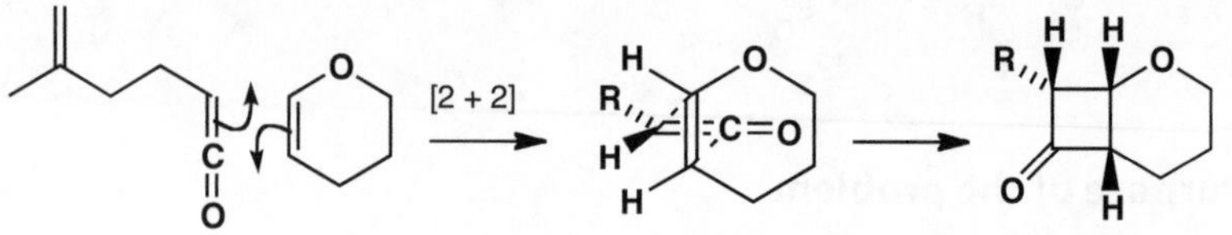

Suggested solutions for Chapter 35

Purpose of the problem

A gentle introduction to an electrocyclic reaction without stereoselectivity.

Suggested solution

Grignard reagents generally prefer direct to conjugate addition, especially with unsaturated aldehydes. MnO₂ specializes in oxidizing allylic alcohols and is the gentle oxidant we need to produce the unstable enone.

The pericyclic process comes next and it is a Nazarov reaction (p. 927 of the textbook), a conrotatory electrocyclic closure of a pentadienyl cation to give a cyclopentenyl cation. There is no stereochemistry and the only regiochemistry is the position of the alkene at the end of the reaction. It prefers the more substituted side of the ring.

The final cuprate addition goes in a conjugate fashion as we should expect as this is what Cu(I) cuprates do. The *cis* 5,5 ring junction is much preferred to *trans* and can equilibrate on work-up by enolization.

PROBLEM 2

Predict the product of this reaction.

Purpose of the problem

A gentle introduction to a sigmatropic reaction without stereo- or regio-selectivity.

Suggested solution

This is a classic Claisen [3,3]-sigmatropic rearrangement sequence starting with an allylic alcohol and forming a vinyl ether by acetal (or in this case orthoester) exchange. The reaction is very *trans* selective.

■ This product was used in a synthesis of chrysanthemic acid by Jacqueline Ficini and Jean d'Angelo: *Tetrahedron Lett.*, 1976, 2441.

PROBLEM 3

Give mechanisms for this alternative synthesis of two fused five-membered rings.

Purpose of the problem

Exploring another aspect of the Nazarov cyclization.

Suggested solution

The first stage is an aliphatic Friedel-Crafts reaction with an acylium ion attacking the alkene.

Next, a Nazarov reaction catalysed by a different Lewis acid closes the five-membered ring and puts the alkene in the only place it can go. The electrocyclic step is conrotatory but that has no meaning with this achiral product.

■ W. Oppolzer and K. Bättig., *Helv. Chim. Acta*, 1981, **64**, 2489.

PROBLEM 4

Explain what is going on here.

Purpose of the problem

Two pericyclic reactions: a sigmatropic shift and a cycloaddition in one reaction scheme.

Suggested solution

The aromatic anion of cyclopentenone displaces tosylate from the alkyl group and then a [1,5] hydrogen shift gives the first product. Such a shift is allowed suprafacially on the ring.

■ This sort of Diels-Alder reaction was used in a synthesis of cedrol by E. G. Breitholler and A. G. Fallis, *J. Org. Chem.*, 1978, **43**, 1964.

Now there is an intramolecular Diels-Alder reaction requiring a high temperature because the dienophile is not activated. The stereochemistry is not obvious but there is no *endo* effect so the molecule folds to give the new five-membered ring a *cis* junction with the old.

PROBLEM 5

A tricyclic hydroxyketone was made by hydrolysis of a *bis* silyl ether. Further reaction gave a new compound. Explain these reactions including the stereochemistry. The diene has the proton NMR spectrum: δ_H 6.06 (1H, dd, J 10.3, 12.1), 6.23 (1H, dd, J 10.3, 14.7), 6.31 (1H, d, J 14.7), and 7.32 (1H, d, J 12.1). Does this agree with the structure given?

Purpose of the problem

Relating the material of this chapter to that of previous chapters with some revision of basic mechanisms.

Suggested solution

The first sequence of reactions is simple. Protonation of the enol ether occurs on the convex face so the OH group is pushed into the *endo* side. Hydrolysis gives the hydroxy-ketone and the tosylate.

The tosylate is displaced with inversion by the excellent S_N2 nucleophile PhS^- and reduction of the ketone from the *exo* face followed by acetylation gives the key intermediate.

Heating this product leads to a retro Diels-Alder reaction: cyclopentadiene is released and a cyclobutene is formed stereospecifically *trans*. This now decomposes by a four-electron conrotatory electrocyclic reaction that could give either the *E,E*- or the *Z,Z*- diene.

The NMR spectrum clearly shows that the *E,E*-diene is formed. The coupling constants for the simple doublets must be for the terminal hydrogens and 14.7 Hz is definitely a *trans* coupling. You might think 12.1 is a bit small for the other *trans* coupling as it is on the low side but the alkene has an electronegative substituent (OAc) and this reduces *J*.

■ It's worth noting for future reference that enol ethers (and enol esters) often have surprisingly small alkene coupling constants.

PROBLEM 6

Treatment of this imine with base followed by an acidic work-up gives a cyclic product with two phenyl groups *cis* to one another. Why is this?

Purpose of the problem

An unusual example of an electrocyclic reaction on an anion.

Suggested solution

The proton from the middle of the molecule is removed to give an anion stabilized by two nitrogens and three phenyl groups. A six-electron electrocyclic reaction closes the five-membered ring and this must be disrotatory, moving both phenyl groups up (or down).

PROBLEM 7

This problem concerns the structure and chemistry of an unsaturated nine-membered ring. Comment on the structure. Explain its different behaviour under thermal or photochemical conditions.

Purpose of the problem

Revision of aromaticity and two alternative electrocyclic reactions.

Suggested solution

The amine has eight electrons in alkenes and two on the nitrogen atom making ten in all. It could be aromatic with $4n + 2$ electrons ($n = 2$). The two reactions are clearly electrocyclic and must be disrotatory to get *cis* ring junctions, the only possible arrangement for two flat rings. Thermally this means a six electron process, but photochemically an eight electron process is all right. The nitrogen does not appear to be involved in either reaction.

■ This was an investigation into the aromaticity of the starting material by A. G. Anastassiou and J. H. Gebrian, *Tetrahedron Lett.*, 1969, 5239.

PROBLEM 8

Propose a mechanism for this reaction that accounts for the stereochemistry of the product.

Purpose of the problem

Another electrocyclic/cycloaddition combination for you to work out.

Suggested solution

The three-membered ring opens using the lone pair on nitrogen in a four-electron conrotatory electrocyclic process. One phenyl group must rotate inwards and the other outwards. Then a cycloaddition of the four-electron 1,3-dipole onto the two-electron dienophile goes without change of stereochemistry. The ester groups remain *cis* and the phenyls must be one up and one down.

■ This extensive study of the opening of three-membered heterocyclic rings came from Huisgen's group in Munich (*J. Chem. Soc., Chem. Commun.*, 1971, 1187, 1188, 1190, and 1192).

PROBLEM 9

Treatment of this amine with base at low temperature gives an unstable anion that isomerizes to another anion above −35 °C. Aqueous work-up gives a bicyclic amine. What are the two anions? Explain the stereochemistry of the product. In the NMR spectrum of the product the two protons in the grey box appear as an ABX system with J_{AB} 15.4 Hz. Comment.

Purpose of the problem

An unusual electrocyclic reaction on an anion with stereochemistry and NMR revision.

Suggested solution

The first anion **A** is formed by removal of the only possible proton: one from the NCH$_2$ group. This anion might be considered aromatic (six electrons from the three alkenes, two from N and two from the anion) but it is clearly unstable as it closes in an electrocyclic reaction at > –35 °C. This is a six-electron process and must therefore be disrotatory. The rotating hydrogens are shown on the structure of **A**. It is essential that the 5,5 ring closure must be *cis* and that demands a disrotatory reaction. Both anions **A** and **B** are extensively delocalized and it is a matter of choice where you draw the anion.

Anion **B** is protonated by water with preservation of the right hand aromatic ring. The final product is a chiral molecule having no plane of symmetry so the boxed CH$_2$ group is diastereotopic with J_{AB} 15.4 Hz. This is larger than usual because of the π-contribution: a neighbouring π-bond increases ^2J by about 2 Hz.

■ This study was originally aimed at finding out the nature of of the starting material, A. G. Anastassiou and group, *J. Chem. Soc., Chem. Commun.*, 1981, 647.

PROBLEM 10

How would you make the starting material for these reactions? Treatment of the anhydride with butanol gives an ester that in turn gives two inseparable compounds on heating. On treatment with an amine, an easily separable mixture of an acidic and a neutral compound is formed. What are the components of the first mixture and how are they formed?

Purpose of the problem

Exploration of alternative conrotatory openings of a cyclobutene.

Suggested solution

The starting material is made by a photochemical [2 + 2] cycloaddition of acetylene and maleic anhydride. Treatment with butanol and base gives the monoester because, after butanol has attacked once, the product is the anion of a carboxylic acid and cannot be attacked again by the nucleophile.

Heat opens the cyclobutene in a conrotatory four-electron electrocyclic reaction. As the two groups are *cis* on the cyclobutene, one must rotate outwards and one inwards. The two groups are similar but not the same so there is little selection and both products are formed.

Treatment with the tertiary amine forms the anions of the carboxylic acids. The one from **B** can do a conjugate addition to the unsaturated ester and form a lactone but that of **A** is too far away and cannot react.

PROBLEM 11

Treatment of this keto-aldehyde (which exists largely as an enol) with the oxidizing agent DDQ (a quinone—see p. 764 of the textbook) gives an unstable compound that turns into the product shown. Explain the reactions and comment on the stereochemistry.

Purpose of the problem

Exploration of a less well defined pericyclic sequence.

Suggested solution

DDQ oxidizes the position between the two carbonyl groups to insert an alkene conjugated with both. We can now put in some stereochemistry as the three-membered ring must be *cis* fused to both six-membered rings. The diene undergoes electrocyclic ring opening to form a seven-membered ring. This is a six-electron and therefore disrotatory reaction and the two bonds to the old three-membered ring are therefore allowed to rotate inwards—the only rotation that can give the product.

■ This observation was vital in developing a synthesis of varucarin A, a natural product with antitumour activity. B. M. Trost and P. G. McDougal, *J. Org. Chem.*, 1984, **49**, 458.

PROBLEM 12

Explain the following observations. Heating this phenol brings it into rapid equilibrium with a bicyclic compound that does not spontaneously give the final product unless treated with acid.

Purpose of the problem

A sigmatropic rearrangement involving a three-membered ring. The σ-bonds in three-membered rings are strained and more reactive than normal σ-bonds and take part readily in pericyclic reactions.

Suggested solution

The first step is a Cope rearrangement—a [3,3]-sigmatropic rearrangement made favourable in this case because the σ-bond that is broken is in a three-membered ring. The product cannot go directly to an aromatic compound as that would require a [1,3] (or a [1,7] depending on how you count) hydrogen shift. Such a shift would have to be antarafacial on the π-system and that is impossible in such a rigid structure.

The aromatization can happen instead by an ionic mechanism. If the extended enol is protonated at the remote end, it can lose a proton from the ring junction to reform the phenol.

■ This reaction was carried out as part of a mechanistic study by E. N. Marvell and S. W. Almond, *Tetrahedron Lett.*, 1979, 2777.

PROBLEM 13

Treatment of cyclohexa-1,3-dione with this acetylenic amine gives a stable enamine in good yield. Refluxing the enamine in nitrobenzene gives a pyridine after a remarkable series of reactions. Fill in the details, give mechanisms for the reaction, structures for the intermediates, and suitable explanations for each pericyclic step. A mechanism is not required for the last step as nitrobenzene simply acts as an oxidant.

Purpose of the problem

Practice in unravelling complicated reaction sequences involving pericyclic steps.

Suggested solution

The formation of the enamine requires only patient adding and subtracting of protons.

■ This enamine is unusually stable and easy to form because it is a vinylogous amide (see p. 512 of the textbook).

The cascade of reactions in hot nitrobenzene starts with a [3,3]-sigmatropic rearrangement that is unusual in that it forms an allene but is otherwise straightforward. To get to the next intermediate, we must go from the ketone to the enol and back again, but with the alkene now in conjugation.

Now we can transfer a proton from nitrogen to the middle of the allene. This is formally a [1,5]-H shift and is, of course, allowed, but it may be an ionic reaction as nitrogen is involved. This gives a diene that can twist round for a six-electron electrocyclic reaction. This is no doubt disrotatory but we can't tell as no stereochemistry is involved.

■ K. Berg-Nielsen and L. Skattebøl, *Acta Chem. Scand.*, 1978, **B32B**, 553.

PROBLEM 1

Rearrangements by numbers: just draw a mechanism for each reaction.

Purpose of the problem

This problem is just to help you acquire the skill of tracking down rearrangements by numbering (arbitrarily) the atoms in the starting material and working out where they've gone in the product.

Suggested solution

The first reaction is the preparation of Corey's 'OBO' protecting group for carboxylic acids. The Lewis acid complexes one of the oxygen atoms and all the atoms of the starting material survive in the product. Atoms 3 and 5 are easy to identify in the product and it doesn't much matter which of the CH_2 groups you label 1, 2, and 4. Of course you may use a completely different numbering system and that's fine. The dotted lines show which new bonds are made and which old bonds are broken.

There is more than one reasonable mechanism: here are two possibilities, the second being perhaps the better.

The second reaction is even easier to work out. Atoms 2 and 3 are easy to find and they identify 1 and 4 in the product.

■ Revisit p. 338-9 of the textbook if you need reminding why.

As the compounds are acetals we must use oxonium ions and not S_N2 reactions. Loss of BF_3 and rotation of the last intermediate gives the product.

The third reaction involves a cyclization. Atoms 1 and 7 clearly make the new bond and the rest of the atoms fit into place except that the bromine has gone and the alkene has moved from 7/8 to 8/9. Zinc inserts oxidatively into the C–Br bond and the mechanism follows from the nucleophilic nature of the organometallic compound.

PROBLEM 2

Explain this series of reactions.

Purpose of the problem

Working out the stereochemistry and mechanism of the Beckmann rearrangement.

Suggested solution

The first reaction forms the oxime by the usual mechanism (chapter 11). This reaction is under thermodynamic control so the OH group will bend away from the aryl substituent. Then we have the Beckmann rearrangement itself (p. 958 of the textbook). The group *anti* to the OH group migrates from C to N and that gives the product after rehydration and adjustment of protons.

■ This example comes from a general investigation into the Beckmann and the related Schmidt rearrangements by R. H. Prager *et al., Aust. J. Chem.,* 1978, **31**, 1989.

PROBLEM 3

Draw mechanisms for the reactions and structures for the intermediates. Explain the stereochemistry, especially of the reactions involving boron. Why was 9-BBN chosen as the hydroborating agent?

Purpose of the problem

Rearrangements involving boron and a ring-closing rearrangement of sorts plus stereochemistry.

Suggested solution

The starting material is symmetrical so it doesn't matter which face of which alkene you attack. The only important things are that boron binds to the more nucleophilic end of the alkene and that R_2B and H are added *cis*. Alkaline H_2O_2 makes the hydroperoxide anion (HOO^-) which attacks boron.

■ The structure of 9-BBN and the mechanism of the oxidation are described on p. 446 of the textbook: here we represent 9-BBN simply as R_2BH.

9-BBN

The mesylate cyclizes in aqueous base. The more nucleophilic end of the remaining alkene displaces the mesylate with inversion to make the *cis* ring junction much preferred by the 5,5 fused system. Water adds to the tertiary cation to give the next intermediate.

Elimination of the alcohol (E1 of course as it is tertiary) gives the alkene and a repeat of the hydroboration from the outside (convex face) of the folded molecule gives the final alcohol with five new stereogenic centres.

9-BBN was chosen because it is very large and reinforces the natural electronic preference of boron to bind to the less substituted end of the alkene with an extra steric effect. It also has bridgehead atoms bound to boron and they make poor migrating groups, forcing the migration of the third B substituent.

■ The final product of these reactions was used as a foundation for the synthesis of the 'iridoid' terpenes by R. S. Matthews and J. J. Whitesell, *J. Org. Chem.*, 1975, **40**, 3313.

PROBLEM 4

It is very difficult to prepare three-membered lactones. One attempted preparation, by the epoxidation of di-*t*-butyl ketone, gave an unstable compound with an IR stretch at 1900 cm^{-1}. This compound decomposed rapidly to a four-membered ring lactone that could be securely identified. Do you think they made the three-membered ring?

Purpose of the problem

Rearrangements as a proof of structure?

Suggested solution

The expected three-membered lactone would have a very high carbonyl stretching frequency because of ring strain. Three-membered cyclic ketones have carbonyl stretches at about 1815 cm^{-1} and lactones have higher frequencies than ketones. So it might be the lactone. If it is, we should find a mechanism for the ring expansion to the four-membered lactone isolated. There is a good mechanism involving migration of a methyl group from one of the *t*-butyl groups. The general conclusion is that R. Wheeland and P. D. Bartlett did indeed make the first α-lactone.

■ See *J. Am. Chem. Soc.*, 1970, **92**, 6057 and also J. K. Crandall and S. A. Sojka, *Tetrahedron Lett.*, 1972, 1641.

possible structure

PROBLEM 5

Suggest a mechanism for this rearrangement.

Purpose of the problem

Working out the mechanism of a new rearrangement.

Suggested solution

The starting material is an enamine and will react with bromine in the manner of an enol. Addition of hydroxide gives the starting material for the rearrangement. Notice that the nitrogen atom migrates rather than the carbon atom and this suggests that it does so by participation. If you numbered the atoms you would have found that the *gem*-dimethyl group and the nitrogen atom give the answer away immediately.

■ This reaction was discovered by L. Duhamel and J.-M. Poirier, *J. Org. Chem.*, 1979, **44**, 3576.

PROBLEM 6

A single enantiomer of the epoxide below rearranges with Lewis acid catalysis to give a single enantiomer of the product. Suggest a mechanism and comment on the stereochemistry.

Purpose of the problem

An unusual group migrates and stereochemistry gives a clue to mechanism.

Suggested solution

The mechanism for the reaction must involve Lewis acid complexation of the epoxide oxygen atom, cation formation, and migration of CO_2Et. This last point may surprise you but inspection of the product shows that CO_2Et is indeed bonded to the other carbon of what was the epoxide.

Although something like this must happen, our mechanism raises as many questions as it answers:

- Why does that bond of the epoxide open? *Answer.* Because the tertiary benzylic cation is much more stable than a secondary cation with a CO_2Et substituent.
- Why does CO_2Et migrate rather than the H atom? *Answer.* For the same reason! If the H atom migrates, the product would be a cation (or at least a partial positive charge would appear in the transition state) next to the CO_2Et group.

• Surely the carbocation intermediate is planar and the product would be racemic? *Answer.* This was the purpose of the investigation. One chiral centre is lost in the reaction so only absolute stereochemistry is relevant. One explanation is that the cation is short-lived and that bond rotation is fast in the direction shown (the CO_2Et group is already down and has to rotate by only 30° to get to the right position for migration). The other is that migration is concerted with epoxide opening. This looks unlikely as the overlap is poor.

■ This mechanism was investigated by R. D. Bach and co-workers; *J. Am. Chem. Soc.*, 1976, **98**, 1975 and 1978, **100**, 1605.

PROBLEM 7

The 'pinacol' dimer of cyclobutanone rearranges with expansion of one of the rings in acid solution to give a cyclopentanone fused *spiro* to the remaining four-membered ring. Draw a mechanism for this reaction. Reduction of the ketone gives an alcohol that rearranges to a bicyclic alkene also in acid. Suggest a mechanism for this reaction and suggest why the rearrangements happen.

Purpose of the problem

An illustration of the easy rearrangement of four-membered rings to form five-membered rings.

Suggested solution

The first reaction is a simple pinacol rearrangement. The diol is symmetrical so protonation of either alcohol and migration of either C–C bond give the product.

Reduction to the alcohol is trivial and then acid treatment allows the loss of water and ring expansion of the remaining four-membered ring. You may well have drawn this as a stepwise process. Elimination gives the most substituted alkene. Both rearrangements occur very easily because of the relief of strain in going from a four- to a five-membered ring.

PROBLEM 8

Give the products of Baeyer-Villiger rearrangements on these compounds, with reasons.

(enantiomerically pure)

Purpose of the problem

Prediction in rearrangements is as important as elsewhere and the Baeyer-Villiger is one of the more predictable rearrangements.

Suggested solution

There are a few minor traps here that we're sure you've avoided. The first compound has two carbonyl groups but esters don't do the Baeyer-Villiger rearrangement so only the ketone reacts. The more substituted carbon migrates with retention of configuration. The aldehyde rearranges with migration of the benzene ring in preference to the hydrogen atom. The last compound is C_2 symmetric so it doesn't matter which group you migrate as long as you ensure retention of configuration. Take care when drawing the product as the migrating group has to be drawn the other way up.

(enantiomerically pure)

PROBLEM 9

Suggest mechanisms for these rearrangements, explaining the stereochemistry in the second reaction.

Purpose of the problem

Unravelling one rearrangement after another with some stereochemistry.

Suggested solution

The first reaction is a simple ring expansion. The amine is not involved, presumably because it is fully protonated. The final loss of proton might be concerted with the migration as this would help explain the position of the alkene in the product.

The second reaction starts with bromination of the alkene and interception of the bromonium ion by the amine. Only when bromine adds to the opposite face of the alkene can the amine cyclize so this reaction resembles iodolactonization. Probably the bromination is reversible.

Finally, the weak base bicarbonate (HCO_3^-) is enough to remove a proton from the nitrogen atom and allow participation in nitrogen migration by displacement of bromide. This alkene is formed because the $C–N^+$ bond to tertiary carbon is broken preferentially.

■ This work allowed the synthesis and trial of some early morphine analogues as painkillers: L Moncovic et al., *J. Am. Chem. Soc.*, 1973, **95**, 647.

PROBLEM 10

Give mechanisms for these reactions that explain any selectivity.

Purpose of the problem

To show that ring expansion from three- to four-membered rings and ring contraction the other way are about as easy.

Suggested solution

The first mechanism is a pinacol rearrangement and the compound is symmetrical so it doesn't matter which alcohol is protonated. Both three- and four-membered rings are strained and the σ-bonds are more reactive than normal (they have a high energy HOMO). This makes ring contraction an easy reaction even though the strain is not relieved.

■ J. M. Conia and J. P. Barnier, *Tetrahedron Lett.*, 1971, 4981.

The second example looks at first to be a similar pinacol rearrangement. But the resulting ketone cannot easily be transformed into the product.

■ J. M. Denis and J. M. Conia, *Tetrahedron Lett.*, 1972, 4593.

Breaking open one of the three-membered rings gets us off to a better start. This gives a hydroxy-ketone that can rearrange in a pinacol fashion with ring expansion of the remaining cyclopropane.

PROBLEM 11

Attempts to produce the acid chloride from this unusual amino acid by treatment with $SOCl_2$ gave instead a β-lactam. What has happened?

Purpose of the problem

To show that ring expansion in small rings is even easier in heterocycles because of participation.

Suggested solution

The formation of the acid chloride might go to completion or it might be that some intermediate on the way to the acid chloride rearranges. We shall use an intermediate. Whichever you use, it is participation by nitrogen that starts the ring expansion going, though the next intermediate is very unstable. When chloride attacks the bicyclic cation, it cleaves the most strained bond, the one common to two three-membered rings.

■ This surprising reaction is one way to make the important β-lactams present in penicillins and other antibiotics. J. A. Deyrup and S. C. Clough, *J. Am. Chem Soc.*, 1969, **91**, 4590.

PROBLEM 12

Treatment of this hydroxy-ketone with base followed by acid gives the enone shown. What is the structure of intermediate **A**, how is it formed, and what is the mechanism of its conversion to the final product?

Purpose of the problem

Fragmentation may be followed by another reaction.

Suggested solution

Removal of the hydroxyl proton by the base promotes a fragmentation that is a reverse aldol reaction. It works because the C–C bond being broken is in a four-membered ring. Then an acid catalysed aldol reaction in the normal direction and elimination via the enol (E1cB) allows the formation of the much more stable six-membered ring.

PROBLEM 13

Just to check your skill at finding fragmentations by numbers, draw the mechanism for each of these one-step fragmentations in basic solution with acidic work-up.

Purpose of the problem

As the problem says, to help you unravel simple fragmentations.

Suggested solution

We can identify the six-membered ring in both compounds—the sequence 1–6 is clearly the same in both with a side chain at C3. The fragmentation is easy enough too—the OH proton is removed and the mesylate must be the leaving group so the groups doing the 'pushing' and 'pulling' are clear from the start.

The two CO_2H groups in the second product might cause a moment's concern but one is on a $-CH_2-CH_2-$ side chain and the other is at a branch point and we can soon fill in the rest of the numbers.

Clearly the OH proton is removed and one of the carboxyls is a leaving group. The stereochemistry disappears in the fragmentation but it is important, as the conformational drawing shows. One lone pair on the O^-, the bond being fragmented, and the bond to the leaving group are all parallel (shown in thick lines).

PROBLEM 14

Explain why both these tricyclic ketones fragment to the same diastereoisomer of the same cyclo-octane.

Purpose of the problem

Fragmentations linked to ester hydrolysis.

Suggested solution

It is obvious from the reactions that two features have disappeared from the starting materials: an ester group (OAc) and a four-membered ring. The ester can be hydrolysed by KOH and the four-membered ring disappears in the fragmentation. As usual, draw the mechanism first and worry about the stereochemistry later. For the first compound, this sequence gives the enolate of a diketone and hence the diketone itself.

The second compound follows the same sequence and a different enolate emerges, but it is simply another enolate of the same ketone. Both compounds give the same basic structure.

KOH
EtOH

But what about stereochemistry? We are not told the stereochemistry of the starting materials but know that 5,4 fused rings must have a *cis* ring junction. This junction survives in the first compound so the stereochemistry must have changed. The second compound gives us the clue as to how. When it tautomerizes to the ketone it will select the more stable *trans* 8,5 ring junction. In the same way, the enolate from the first sequence is in equilibrium under the reaction conditions with all the other enolates of the same ketone, including those at ring junctions. This is a stereo*selective* reaction.

first product from the
second fragmentation

PROBLEM 15

Suggest a mechanism for this fragmentation and explain the stereochemistry of the alkenes in the product. This is a tricky problem, but find the mechanism and the stereochemistry will follow.

Purpose of the problem

Probably the most beautiful application of fragmentation yet by a true genius of chemistry, Albert Eschenmoser.

Suggested solution

The tosylate is obviously the leaving group, the two oxygens in the ring must become the ester group, and the CO_2^- must leave as CO_2. All that remains is to trace a pathway from CO_2^- to OTs via one of the ring oxygens using parallel bonds. Though you could draw a mechanism for this double fragmentation, it is not convincing. The only electrons *anti*-parallel to the C–OTs bond are those in the ring junction bond and the equatorial lone pair on one of the ring oxygens. Marking these with heavy lines, we carry out the first fragmentation. We've also drawn in the hydrogen that ends up on the alkene so you can see clearly where the *trans* geometry comes from

■ *Angew. Chem. Int. Ed. Engl.*, 1979, **18**, 634, 636.

The second fragmentation is easier to see if we redraw the intermediate so that we can see which groups are antiparallel. A conformational drawing also reveals the correct alkene geometry.

PROBLEM 16

Suggest a mechanism for this reaction and explain why the molecule is prepared to abandon a stable six-membered ring for a larger ring.

Purpose of the problem

A simple example of fragmentation used to create a medium size (11-membered) ring.

Suggested solution

The strong base removes the proton from the OH group and the oxyanion attacks one of the carbonyl groups (they are the same). This intermediate might decompose back to starting materials but it can also fragment with the loss of an enolate. The product is then an ester, and protonation of the enolate completes the reaction. The eleven-membered ring is more stable than usual because of the benzene ring (see problem 2, chapter 34), and because the ester does not suffer from cross-ring interactions in its favoured s-*trans* conformation.

■ J. R. Mahajan and H. de Carvalho, *Synthesis*, 1979, 518.

PROBLEM 17

Give mechanisms for these reactions, commenting on the fragmentation.

Purpose of the problem

Revision of conjugate addition of enols, another ring expansion with an enolate as leaving group and an interesting piece of stereochemistry.

Suggested solution

The first step is enamine formation and the second is conjugate addition. This appears to lead to a dead end as we cannot find a way to make the intermediate from the product.

The answer is to exchange the enamine of the ketone with the enamine of the aldehyde. Under the conditions, enamine formation is reversible and there are various ways you could draw details. Cyclization of this compound now gives the intermediate we are looking for.

The last two diagrams show where the stereochemistry comes from. The final product has a chair six-membered ring. The 1,3-bridge on the bottom of this ring must be diaxial or it cannot reach round. The pyrrolidine is equatorial and the five-membered ring must be *cis* fused. No doubt the stereochemistry as well as the intermediates are under thermodynamic control.

Finally the fragmentation itself. Methylation of the nitrogen makes it into a leaving group and addition of hydroxide to the ketone provides the electronic push. Notice that the C–N$^+$ bond, the C–C bond being fragmented, and a lone pair on the O$^-$ group are all parallel. The stereochemistry is already there in the intermediate.

PROBLEM 18

Suggest mechanisms for these reactions, explaining the alkene geometry in the first case. Do you consider that they are fragmentations?

Purpose of the problem

Simple fragmentations involving the opening of three-membered rings.

Suggested solution

The first reaction is a fragmentation without any 'push' but that is all right because the bond that is being broken is in a three-membered ring. You may have drawn a concerted mechanism or a stepwise one with a cation as intermediate. Either may be correct. The stereochemistry of the alkene is thermodynamically controlled.

■ W. S. Johnson *et al., J. Am. Chem. Soc.*, 1978, **100**, 4268.

The second reaction is base-catalysed and starts with the hydrolysis of the ester by NaOH. This fragmentation also needs 'push', though only a three-membered ring is being broken, because the leaving group is an enolate, nowhere near as electron-withdrawing as the water molecule or the carbocation in the first example. Are they fragmentations? In both cases a C–C bond is being broken but we would understand if you felt the first was not strictly a fragmentation, particularly if it goes stepwise. Neither reaction breaks the molecule into three pieces and the terminology is merely a matter of opinion.

■ K. Kondo *et al., Tetrahedron*, 1978, 907.

PROBLEM 19

What steps would be necessary to carry out an Eschenmoser fragmentation on this ketone, and what products would be formed?

Purpose of the problem

Revision of an important and complex reaction involving fragmentation.

Suggested solution

The Eschenmoser fragmentation (p. 965 of the textbook) uses the tosylhydrazone of an α,β-epoxy-ketone. The epoxide can be made with alkaline hydrogen peroxide and the tosylhydrazone needs just tosylhydrazine to form what is essentially an imine. Then the fun can begin. The stereochemistry doesn't matter for once.

The fragmentation is initiated with base that removes the proton from the NHTs group. This anion fragments the molecule one way and then the oxyanion fragments it the other way with nitrogen gas and Ts⁻ as leaving groups. The product is an acetylenic aldehyde or ketone.

PROBLEM 20

Revision content. Suggest mechanisms for these reactions to explain the stereochemistry.

Purpose of the problem

Rearrangements and a fragmentation.

Suggested solution

The ring opening and the rearrangement cannot be concerted because the group on the 'wrong' side of the molecule migrates. There must be a cationic intermediate. In contrast, attack of bromide occurs stereospecifically from the side opposite the migrating group, so this is presumably concerted with the rearrangement.

The second reaction is a fragmentation. Silver(I) is an excellent Lewis acid for halogens and probably produces a secondary carbocation intermediate. Push from the OH group completes the fragmentation.

■ As it happens the starting epoxide is that of natural α-pinene so it and the product are single enantiomers. P. H. Boyle *et al.*, *J. Chem. Soc., Chem. Commun.*, 1971, 395.

Suggested solutions for Chapter 37

PROBLEM 1

Give a mechanism for the formation of this silylated ene-diol and explain why the Me₃SiCl is necessary.

Purpose of the problem

Reminder of an important radical reaction.

Suggested solution

This is an acyloin condensation linking radicals derived from esters by electron donation from a dissolving metal (here sodium). If the esters can form enolates, the addition of Me₃SiCl protects against that problem by removing the MeO⁻ by-product.

The first product is a very electrophilic 1,2-dione and it accepts electrons from sodium atoms even more readily than do the original esters. The product is an ene diolate that is also silylated under the reaction conditions.

■ Details from B. M. Trost and group, *J. Org. Chem.*, 1978, **43**, 4559.

PROBLEM 2

Heating the diazonium salt below in the presence of methyl acrylate gives a reasonable yield of a chloroacid. Why is this unlikely to be nucleophilic aromatic substitution by the S_N1 mechanism (p. 520 of the textbook)? Suggest an alternative mechanism that explains the regioselectivity.

Purpose of the problem

Revision of nucleophilic aromatic substitution with diazonium salts and contrasting cations and radicals.

Suggested solution

The cation mechanism is perfectly reasonable as far as the diazonium salt is concerned but it will not do for the alkene. Conjugated esters are electrophilic and not nucleophilic alkenes. Even if it were to attack the aryl cation, we should find the reverse regioselectivity.

The only way to produce the observed product is to decompose the diazonium salt homolytically. To do this we can draw the salt as a covalent compound or transfer one electron from the chloride ion to the diazonium salt. The other product would be a chlorine radical. Addition to the alkene gives the more stable radical which abstracts chlorine from the diazonium salt and keeps the chain going.

■ Notice that in the last step we have put in only half the mechanism—we shall generally do this from now on as it is clearer. There is nothing wrong with putting in another chain of half-headed arrows going in the other direction.

PROBLEM 3

Suggest a mechanism for this reaction and comment on the ring size formed. What is the minor product likely to be?

Purpose of the problem

Activated alkenes are not necessary in radical cyclizations.

Suggested solution

The peroxide is a source of benzoyloxy radicals (PhCO$_2$·) and these capture hydrogen atoms to give the most stable radical. The best one here is stablized by both CN and CO$_2$Et. Cyclization onto the alkene gives mainly a secondary radical on a six-membered ring and this abstracts a hydrogen from starting material to complete the cycle.

The alternative is to add to the more substituted end of the alkene. This gives a less stable primary radical, but this '5-*exo*' ring closure is often preferred because the orbital alignment is better. The minor product has a five-membered ring.

PROBLEM 4

Treatment of this aromatic heterocycle with NBS (*N*-bromosuccinimide) and AIBN gives mainly one product but this is difficult to purify from minor impurities containing one or three bromine atoms. Further treatment with 10% aqueous NaOH gives one easily separable product in modest yield (50%). What are the mechanisms for the reactions?

Purpose of the problem

An important radical reaction: bromination at benzylic and allylic positions by NBS, and an application.

Suggested solution

Two preliminary reactions need to take place: NBS is a source of a low concentration of bromine molecules and AIBN initiates the radical chain by forming a nitrile-stabilized tertiary radical.

The new radical abstracts hydrogen atoms from the benzylic positions to make stable delocalized radicals. These react with bromine to give the benzylic bromide and release a bromine atom.

All subsequent hydrogen abstractions are carried out by bromine atoms, either of the kind we have just seen or to remove a hydrogen atom from the other methyl group. This reaction provides the HBr that generates more bromine from NBS.

Finally the dibromide reacts with NaOH to give the new heterocycle. Both S$_N$2 displacements are very easy at a benzylic centre and the second is intramolecular.

■ This product was used to make constrained amino acids by S. Kotha and co-workers, *Tetrahedron Lett.*, 1997, **38**, 9031.

PROBLEM 5

Propose a mechanism for this reaction accounting for the selectivity. Include a conformational drawing of the product.

Purpose of the problem

Another important radical reaction: cyclization of alkyl bromides onto alkenes.

Suggested solution

This time AIBN abstracts the hydrogen from Bu$_3$SnH and the tin radicals carry the chain along. First they remove the bromine atom from the starting material to make a vinyl radical that cyclizes onto the unsaturated ketone to give a radical stabilized by conjugation with the carbonyl group. The chain is completed by abstraction of hydrogen from another molecule of Bu$_3$SnH, the tin radical formed then allowing the cycle to restart.

The stereochemistry of the product comes from the requirement of a 1,3-bridge to be diaxial as this is the only way the bridge can reach across the ring. At the moment of cyclization, the vinyl radical side chain must be in an axial position.

PROBLEM 6

An ICI process for the manufacture of the diene used to make pyrethroid insecticides involved heating these compounds to 500 °C in a flow system. Propose a radical chain mechanism for the reaction.

Purpose of the problem

Learning how to avoid a trap in writing radical reactions and to show you that radical reactions can be useful.

Suggested solution

The most likely initiation at 500 °C is the homolytic cleavage of the C–Cl bond to release allyl and chloride radicals. The chloride radicals then attack the alkene and abstract a hydrogen atom to give more of the same allylic radical.

The trap is to form the product by dimerizing the allylic radical. Dimerizing radicals does sometimes occur (in the acyloin reaction for example) but it is a rare process.

Much more likely is a chain reaction. If we add the allylic radical to the alkene part of the allylic chloride we make a stable tertiary radical that can lose chloride radical and propagate the chain.

■ The original workers at ICI suggested a different mechanism: D. Holland and D. J. Milner, *Chem. and Ind. (London)*, 1979, 707.

PROBLEM 7

Heating this compound to 560 °C gives two products with the spectroscopic data shown below. What are they and how are they formed?

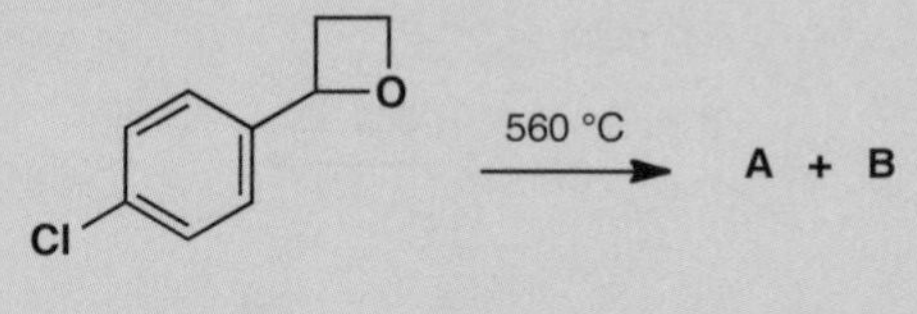

A has IR 1640 cm^{-1}; *m/z* 138 (100%) and 140 (33%), δ_H (ppm) 7.1 (4H, s), 6.5 (1H, dd, *J* 17, 11 Hz), 5.5 (1H, dd, *J* 17, 2 Hz), and 5.1 (1H, dd, *J* 11, 2 Hz).

B has IR 1700 cm^{-1}; *m/z* 111 (45%), 113 (15%), 139 (60%), 140 (100%), 141 (20%), and 142 (33%), δ_H (ppm) 9.9 (1H, s), 7.75 (2H, d, *J* 9 Hz), and 7.43 (2H, d, *J* 9 Hz).

Purpose of the problem

Revision of structure determination and a radical reaction with a difference.

Suggested solution

Compound **A** contains chlorine (*m/z* 138/140, 3:1) and that fits C_8H_7Cl. It still has the 1,4-disubstituted benzene ring (four aromatic Hs) and it is an alkene (IR 1640) with three hydrogens on it with characteristic coupling. We can write the structure immediately as there is no choice. The four aromatic hydrogens evidently have the same chemical shift.

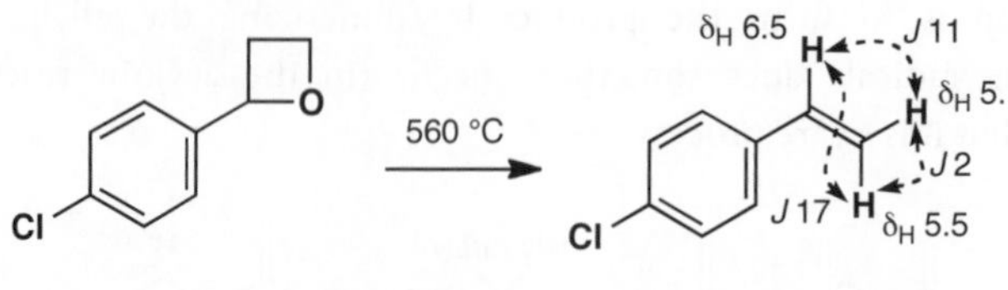

Compound **B** has *m/z* 140/142, 3:1 and a carbonyl group (at 1700 cm^{-1}) which fits C_7H_5ClO and looks like an aldehyde (δ_H 9.9). It still has the disubstituted benzene. The structure is even easier this time!

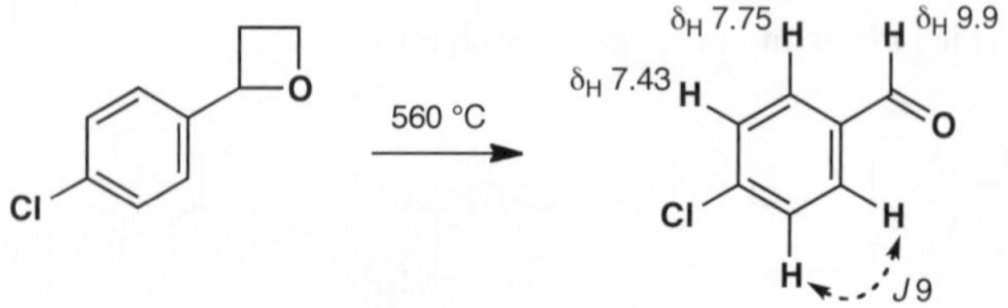

So how are these products formed? At such high temperatures, σ-bonds break and the weakest bonds in the molecule are the C–C and C–O bonds in the four-membered ring next to the benzene ring. Breaking these bonds releases strain and allows one of the radical products to be secondary and delocalized.

PROBLEM 8

Treatment of methylcyclopropane with peroxides at very low temperature (–150 °C) gives an unstable species whose ESR spectrum consists of a triplet with coupling of 20.7 gauss and fine splitting showing dtt coupling of 2.0, 2.6, and 3.0 gauss. Warming to a mere –90 °C gives a new species whose ESR spectrum consists of a triplet of triplets with coupling 22.2 and 28.5 gauss and fine splitting showing small ddd coupling of less than 1 gauss.

If methylcyclopropane is treated with *t*-BuOCl, various products are obtained but the two major products are **C** and **D**. At lower temperatures more of **C** is formed and at higher temperatures more of **D**.

Treatment of the more substituted cyclopropane below with PhSH and AIBN gives a single product in quantitative yield. Account for all these reactions, identifying **A** and **B** and explaining the differences between the various experiments.

Purpose of the problem

Working out the consequences of an important substituent effect on radical reactions: the cyclopropyl group.

Suggested solution

The peroxide is a source of *t*-BuO· radicals and these abstract a hydrogen from the methyl group of the hydrocarbon. The first spectrum is that of the cyclopropylmethyl radical. The odd electron is in a p orbital represented by a circle and the planar CH_2· group is orthogonal to the plane of the ring but the two Hᵃs are the same because of rapid rotation. The odd electron has a large coupling to the two hydrogens (Hᵃ) on the same carbon, a smaller doublet coupling to Hᵇ, and small couplings to the two Hᶜs and two Hᵈs. The coupling to Hᵇ is small because the p orbital containing the odd electron is orthogonal to the C–Hᵇ bond.

Warming to –90 °C causes decomposition to an open-chain radical. The odd electron is coupled to the two hydrogens on its own carbon (Hᵃ) and those on the next carbon (Hᵇ) each giving a triplet (22.2 and 28.5). Coupling to the more remote hydrogens is small.

Decomposition of the same hydrocarbon with *t*-BuOCl produces the same sequence of radicals but they can now be intercepted by the chlorine atom of the reagent, releasing more *t*-BuO· radicals and a radical chain is started. At lower temperatures the ring opening is slower so more of the cyclopropane is captured.

■ C. S. Walling and P. S. Fredericks, *J. Am. Chem. Soc.*, 1962, **91**, 1877.

The last example also produces a radical next to a cyclopropane ring but this time it can decompose very easily to give a stable secondary benzylic radical. This captures a hydrogen atom from PhSH releasing PhS· and maintaining an efficient radical chain. Ring opening of cyclopropanes is now a standard way of detecting radicals.

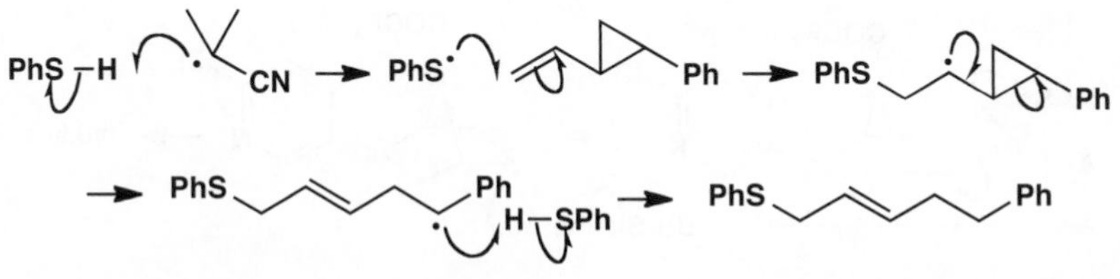

PROBLEM 9

The last few stages of Corey's epibatidine synthesis are shown here. Give mechanisms for the first two reactions and suggest a reagent for the last step.

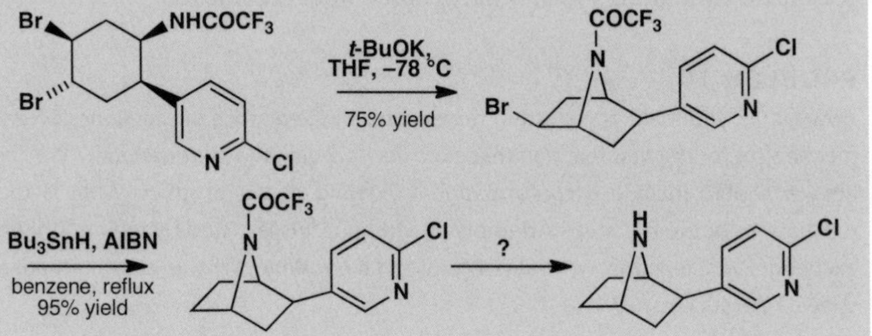

Purpose of the problem

Application of radical reactions in an important sequence plus revision of conformation and stereochemistry.

Suggested solution

The first step involves deprotonation of the rather acidic amide (the CF_3 group helps) and the displacement of the only possible bromide—the one on the opposite face of the six-membered ring as the S_N2 reaction must take place with inversion.

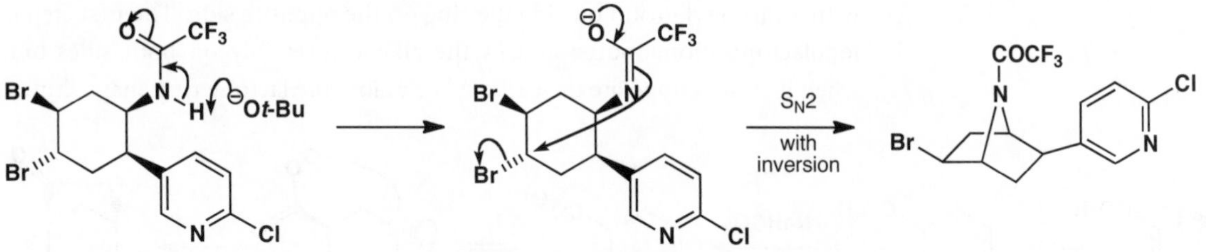

The second step is a standard dehalogenation by Bu_3SnH. AIBN generates $Bu_3Sn^\bullet$ by hydrogen abstraction from the reagent and this removes the bromine. Make sure you complete the chain and do not use $H^\bullet$ at any point.

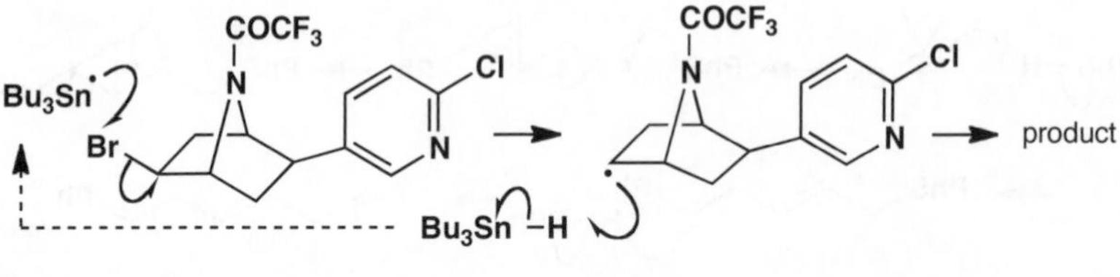

Finally we need to hydrolyse the amide. This normally requires strong acid or alkali but the CF$_3$ group makes this amide significantly more electrophilic than most and milder conditions can be used. Corey actually used NaOMe in methanol at 13 °C for two hours and got a yield of 96%. Any reasonable conditions you may have chosen would be fine too.

■ E. J. Corey and group, *J. Org. Chem.*, 1993, **58**, 5600.

PROBLEM 10

How would you make the starting material for this sequence of reactions? Give a mechanism for the first reaction that explains its regio- and stereoselectivity. Your answer should include a conformational drawing of the product. What is the mechanism of the last step? Attempts to carry out this last step by iodine/lithium exchange and reaction with allyl bromide failed. Why? Why is the alternative shown here successful?

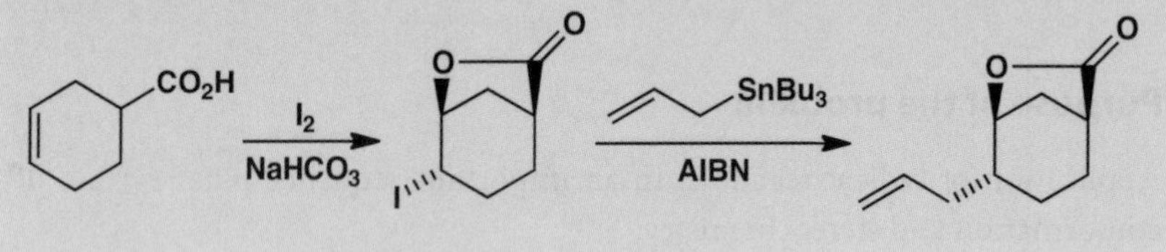

Purpose of the problem

Application of radical reactions when the alternative ionic reactions fail.

Suggested solution

The starting material is an obvious Diels-Alder product as it is a cyclohexene with a carbonyl group outside the ring on the opposite side. The first step is iodolactonization. Iodine attacks the alkene reversibly on both sides but, when it attacks opposite the carboxylate anion, the lactone ring snaps shut.

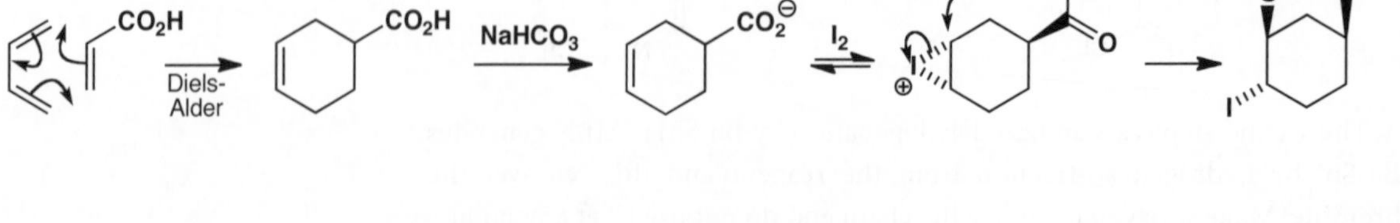

The problem asks for a conformational drawing of the product and indeed that is necessary. The 1,3-lactone bridge must be diaxial as that is the only way for the carboxylate to reach across and therefore it must attack from an axial direction too.

The last step is initiated by AIBN which removes the iodine atom from the compound to make a secondary radical. This attacks the allyl stannane and the intermediate loses $Bu_3Sn^\bullet$ and that takes over the job of removing iodine atoms to keep the chain going. The radical intermediate has no stereochemistry at the planar radical carbon and attack occurs from the bottom face to avoid the blocking lactone bridge.

■ You can regard this as the formation of a diaxial product as in the opening of a cyclohexene oxide with a nucleophile (p. 836 of the textbook).

Anionic reactions cannot be used for this allylation. If the iodine were metallated, the organometallic compound would immediately expel the lactone bridge as carboxylate ion is a good leaving group. The radical is stable because the C–O bond is strong and not easily cleaved in radical reactions.

PROBLEM 11

Suggest a mechanism for this reaction explaining why a mixture of diastereoisomers of the starting material gives a single diastereoisomer of the product. Is there any other form of selectivity?

Purpose of the problem

A radical ring-closing reaction with a curious stereochemical outcome.

Suggested solution

The abstraction of bromine, at first by AIBN and thereafter by Bu$_3$Sn˙ produces a radical that again does not eliminate but adds to an alkene. A five-membered ring is formed (this is usually the more favourable closure) by attack on the alkene on the opposite side from that occupied by the *i*-Pr group. The product is a mixture of diastereoisomers as no change occurs at the acetal centre.

Acid-catalysed oxidation first hydrolyses the acetal and then oxidizes either the hemiacetal or the aldehyde to the lactone. Now the molecule is one diastereoisomer as the ambiguous centre is planar. The other form of selectivity is the ring size (see the textbook, p. 1000).

PROBLEM 12

Reaction of this carboxylic acid $(C_5H_8O_2)$ with bromine in the presence of dibenzoyl peroxide gives an unstable compound **A** $(C_5H_6Br_2O_2)$ that gives a stable compound **B** $(C_5H_5BrO_2)$ on treatment with base. Compound **B** has IR 1735 and 1645 cm^{-1} and NMR δ_H 6.18 (1H, s), 5.00 (2H, s) and 4.18 (2H, s). What is the structure of the stable product **B**? Deduce the structure of the unstable compound **A** and mechanisms for the reactions.

Purpose of the problem

Revision of structural analysis in combination with an important radical functionalization.

Suggested solution

The starting material is $C_5H_8O_2$ so the stable compound **B** has gained a bromine and lost three hydrogens. There must be an extra double bond equivalent (DBE) somewhere in **B**. The IR spectrum shows that the OH has gone and suggests a carbonyl group, possibly an ester because of the high frequency, and an alkene. The NMR shows that both methyl groups have gone and have been replaced by CH_2 groups. The bromine must be on one of them and the ester oxygen on the other. The extra DBE is a ring.

Since both methyl groups are functionalized, unstable **A** must have one Br on each methyl group. The peroxide produces benzoyl radicals that abstract protons from both allylic positions to give stabilized radicals that sttack bromine molecules to give bromide radicals to continue the chain reaction. In base the carboxylate cyclizes onto the *cis* CH_2Br group.

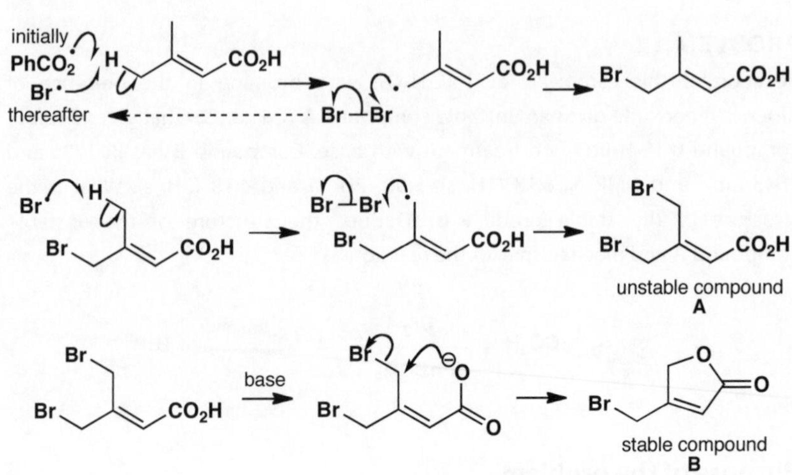

initially PhCO$_2$•
Br•
thereafter

Br$-$Br

unstable compound
A

base

stable compound
B

PROBLEM 1

Suggest mechanisms for these reactions.

Purpose of the problem

Two simple carbene reactions initiated by base.

Suggested solution

Going to the right we must remove the rather acidic proton from $CHBr_3$ to give the carbanion. This loses bromide to give dibromocarbene and insertion into cyclohexene gives the product.

The second reaction is very similar. α-Elimination of HCl gives a carbene that inserts into an alkene. These are the simplest reactions of carbenes and are very common.

PROBLEM 2

Suggest a mechanism for this reaction and explain the stereochemistry.

Purpose of the problem

Another important carbene method used in the synthesis of a natural antibiotic.

Suggested solution

■ This reaction established the skeleton of cycloeudesmol and was carried out by E. Y. Chen, *Tetrahedron Lett.*, 1982, **23**, 4769, at Sandoz.

cycloeudesmol

The diazo compound decomposes to gaseous nitrogen and a carbene under catalysis by Cu(II). Insertion into the exposed alkene gives the three-membered ring. The stereochemistry partly comes from the 'tether'—the linkage between the carbene and the rest of the molecule that delivers the carbene to the bottom face of the alkene. The rest comes from the inevitable *cis* fusion between the five- and three-membered rings.

PROBLEM 3

Comment on the selectivity shown in these reactions.

Purpose of the problem

A study in chemoselectivity during carbene insertion into alkenes.

Suggested solution

The first reaction is a variation on Simmons-Smith cyclopropanation. Though strictly a carbenoid rather than a carbene, it delivers a CH_2 group from an organozinc compound bound to an oxygen atom, in this case the OMe group. Only that alkene reacts.

The second cyclopropanation occurs at the only remaining alkene with a carbene generated from a diazoester. The stereoselectivity comes from attack on the opposite side of the ring from the already established cyclopropane.

■ There is little selectivity for the stereochemistry of the CO_2Et group but this fortunately did not matter in the synthesis of a natural defence substance from a sponge by G. A. Schieser and J. D. White, *J. Org. Chem.*, 1980, **45**, 1864.

PROBLEM 4

Suggest a mechanism for this ring contraction.

Purpose of the problem

Drawing mechanisms for a rearrangement involving a carbene formed photochemically.

Suggested solution

■ Reaction used by J. Froborg and G. Magnusson, *J. Am. Chem. Soc.*, 1978, **100**, 6728.

The carbene formed by loss of nitrogen from the diazoketone rearranges with the migration of either C–C bond to give a ketene picked up by methanol.

PROBLEM 5

Suggest a mechanism for the formation of this cyclopropane.

Purpose of the problem

An unusual type of carbene but it behaves normally.

Suggested solution

There is no doubt that *t*-BuO⁻ is a base, but which proton does it remove? The OH proton perhaps, but that doesn't lead to a carbene. The proton on the alkyne? That molecule has a leaving group, but is it too far away?

can be thought of as

Not if you push the electrons through the molecule in a γ-elimination. Normal elimination is β-elimination: both α- and γ-elimination can produce carbenes. The arrows are easy to make sense of if you think of a carbene as a carbon with both a + and a – charge. The carbene is an allenyl carbene with no substituent at the carbene centre. It inserts into the alkene in the other molecule.

PROBLEM 6

Decomposition of this diazo compound in methanol gives an alkene **A** ($C_8H_{14}O$) whose NMR spectrum contains two signals in the alkene region: δ_H 3.50 (3H, s), 5.50 (1H, dd, J 17.9, 7.9), 5.80 (1H, ddd, J 17.9, 9.2, and 4.3), 4.20 (1H, m) and 1.3–2.7 (8H, m). What is its structure and geometry?

When you have done that, suggest a mechanism for the reaction using this extra information: Compound **A** is unstable and even at 20 °C isomerizes to **B**. If the diazo compound is decomposed in methanol containing a diene, compound **A** is trapped as the adduct shown. Account for all these reactions.

Purpose of the problem

Revision of structural analysis, alkene geometry, and cycloadditions with carbenes as a mechanistic link.

Suggested solution

The starting material is $C_7H_{10}N_2$ so it has lost nitrogen and gained CH_4O—one molecule of methanol. We can see the MeO group at δ_H 3.50 and the four CH_2 groups in the ring are still there (8H m at 1.3–2.7). All that is left is a multiplet at δ_H 4.2, obviously next to OMe, and a pair of alkene protons at δ_H 5.5 and 5.8, coupled with J 17.9—obviously a *trans* alkene. That at δ_H 5.5 is coupled to one proton and the one at 5.8 is coupled to two. We now have these fragments:

C_4H_8 C_4H_5 C_2H_4O $= C_{10}H_{12}O$

But these add up to C_2H_3 too much! Clearly the CH attached to OMe and the CH attached to the alkene are the same atom and the CH_2 at the other end of the alkene must be one end of the chain of four CH_2s. We now have a structure but it doesn't join up!

$= C_8H_{14}O$

This is the test of your belief in spectroscopy—the dotted ends must join up to give **A**. Yes, this does put an *E*-alkene in a seven-membered ring, and it is difficult to draw, but you were warned that A is unstable. The CH_2 group next to the CHOMe group is diastereotopic so the coupling constants are different.

$= $ $= C_8H_{14}O$

■ This was the discovery of H. Jendralla, *Angew. Chem. Int. Ed. Engl.*, 1980, **19**, 1032. If you were really on the ball, you'll have noticed that a *trans*-cycloheptene is chiral, so this compound must be a single diastereoisomer though we don't know which.

Now that we know the structure of **A**, it is easy enough to find a mechanism. Loss of nitrogen produces a carbene that gives an allene in a pericyclic process and this twisted compound (the two alkenes are at 90° to each other) and protonation gives the *trans* alkene as a cation that reacts with methanol to give **A**.

The twisted alkene is unstable and rotates to the much more stable *cis* alkene even at 20 °C. It can rotate because the overlap between the p orbitals is weak as they are not parallel. Trapping in a Diels-Alder reaction preserves the *trans* stereochemistry.

PROBLEM 7

Give a mechanism for the formation of the three-membered ring in the first of these reactions and suggest how the ester might be converted into the amine with retention of configuration

Ph— + N_2—CO_2Et →[Cu(I)] Ph—△—CO_2Et →[?] Ph—△—NH_2

Purpose of the problem

A routine carbene insertion and a reminder of nitrenes as analogues of carbenes.

Suggested solution

The diazoester gives the carbene under Cu(I) catalysis and insertion into the alkene follows its usual course. The only extra is stereoselectivity: the insertion happens more easily if the two large groups (Ph and CO_2Et) keep as far apart as possible.

N_2—CO_2Et →[Cu(I)] Ph⋯:CH—CO_2Et → Ph—△—CO_2Et

Conversion of acid derivatives into amines with the loss of the carbonyl group can be done in various ways. In chapter 36 we recommended the Curtius and the Hofmann. The Hofmann degradation is the easier if we start with an ester, converting into the amide with ammonia and then treating with bromine in basic solution. The N-bromo amide undergoes α-elimination to a nitrene that rearranges to an isocyanate.

■ The amine product is an antidepressant discovered by A. Burger and W. L. Yost, *J. Am. Chem. Soc.*, 1948, **70**, 2198.

Ph—△—CO_2Et →[NH_3] Ph—△—$CONH_2$ →[Br_2][NaOH] Ph—△—C(O)N(H)Br →[NaOH]

Ph—△—C(=O)N⁻Br → Ph—△—C(=O)N: → Ph—△—N=C=O → amine

PROBLEM 8

Explain how this highly strained ketone is formed, albeit in very low yield, by these reactions. How would you attempt to make the starting material?

Purpose of the problem

To show that intramolecular carbene insertion is a powerful way to make cage compounds.

Suggested solution

Oxalyl chloride makes the acid chloride, and diazomethane converts this into the diazoketone.

Now the carbene chemistry. Treatment with Cu(I) removes nitrogen and forms the carbene. Remarkably, this is able to reach across the molecule and insert into the alkene, thus forming one three- and two new four-membered rings in one step. You will not be surprised at the yield: 10%.

■ This very strained ketone was used in vain while attempting to make tetrahedrane by G. Maier and group (*Angew. Chem. Int. Ed. Engl.*, 1983, **22**, 990).

How would you attempt to make the starting material? The original workers used another carbene reaction—the Cu(I) catalysed insertion of a diazoester into *bis*-trimethylsilyl acetylene.

PROBLEM 9

Attempts to prepare compound **A** by phase-transfer catalysed cyclization required a solvent immiscible with water. When chloroform ($CHCl_3$) was used, compound **B** was formed instead and it was necessary to use the more toxic CCl_4 for success. What went wrong?

Purpose of the problem

Carbene chemistry is not always what is wanted: how do you avoid it?

Suggested solution

Product **B** is clearly the adduct of product **A** and dichlorocarbene which must have come from the chloroform and base. The good news is that product **A** was evidently formed in the basic reaction mixture so, if we simply avoid a solvent that is also a carbene source, all is well.

■ This chemistry was used to make new β-lactams at ICI by S. R. Fletcher and L. T. Kay, *J. Chem. Soc., Chem. Commun.*, 1978, 903.

PROBLEM 10

Revision content. How would you carry out the first step in this sequence? Propose mechanisms for the remaining steps explaining any selectivity.

Purpose of the problem

Revision of specific enol formation, rearrangement reactions, electrocyclic reactions and conjugate addition plus some carbene chemistry.

Suggested solution

The first step requires a specific enol from an enone. Treatment with LDA achieves kinetic enolate formation by removing one of the more acidic hydrogens immediately next to the carbonyl group. The lithium enolate is trapped with Me_3SiCl to give the silyl enol ether.

The next step is dichlorocarbene insertion into the more nucleophilic of the two akenes. Dichlorocarbene is an electrophilic carbene so the main interaction is between the HOMO (π) of the alkene and the empty p orbital of the carbene. The carbene is formed by decarboxylation, a process that needs no strong base.

You can draw the ring expansion in a number of ways. All start with the removal of the Me₃Si group with water. You might then simply use a one-step mechanism (a) but an electrocyclic process via the cyclopropyl cation (b) might be better. This is allowed since the inevitable *cis* ring junction requires H and OH to rotate outwards.

Finally, a double conjugate addition of $MeNH_2$ to the dienone forms the bicyclic amine. Conjugate addition probably occurs first on the more electrophilic chloroenone, though it doesn't much matter. There is some stereoselectivity in that the remaining chlorine prefers the equatorial position on the new six-membered ring but this is thermodyanmic control as that position is easily enolized.

■ The product has the skeleton of the tropane alkaloids and this chemistry allowed T. L. Macdonald and R. Dolan (*J. Org. Chem.*,1979, **44**, 4973) to make a number of these natural products.

PROBLEM 11

How would you attempt to make these alkenes by metathesis?

Purpose of the problem

Applications of this important and powerful method.

Suggested solution

Metathesis is usually *E*-selective and these are both *E*-alkenes so prospects are good. We must disconnect each compound at the alkene and add something to the end of each, probably just CH_2 as the by-product will then be volatile ethylene.

Each starting material must now be made. The stereochemistry of the first tells us that we should add an allyl metal compound to an epoxide. The metathesis catalyst will be one of those mentioned in the chapter.

The second molecule is not symmetrical but this is all right as it will be an intramolecular (ring-closing) metathesis so we can expect few cross-products. There are many ways to make the starting material: alkylation of a ketone is probably the simplest though conjugate addition would have its advantages. The same catalyst can be used and very little would be needed.

PROBLEM 12

Heating this acyl azide in dry toluene under reflux for three hours gives a 90% yield of a heterocycle. Suggest a mechanism, emphasizing the role of any reactive intermediates.

Purpose of the problem

Demonstrating the practical nature of nitrene chemistry in the context of heterocyclic synthesis.

Suggested solution

Heating an azide liberates nitrogen gas and forms a nitrene. In this case, rearrangement to an isocyanate is followed by intramolecular nucleophilic attack by the *ortho* amino group.

PROBLEM 13

Give mechanisms for the steps in this conversion of a five- into a six-membered aromatic heterocycle.

Purpose of the problem

It is the turn of carbene chemistry to show its usefulness in that most practical of all subjects: heterocyclic synthesis.

Suggested solution

Decomposition of trichloroacetate ion releases the Cl_3C^- carbanion. Loss of chloride gives dichlorocarbene and addition to one of the double bonds in the pyrrole gives a bicyclic intermediate.

■ A. O. Fitton and R. K. Smalley, *Practical heterocyclic chemistry*, Academic Press, London, 1968, p. 17.

Ring expansion can be drawn in various ways. There is a direct route from the neutral amine, or its anion, that doesn't look very convincing, or you can ionize one of the chlorides first and open the cyclopropyl cation in an electrocyclic reaction. However you explain it, this is a good way to make 3-substituted pyridines.

39

Suggested solutions for Chapter 39

PROBLEM 1

Propose three fundamentally different mechanisms (other than variations of the same mechanism with different kinds of catalysis) for this reaction. How would (a) D labelling and (b) ^{18}O labelling help to distinguish the mechanisms? What other experiments would you carry out to rule out some of these mechanisms?

Purpose of the problem

Investigating a reaction where there are several reasonable mechanisms.

Suggested solution

The reaction is an ester hydrolysis so the obvious mechanism is to attack the carbonyl group with hydroxide. Notice that we draw out each stage of the mechanism and do not use any summary or shorthand.

Mechanism 1: Normal ester hydrolysis

But the ester oxygen atom is attached to an aromatic ring with a *para* nitro group. Nucleophilic aromatic substitution would give the same product.

Mechanism 2: Nucleophilic aromatic substitution

Finally, the ester can be transformed into an enolate, using hydroxide as a base. Elimination gives a ketene that can be attacked by hydroxide as a nucleophile to give the product.

Mechanism 3: Enolate elimination to give a ketene

Mechanism 3 requires the exchange of at least one hydrogen atom with the solvent so, if D_2O were used as the solvent, or better deuterated starting material were used, the exchange of one whole deuterium atom would indicate mechanism 3 while no exchange, or only minor amounts from the inevitable enolization, would show mechanisms 1 or 2. In mechanisms 1 and 3, the added OH group ends up in in CO_2H but in mechanism 2 it ends up as the phenol. Using $H_2^{18}O$ as solvent, or better labelling the ester oxygen as ^{18}O would separate mechanisms 1 and 3 from 2.

$\star = {}^{18}O$

Other experiments we could do might include trying to trap the ketene intermediate in a [2 + 2] cycloaddition, studying the reaction by UV, hoping to see the release of *p*-nitrophenolate in mechanism 3, changing the structure of the starting material so that one or other of the mechanisms would be difficult, even measuring the effects of the substituent on the benzene ring on the rate, or looking for a deuterium isotope effect in the labelled lactone.

PROBLEM 2

Explain the stereochemistry and labelling pattern in this reaction.

Purpose of the problem

A combination of labelling and stereochemistry reveals the details of a surprisingly interesting rearrangement.

Suggested solution

The randomization of the label and the racemization suggest that the carboxylate falls off the allyl cation and then comes back on again at either end. While they are detached the distinction between the two ends of both cation and anion disappears as they are delocalized.

The product is racemic because the two intermediates each have a plane of symmetry and are achiral. The retention of *relative* stereochemistry (formation of the *trans* product from *trans* starting material) could result from stereoselective recombination (the two faces of the allyl cation are not the same) or from the two ions sticking together as an ion pair so that the acetate slides across one face of the cation. An alternative [3,3] sigmatropic rearrangement would not randomize the labels in the same way.

■ This question is based on more complex chemistry described by H. L. Goering *et al.*, *J. Am. Chem. Soc.*, 1964, **86**, 1951.

PROBLEM 3

The Hammett ρ value for migrating aryl groups in the acid-catalysed Beckmann rearrangement is –2.0. What does that tell us about the rate-determining step?

Purpose of the problem

The Hammett relationship gives an intimate picture of the Beckmann rearrangement.

Suggested solution

The normal mechanism for the Beckmann rearrangement (pp. 958–960 of the textbook) involves protonation at OH and migration of the group *anti* to the N–O bond: in this case the substituted benzene ring.

If this mechanism is correct here, we should expect the migration itself to be the slow step. The first step is just a proton transfer to oxygen and must be fast. The steps after the migration involve attack of water on a carbocation and proton transfers to O and N and these must all be fast. The migration breaks a C–C bond, forms a C–N bond and creates an unstable cation. But does this agree with the evidence? Starting material and product in the migration step are cations so the transition state must be a cation too. Any contribution to cation stability made by the migrating group should help and we should therefore expect electron-donating groups to migrate faster. This is what we see: a ρ value of –2.0 shows a modest acceleration by electron-donating groups (p. 1041 ff.).

In the Beckmann rearrangement, the *anti* group migrates but in other rearrangements the migrating group is chosen for a very different reason: it is normally the group that is best able to stabilize a positive charge and benzene rings can do this by π participation. This would be the participation mechanism:

The Hammett ρ value of –2.0 gives very definite evidence that participation does not occur. If it did the closure of the unstable three-membered ring would be the slow step and a positive charge would form on the benzene ring itself. This would give a much larger ρ value of something like –5.0. One reason that participation does not occur is that the starting material is planar and the p orbitals in the benzene ring cannot point in the right direction to interact with the σ* orbital of the N–O bond. They are orthogonal to it.

■ This is the early work of D. E. Pearson and group, *J. Org. Chem.,* 1952, **17**, 1511; 1954, **19**, 957.

PROBLEM 4

Between pH 2 and 7 the rate of hydrolysis of this ester is independent of pH. At pH 5 the rate is proportional to the concentration of acetate ion (AcO⁻) in the buffer solution and the reaction goes twice as fast in H_2O as in D_2O. Suggest a mechanism for the pH-independent hydrolysis. Above pH 7 the rate increases with pH. What kind of change is this?

Purpose of the problem

Time for you to try your skill at interpreting pH-rate profiles.

Suggested solution

The second part of the question is easily dealt with. In alkaline solution the rate of hydrolysis simply increases with pH and we have the normal specific base-catalysed reaction in which hydroxide ion attacks the carbonyl group.

■ Nucleophilic attack on this ester is much faster than that on $MeCO_2Et$ but is still the slow step. This is a much simplified description of a series of experiments described by L. R. Fedor and T. C. Bruice, *J. Am. Chem. Soc.,* 1965, **87**, 4138.

But this is no ordinary ester. The leaving group is a thiol (pK_a about 8) not the usual alcohol (pK_a about 16) and so the thiolate anion is a much better leaving group than EtO^-. Also the CF_3 group is very electron-withdrawing so nucleophilic attack on the carbonyl group will be unusually fast. This is why there is a region of pH-independent hydrolysis not found with EtOAc. You might have suggested that acetate is a nucleophile or a general base catalyst but the solvent deuterium isotope effect suggests that it is a general base. The change at pH 7 is a change of mechanism as the faster of two mechanisms applies—a sketch of the pH-rate profile will show you the upward curve.

PROBLEM 5

In acid solution, the hydrolysis of this carbodiimide has a Hammett ρ value of –0.8. What mechanism might account for this?

Purpose of the problem

Interpretation of a small Hammett ρ value.

Suggested solution

The most obvious explanation for a low Hammett ρ value, that the aromatic ring is too far away from the reaction, will not wash here as the aromatic rings are joined directly to the reacting nitrogen atoms of the carbodiimide. The reaction must surely start with the protonation of one of the nitrogens. This cannot be the slow step and it would in any case have a large negative ρ value. The small ρ value observed suggests that the rate-determining step must have a large positive ρ value that nearly cancels out the large negative value for the first step. Attack by water on the protonated carbodiimide looks about right.

The expected equilibrium Hammett ρ value for the protonation would be about –2.5 to –3 so the kinetic Hammett ρ value for the attack of water would have to be about +2 to give a net Hammett ρ value of –0.8. This looks fine. The rest of the mechanism involves proton transfers, hydrolysis of an imide, and decarboxylation.

■ The hydrolysis of carbdiimides in acid and base was studied by S. Hünig *et al.*, *Liebigs Annalen*, 1953, **579**, 87.

PROBLEM 6

Explain the difference between these Hammett ρ values by mechanisms for the two reactions. In both cases the ring marked with the substituent X is varied. When R = H, ρ = –0.3 but when R = Ph, ρ = –5.1.

Purpose of the problem

Interpretation of a variation in Hammett ρ value with another structural variation.

Suggested solution

The reaction is obviously nucleophilic substitution at the benzylic centre so we are immediately expecting S_N1 or S_N2. When R = H, the reaction occurs at a primary alkyl group and S_N2 is expected. When R = Ph, the reaction occurs at a secondary benzylic centre and S_N1 is expected.

Since S$_N$1 produces a cation delocalized round the benzene ring in the slow step, a large negative Hammett ρ value is reasonable. It is not obvious what sign the Hammett ρ value would have in the S$_N$2 reaction but as there is no build-up of negative charge on the carbon atom in the transition state, a small value is reasonable. The actual value (–0.3) is very small indeed but, if we can read anything into it, it suggests a loose S$_N$2 transition state with a small positive charge on carbon.

PROBLEM 7

Explain how chloride catalyses this reaction.

Purpose of the problem

An extreme example of surprising catalysis.

Suggested solution

At first you might ask how chloride can catalyse anything at all. It is a weak base and not a very good nucleophile for the carbonyl group. However, in polar aprotic solvents like acetonitrile (MeCN), chloride is not solvated and is both more basic and more nucleophilic. In this reaction it cannot be a nucleophilic catalyst as attack on the carbonyl group simply regenerates starting material. It cannot be a specific base as it is too weak, even in acetonitrile, to remove a proton from methanol. But it can act as a general base. As methanol attacks the carbonyl group its proton becomes more acidic and, in the transition state, chloride is at last able to act.

PROBLEM 8

The hydrolysis of this oxaziridine in 0.1M sulfuric acid has $k(H_2O)/k(D_2O) = 0.7$ and an entropy of activation of $\Delta S = -76$ J mol^{-1} K^{-1}. Suggest a mechanism.

Purpose of the problem

Deducing a mechanism from isotope effects and entropy of activation.

Suggested solution

The inverse solvent deuterium isotope effect indicates specific acid catalysis and the modest negative entropy of activation suggests some bimolecular involvement. There are various mechanisms you might have proposed and a likely one involves cleavage of the three-membered ring in the protonated amine. The second or possibly the third step could be rate-determining.

Once the three-membered ring is opened, the rest of the mechanism amounts to acid-catalysed hemiacetal hydrolysis. The original workers favoured an alternative mechanism that starts with protonation of the oxygen atom and ends up with the hydrolysis of an imine. Again, the second or third step could be rate-determining.

■ The original work was by J. H. Fendler and group, *J. Chem. Soc., Perkin Trans 2*, 1973, 1744. You might also have considered an electrocyclic opening of the three-membered ring.

PROBLEM 9

Explain how both methyl groups in the product of this reaction come to be labelled. If the starting material is reisolated at 50% reaction, its methyl group is also labelled.

Purpose of the problem

Exploring a mechansim through labelling.

Suggested solution

The role of silver ion (Ag^+) is the removal of the halide to give an acylium ion that reacts, not at the carbonyl group, but at the methyl group to give CO_2 and a methylated benzene ring. The simple Friedel-Crafts route cannot be the whole story: it explains how the added methyl group is labelled, but not why it is only partly labelled and how label gets into the other methyl group.

The only way in which we can explain those extra features is to suggest that methylation initially occurs on the oxygen atom and that a methyl group is transferred from there to the benzene ring. We should never have detected this detail without the labelling experiment. Alkylation on oxygen provides an alkylating agent that can transfer either CH_3 or CD_3 and also explains the formation of trideuterotoluene.

■ We hope you didn't suggest a methyl cation as an intermediate.

intermediate can transfer either CH_3 or CD_3

PROBLEM 10

The pK_a values of some protonated pyridines are as follows:

X	H	3-Cl	3-Me	4-Me	3-MeO	4-MeO	3-NO$_2$
pK_a	5.2	2.84	5.68	6.02	4.88	6.62	0.81

Can the Hammett correlation be applied to pyridines using the σ values for benzene? What equilibrium ρ value does it give and how do you interpret it? Why are no 2-substituted pyridines included in the list?

Purpose of the problem

Making sure you understand the ideas behind the Hammett relationship.

Suggested solution

The obvious thing to do is to plot the pK_a values against the σ values for the substituents using the *meta* values for the 3-substituted and *para* values for the 4-substituted compounds (see table on p. 1042 of the textbook). This gives quite a good straight line and we get a slope (Hammett ρ value) of +5.9. The sign is of course positive as the same electronic effects that make benzoic acids more acidic will also make pyridinium ions more acidic. The large ρ value may have surprised you, but reflect: ionization of benzoic acids occurs outside the ring and the charge isn't delocalized round the ring. Deprotonation of pyridinium ions occurs on the ring and the charge (positive this time) is delocalized round the ring.

■ This work was done to apply the Hammett relationship to reactions of pyridines with acid chlorides. R. B. Moody and group, *J. Chem. Soc., Perkin Trans 2*, 1976, 68.

There are no 2-substituted pyridines on the list since, like *ortho*-substituted benzenes, they cannot be expected to give a good correlation because of steric effects.

PROBLEM 11

These two reactions of diazo compounds with carboxylic acids give gaseous nitrogen and esters as products. In both cases the rate of reaction is proportional to [diazo compound][RCO_2H]. Use the data for each reaction to suggest mechanisms and comment on the difference between them.

$\rho = 1.6$
$k(RCO_2H)/\, k(RCO_2D) = 3.5$

$k(RCO_2D)/\, k(RCO_2H) = 2.9$

Purpose of the problem

Application of contrasting isotope effects to detailed mechanistic analysis.

Suggested solution

The first reaction has a normal kinetic isotope effect (RCO_2H reacts *faster* than RCO_2D) while the second has an inverse deuterium isotope effect (RCO_2H reacts *slower* than RCO_2D). This suggests that there is a rate-determining proton transfer in the first reaction but specific acid catalysis in the second (i.e. fast equilibrium proton transfer followed by slow reaction of the protonated species). Protonation occurs at carbon in both reactions, and this can be a slow step.

The second reaction follows much the same pathway except that loss of nitrogen is now difficult because the cation would be very unstable (primary and next to a CO_2Et group) so the second step is S_N2 and rate determining. .

PROBLEM 12

Suggest mechanisms for these reactions and comment on their relevance to the Favorskii family of mechanisms.

1. Br_2
2. $EtO^\ominus$, EtOH

$MeO^\ominus$, MeOH
bromoketone added to base

$MeO^\ominus$, MeOH
base added to bromoketone

Purpose of the problem

Extension of a section of the chapter (pp. 1061–3 of the textbook) into new reactions with internal trapping of intermediates.

Suggested solution

In the first reaction the bromination must occur on the alkene to give a dibromide. We cannot suggest stereochemistry at this stage and it is better to continue with the standard Favorskii mechanism and see what happens. Everything follows until the very last step when the opening of the cyclopropane provides electrons in just the right place to eliminate the second bromide and put the alkene back where it was. This alternative behaviour of a proposed intermediate gives us confidence that the intermediate really is involved.

The stereochemistry of the initial bromination turns out to be irrelevant as it disappears when the oxyallyl cation is formed. We know the stereochemistry of the final product so we know the stereochemistry of the cyclopropanone: it must be on the opposite face of the five-membered ring to the methyl group. The disrotatory closure of the oxyallyl cation evidently goes preferentially one way with the H and the CMe₂Br substituents going upwards and the carbonyl group going down.

The second reaction to the right is a normal Favorskii. The only point of interest is the way the three-membered ring breaks up. The more stable carbanion is the doubly benzylic one so that leaves.

The reaction with excess bromoketone starts the same way but the oxyallyl cation is intercepted by one of the benzene rings in a four-electron conrotatory electrocyclic reaction like the Nazarov reaction (p. 927 of the textbook).

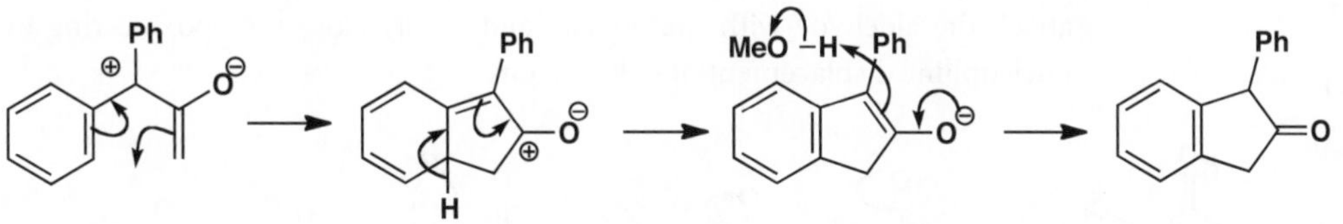

You may wonder how excess MeO⁻ stops this from happening. It doesn't. The oxyallyl cation and the cyclopropanone are in equilibrium and excess MeO⁻ captures the cyclopropanone and drives the normal Favorskii onwards. If there is no excess MeO⁻ the oxyallyl cation lasts long enough for the five-membered ring to be the main product.

■ This work was part of a thorough investigation into the mechanism of the Favorskii rearrangement by F. G. Bordwell and group, *J. Am. Chem. Soc.*, 1970, **92**, 2172.

PROBLEM 13

A typical Darzens reaction involves the base-catalysed formation of an epoxide from an α-haloketone and an aldehyde. Suggest a mechanism consistent with the data below.

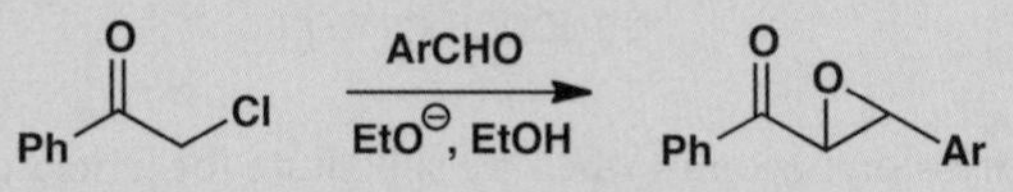

(a) The rate expression is: rate = k_3[PhCOCH$_2$Cl][ArCHO][EtO⁻]

(b) When Ar is varied, the Hammett ρ value is +2.5.

(c) The following attempted Darzens reactions produced unexpected results:

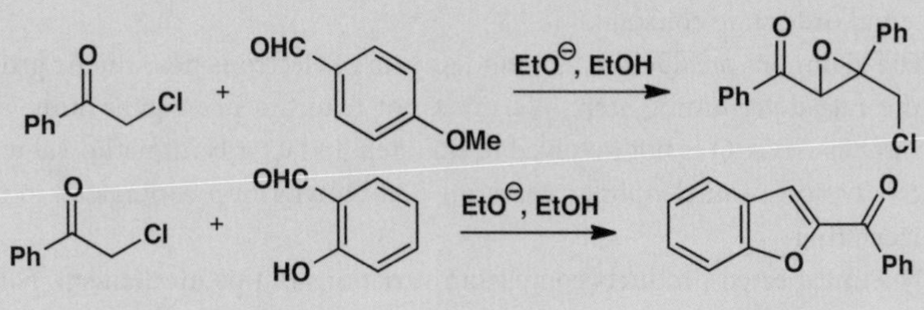

Purpose of the problem

Trying to get a complete picture of a reaction using physical data and structural variation.

Suggested solution

The ethoxide is not incorporated into the product but appears in the rate expression. Its role must be as a base and there is only one set of enolizable protons. We start by making the enolate of the chloroketone. This cannot be the slow step as the aldehyde appears in the rate expression. Then we can attack the aldehyde with the enolate and finally close the epoxide ring by nucleophilic displacement of chloride ion.

If this mechanism is right, the kinetic data show that the second step is rate-determining (a reasonable deduction as it is a bimolecular step) and that the first step is a pre-equilibrium. We can write:

$$\text{rate} = k_2[\text{enolate}][\text{ArCHO}]$$

And we know from the pre-equilibrium that

$$K_1 = \frac{[\text{enolate}]}{[\text{PhCOCH}_2\text{Cl}][\text{EtO}^-]}$$

So the rate expression becomes when we substitute for [enolate]:

$$\text{rate} = K_1 k_2[\text{PhCOCH}_2\text{Cl}][\text{EtO}^-][\text{ArCHO}]$$

and this matches the observed rate expression though the apparently third order rate constant is revealed as the product of an equilibrium constant and a second order rate constant.

The Hammett ρ value shows a modest gain of electrons near the Ar group in the rate-determining step. We must not take the pre-equilibrium into account as ArCHO is not involved in this step. In fact a Hammett ρ value of +2.5 is typical of nucleophilic attack on a carbonyl group conjugated to the benzene ring.

The unexpected products come from variations in this mechanism. *para*-Methoxybenzaldehyde is conjugated and unreactive so the enolate ignores it and reacts with the unenolized version of itself.

With salicylaldehyde, the second example, the phenolic OH group will exist as an anion under the reaction conditions. Alkylation by the chloroketone allows enolate formation leading to an intramolecular aldol reaction.

PROBLEM 14

If you believed that this reaction went by elimination followed by conjugate addition, what experiments would you carry out to try and prove that the enone is an intermediate?

Purpose of the problem

Turning the usual question backwards: what evidence do you want, rather than how to interpret what you are given.

Suggested solution

The suggested mechanism of elimination followed by conjugate addition might be contrasted with direct S_N2 to see what evidence is needed.

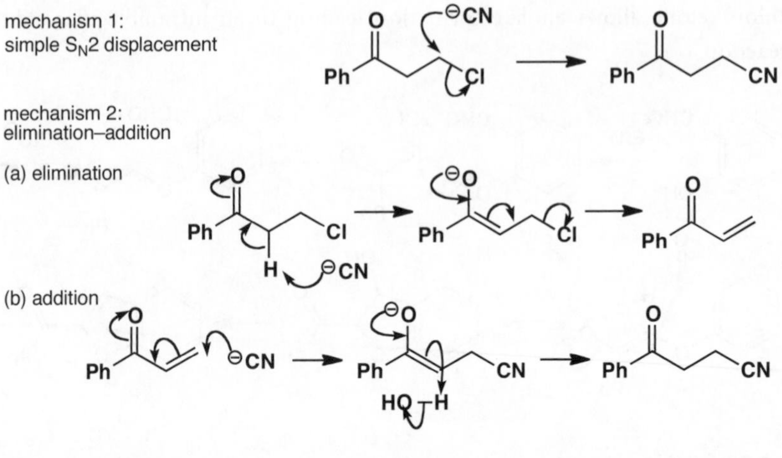

mechanism 1:
simple S$_N$2 displacement

mechanism 2:
elimination–addition

(a) elimination

(b) addition

There are many types of evidence you might suggest: here are some of them.

- Exchange of protons in D$_2$O/EtOD would suggest elimination/addition.
- Kinetic evidence (tricky as you cannot be sure which is the slow step.
- A Hammett plot with substituted benzene rings. The S$_N$2 mechanism would have a small ρ as the benzene ring is a long way from the action.
- Base catalysis: mechanism 2 is base catalysed, mechanism 1 isn't.
- Kinetic isotope effect might be found in mechanism 2.
- Stereochemistry. If a substituent were added to make the terminal carbon chiral, inversion would be expected for mechanism 1 and racemization for mechanism 2. But choose a small substituent otherwise it would be a very different compound.

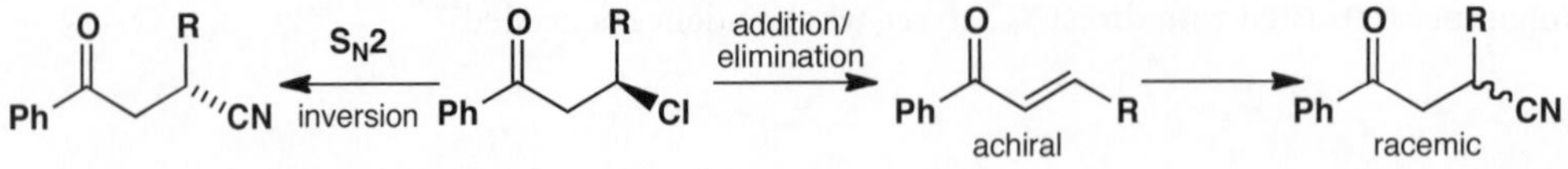

Suggested solutions for Chapter 40

40

PROBLEM 1

Suggest mechanisms for these reactions, explaining the role of palladium in the first step.

Purpose of the problem

Revision of enol ethers and bromination, the Wittig reaction, and, of course, first steps in palladium chemistry.

Suggested solution

The first step is a reaction of an enol with an allylic acetate catalysed by palladium(0) via an η^3 allyl cation. There is no regiochemistry to worry about as the diketone and allylic acetate are both symmetrical.

■ You might have drawn the η^3 allyl cation complex in various satisfactory ways—some are mentioned on p. 1089 of the textbook.

NBS in aqueous solution is a polar brominating agent, ideal for reaction with an enol ether. The intermediate is hydrolysed to the ketone by the usual acetal style mechanism.

Finally, an intramolecular Wittig reaction. This is a slightly unusual way to do what amounts to an aldol reaction but the 5,5 fused enone system is strained and the Wittig went under very mild conditions (K_2CO_3 in aqueous solution). The stereochemistry of the new double bond is the only one possible and Wittig reactions with stabilized ylids generally give the most stable of the possible alkene.

■ This process is a general way to make 5,5 fused systems devised by B. M. Trost and D. P. Curran, *J. Am. Chem. Soc.*, 1980, **102**, 5699.

PROBLEM 2

This Heck-style reaction does not lead to regeneration of the alkene. Why not? What is the purpose of the formic acid (HCO_2H) in the reaction mixture?

Purpose of the problem

Making sure you understand the steps in the mechanism of the Heck reaction.

Suggested solution

The reaction must start with the oxidative addition of Pd(0) into the Ph–I bond. The reagent added is Pd(II) so one of the reduction methods on page 1081 of the textbook must provide enough Pd(0) to start the reaction going. The oxidative addition gives PhPdI and this does the Heck reaction on the alkene. Addition occurs on the less hindered top (*exo-*) face and the phenyl group is transferred to the same face.

Normally now the alkyl palladium(II) species would lose palladium by β-elimination. This is impossible in this example as there is no hydrogen atom *syn* to the PdI group. Instead, an external reducing agent is needed and that is the role of the formate anion: it provides a hydride equivalent by 'transfer hydrogenation' when it loses CO_2.

■ A heterocyclic version of this reaction was part of a synthesis of the natural analgesic epibatidine by S. C. Clayton and A. C. Regan, *Tetrahedron Lett.*, 1993, **34**, 7493.

PROBLEM 3

Cyclization of this unsaturated amine with catalytic Pd(II) under an atmosphere of oxygen gives a cyclic unsaturated amine in 95% yield. How does the reaction work? Why is the atmosphere of oxygen necessary? Explain the stereochemistry and regiochemistry of the reaction. How would you remove the CO_2Bn group from the product?

Purpose of the problem

Introducing you to 'aminopalladation': like oxypalladation, nucleophilic attack on a palladium π-complex.

Suggested solution

The π-complex between the alkene and Pd(II) permits nucleophilic attack by the amide on its nearer end and in a *cis* fashion because the nucleophile is tethered by a short chain of only two carbon atoms. Nucleophilic attack and elimination of Pd(0) occur in the usual way. The removal of the CO_2Bn group would normally be done by hydrogenolysis but in this case ester hydrolysis by, say, HBr would be preferred to avoid reduction of the alkene. The free acid decarboxylates spontaneously.

■ This general synthesis of heterocycles was introduced by J.-E. Bäckvall and group, *Tetrahedron Lett.*, 1995, **36**, 7749.

PROBLEM 4

Suggest a mechanism for this lactone synthesis.

Purpose of the problem

Introducing you to carbonyl insertion into a palladium (II) σ-complex.

Suggested solution

Oxidative insertion into the aryl bromide, carbonylation, and nucleophilic attack on the carbonyl group with elimination of Pd(0) form the catalytic cycle. No doubt the palladium has a number (1 or 2?) of phosphine ligands complexed to it during the reaction and these keep the Pd(0) in solution between cycles.

■ M. Moru *et al.*, *Heterocycles*, 1979, **12**, 921.

PROBLEM 5

Explain why enantiomerically pure lactone gives *syn* but racemic product in this palladium-catalysed reaction.

MeO$_2$C   CO$_2$Me

(Ph$_3$P)$_4$Pd

CO$_2$H

CO$_2$Me

CO$_2$Me

(–)-lactone   *syn* but racemic

Purpose of the problem

Helping you to understand the details of palladium-catalysed allylation.

Suggested solution

Following the usual mechanism, the palladium complexes to the face of the alkene opposite the bridge. The ester leaves to give an allyl cation complex. This is attacked by the malonate anion from the opposite face to the palladium. So the overall result is retention of configuration, the *syn* starting material giving the *syn* product.

CO$_2^\ominus$ CO$_2^\ominus$ CO$_2$H

CO$_2$Me   CO$_2$Me   CO$_2$Me

L$_n$Pd   L$_n$Pd   L$_n$Pd   CO$_2$Me   CO$_2$Me   CO$_2$Me

The racemization comes from the structure of the allyl cation complex. It is symmetrical with a plane of symmetry running vertically through the complex as drawn. Attack by the malonate anion occurs equally at either side of the plane giving the two enantiomers of the *syn* diastererereoisomer in equal amounts.

■ This investigation helped to establish the mechanism of these reactions: B. M. Trost and N. R. Schmuff, *Tetrahedron Lett.*, 1981, **22**, 2999.

CO$_2$H   CO$_2^\ominus$ CO$_2$H

MeO$_2$C   MeO$_2$C   CO$_2$Me

CO$_2$Me   MeO$_2$C   L$_n$Pd   CO$_2$Me   CO$_2$Me

PROBLEM 6

Explain the reactions in this sequence, commenting on the regioselectivity of the organometallic steps.

Purpose of the problem

Revision of allylic Grignard reagents, the synthesis of pyridines, and the mechanism of the Wacker oxidation.

Suggested solution

The allylic Grignard reagent does direct addition from the end remote to the magnesium atom, as often happens. Hydrolysis of the silyl enol ether reveals an aldehyde.

Now the Wacker oxidation, by whatever detailed mechanism you prefer, must involve the addition of water to a Pd(II) π-complex of the alkene and β-elimination of palladium to give Pd(0) which is recycled by oxidation with oxygen mediated by copper.

Finally, the pyridine synthesis is simply a double enamine/imine formation between ammonia and the two carbonyl groups. Probably the aldehyde reacts first.

■ M. A. Tius, *Tetrahedron Lett.*, 1982, **23**, 2819

PROBLEM 7

Give a mechanism for this carbonylation reaction. Comment on the stereochemistry and explain why the yield is higher if the reaction is carried out under a carbon monoxide atmosphere.

Hence explain this synthesis of part of the antifungal compound pyrenophorin.

Purpose of the problem

More carbonylation with a Stille coupling.

Suggested solution

The tin-palladium exchange (transmetallation) occurs with retention of configuration at the alkene. The exchange of the benzyl group for the benzoyl group is necessary to get the reaction started.

Now the coupling can take place on the palladium atom producing the product and Pd(0) which can insert oxidatively into the C–Cl bond. Transmetallation sets up a sustainable cycle of reactions. It is better to have an atmosphere of carbon monoxide because the acyl palladium complex can give off CO and leave a PdPh σ-complex. The atmosphere of CO reverses this reaction.

The second sequence starts with a radical hydrostannylation (chapter 37) giving the *E*-vinyl stannane preferentially if a slight excess of Bu₃SnH is used.

Now the coupling with the acid chloride takes place as before though this time we have an aliphatic carbonyl complex. There is no problem with β-elimination as that would give a ketene. Again, the stereochemistry of the vinyl stannane is retained in the product.

PROBLEM 8

A synthesis of an antifungal drug made use of this palladium-catalysed reaction. Give a mechanism, explaining the regio- and stereochemistry.

Purpose of the problem

A simple example of amine synthesis using palladium.

Suggested solution

The palladium forms the usual allyl cation complex and the nitrogen nucleophile attacks the less hindered end thus also retaining the conjugation. Attack at the triple bond would give an allene. The *E* stereochemistry of the palladium complex is retained in the product.

PROBLEM 9

Work out the structures of the compounds in this sequence and suggest mechanisms for the reactions, explaining any selectivity.

B has IR: 1730, 1710 cm⁻¹, δ_H 9.4 (1H, s), 2.6 (2H, s), 2.0 (3H, s), and 1.0 (6H, s).
C has IR: 1710 cm⁻¹, δ_H 7.3 (1H, d, J 5.5 Hz), 6.8 (1H, d, J 5.5 Hz), 2.1 (2H, s), and 1.15 (6H, s).

Purpose of the problem

An intramolecular aldol reaction (p. 636 of the textbook) and a Wacker oxidation (p. 1096 of the textbook).

Suggested solution

B clearly has aldehyde and ketone functional groups with nothing but singlets in the NMR. On the other hand **C** has a *cis* disubstituted alkene with a small (and therefore *cis*) J value and is a cyclopentenone.

PROBLEM 10

A synthesis of the Bristol-Myers Squibb anti-migraine drug Avitriptan (a 5-HT receptor antagonist) involves this palladium-catalysed indole synthesis. Suggest a mechanism and comment on the regioselectivity of the alkyne attachment.

Purpose of the problem

A new reaction for you to try—a palladium-catalysed indole synthesis.

Suggested solution

Although palladium(II) is added to the solution, the aryl iodide tells you that this is an oxidative insertion of Pd(0) produced by one of the methods described on p. 1081 of the textbook. The resulting Pd(II) species complexes to the alkyne and the amine can now attack the triple bond. This gives a heterocycle with the Pd(II) in the ring. Coupling of the two organic fragments extrudes Pd(0) to start a new cycle. The nitrogen attacks the more hindered end of the alkyne so that the palladium can occupy the less hindered end.

■ This is the Larock indole
synthesis (R. C. Larock and E. K. Yum,
J. Am. Chem. Soc., 1991, **113**, 6689)
and its use in the synthesis of
Avitriptan is described in P. D.
Brodfuehrer *et al.*, *J. Org. Chem.*,
1997, **62**, 9192).

PROBLEM 1

Explain how this synthesis of amino acids, starting with natural proline, works. Explain the stereoselectivity of each step after the first.

Purpose of the problem

A simple exercise in the creation of a new stereogenic centre via a cyclic intermediate.

Suggested solution

Nothing exciting happens until the hydrogenation step. The stereoselectivity of the reaction with ammonia is interesting but not of any consequence as that stereochemistry disappears in the elimination. This gives the E-enone as expected since the alkene and the carbonyl group are in the same plane.

■ This method was invented by B. W. Bycroft and G. R. Lee, *J. Chem. Soc., Chem. Commun.*, 1975, 988.

The new stereogenic centre is created in the hydrogenation step. The molecule is slightly folded and the catalyst interacts best with the outside (convex) face so that it adds hydrogen from the same face as the ring junction hydrogen. All that remains is to hydrolyse the product without racemization. Did you notice that the configuration of the new amino acid (S) is the same as that of the natural amino acids?

PROBLEM 2

This is a synthesis of the racemic drug tazodolene. If the enantiomers of the drug are to be evaluated for biological activity, they must be separated. At which stage would you recommend separating the enantiomers and how would you do it?

Purpose of the problem

First steps in planning an asymmetric synthesis by resolution.

Suggested solution

You need to ask: which is the first chiral intermediate? Can it be conveniently resolved? Will the chirality survive subsequent steps? The first intermediate is chiral but it enolizes very readily and the enol is achiral, so that's no good. The second intermediate is chiral but it has three chiral centres and these are evidently not controlled. We would have to separate the diastereoisomers before resolution and that would be a waste of time and material since all of them give the next intermediate anyway.

The next intermediate, the amino alcohol is ideal: it has only one chiral centre and that is not affected by the last reaction. It has two 'handles' for resolution—the amine and the alcohol. We might make a salt with tartaric acid or an ester of the alcohol with some chiral acid. Alternatively we could resolve tazadolene itself: it still has an amino group and we could form a salt with a suitable acid.

■ This synthesis is from the Upjohn company and is in only the patent literature (*Chem. Abstr.*, 1984, **100**, 6311.

PROBLEM 3

How would you make enantiomerically enriched samples of these compounds (either enantiomer)?

Purpose of the problem

First steps in planning an asymmetric synthesis.

Suggested solution

There are many possible answers here. What we had in mind was some sort of asymmetric Diels-Alder reaction for the first, an asymmetric aldol for the second or else opening an epoxide made by Sharpless epoxidation, asymmetric dihydroxylation for the third, and perhaps asymmetric dihydroxylation of a Z-alkene for the fourth. Of course you might have used resolution or asymmetric hydrogenation, or the chiral pool, or any other strategy from chapter 41.

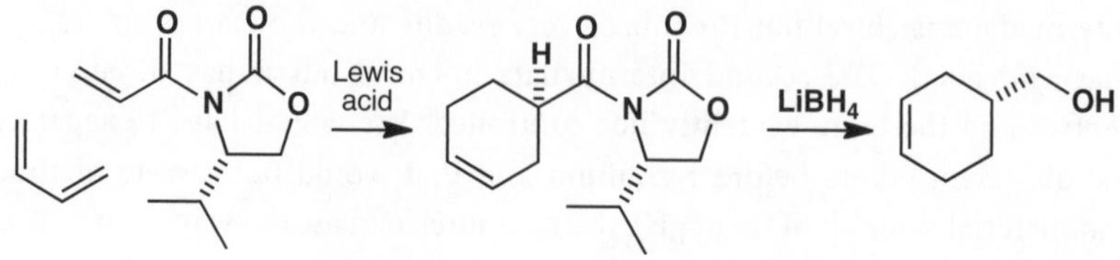

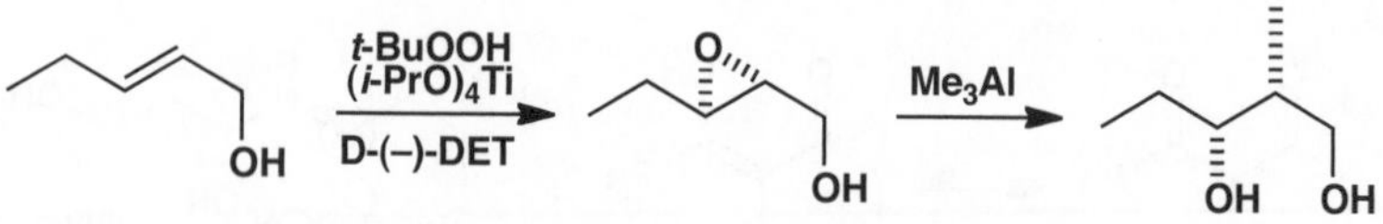

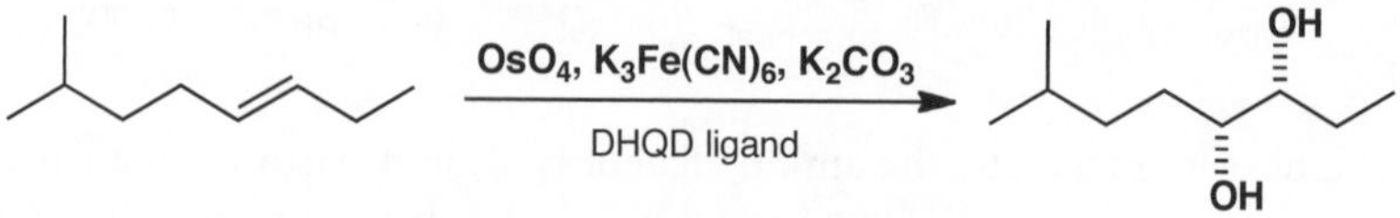

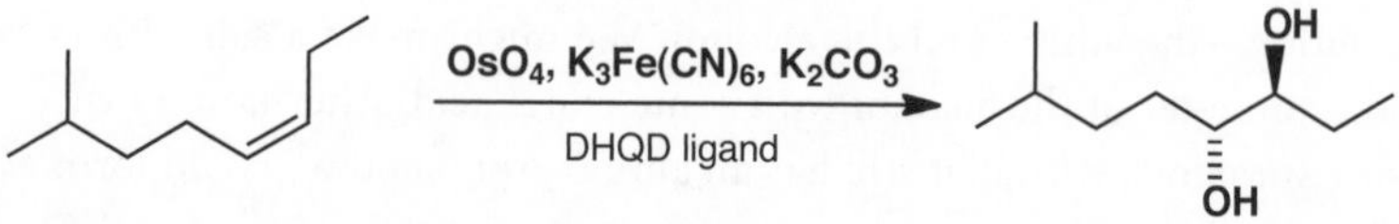

PROBLEM 4

In the following reaction sequence, the stereochemistry of mandelic acid is transmitted to a new hydroxy-acid by stereochemically controlled reactions. Give mechanisms for each reaction and state whether it is stereospecific or stereoselective. Offer some rationalization for the creation of new stereogenic centres in the first and last reactions.

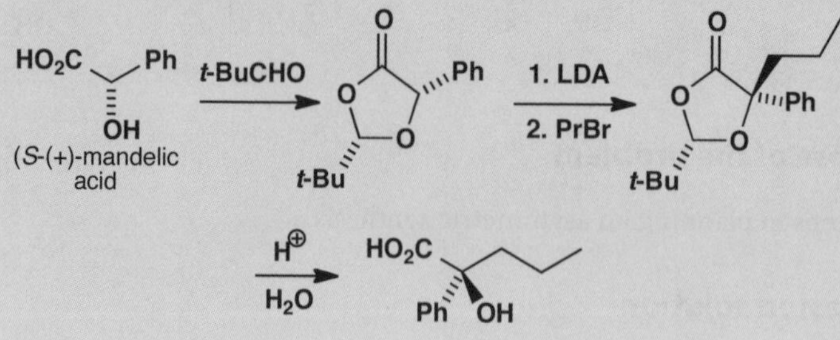

Purpose of the problem

Your chance to examine an ingenious method of asymmetric induction.

Suggested solution

The first reaction amounts to cyclic acetal formation except that one of the 'alcohols' is a carboxylic acid. The reaction is stereospecific (no change) at the original chiral centre and stereoselective at the new one.

The second reaction creates a lithium enolate and alkylates it. It is again stereospecific at the unchanged chiral centre and stereoselective at the new one. Finally, acetal hydrolysis preserves the new quaternary centre unchanged (stereospecific) by a mechanism that is the reverse of the first step.

Now, as far as the rationalization is concerned, the first step takes place through a sequence of reversible reactions and therefore under thermodynamic control so the most stable product will be formed. It may seem surprising that this should be the *cis* compound, but the conformation of this chair-like five-membered ring prefers to have the two substituents pseudoequatorial.

The alkylation is under kinetic control and, as a lithium enolate has more or less a flat ring, the alkyl halide approaches the opposite face to the *t*-Bu group. It has to approach orthogonally to the ring as it must overlap with the p orbital of the enolate.

■ This is Seebach's clever method of preserving the knowledge of a chiral centre while it is destroyed in a reaction. First a temporary centre (at the *t*-butyl group) is created in a stereoselective reaction; the original centre is destroyed by enolization but the temporary centre can be used to re-create it: D. Seebach *et al.*, *J. Am. Chem. Soc.*, 1983, **105**, 5390.

PROBLEM 5

This reaction sequence can be used to make enantiomerically enriched amino acids. Which compound is the origin of the chirality and how is it made? Suggest why this particular enantiomer of the product amino acid might be formed. Suggest reagents for the last stages of the process. Would the enantiomerically enriched starting material be recovered?

Purpose of the problem

Step-by-step discusssion of a simple but useful sequence.

Suggested solution

The amine, phenylethylamine, is the origin of the chirality. It is easily made by resolution, for example by crystallizing the salt of the racemic amine with tartaric acid. This means that both enantiomers are readily available.

■ There is a good example of the application of this method by K. Q. Do *et al.* in *Helv. Chim. Acta*, 1979, **62**, 956.

This particular enantiomer of the amino acid product belongs to the natural (*S*) series. The unnatural (*R*) enantiomer would also be valuable and can be made from the other enantiomer of the starting material. The last stages of the process require cleavage of one C–N bond and hydrolysis of the nitrile. It will be important to do this without racemizing the newly created centre.

The C–N bond can be cleaved reductively by hydrogenation as it is an *N*-benzyl bond. This would also hydrogenate the nitrile so that must first be hydrolysed using acid or base, as weak as possible. The starting material is not recovered and the chirality is lost as the by-product is just ethyl benzene, the nitrogen atom being transferred to the product.

PROBLEM 6

Explain the stereochemistry and mechanism in the synthesis of the chiral auxiliary 8-phenylmenthol from (+)-pulegone. After the reaction with Na in *i*-PrOH, what is the minor (13%) component of the mixture?

Purpose of the problem

A combination of conformational analysis, stereoselective reactions, and resolution to get a single enantiomer.

Suggested solution

The first reaction is a conjugate addition that evidently goes without any worthwhile stereoselectivity. The stereochemistry is not fixed in the addition but in the protonation of the enolate in the work-up. Equilibration of the mixture by reversible enolate formation with KOH in ethanol gives mostly the all-equatorial compound.

Reduction by that smallest of reagents, an electron, gives the all-equatorial product. Since the stereochemical ratio in the product is the same as in the starting materials (87:13), the reduction must be totally stereoselective. The all-equatorial ketone gives 100% all-equatorial alcohol and the minor isomer must give one other diastereoisomer (we cannot say which).

■ Two fractional crystallizations actually give 50% of the required compound according to the details given by O. Ort, *Org. Synth. Coll.*, 1993, **VIII**, 552.

The mixture still has to be separated and, as it is a mixture of diastereoisomers, it can be separated by physical means. The chloroacetate is just a convenient crystalline derivative.

PROBLEM 7

Suggest syntheses for single enantiomers of these compounds.

Purpose of the problem

Devising your own asymmetric syntheses.

Suggested solution

The first compound is an ester derived from a cyclic secondary alcohol that could be made from the corresponding enone by asymmetric reduction.

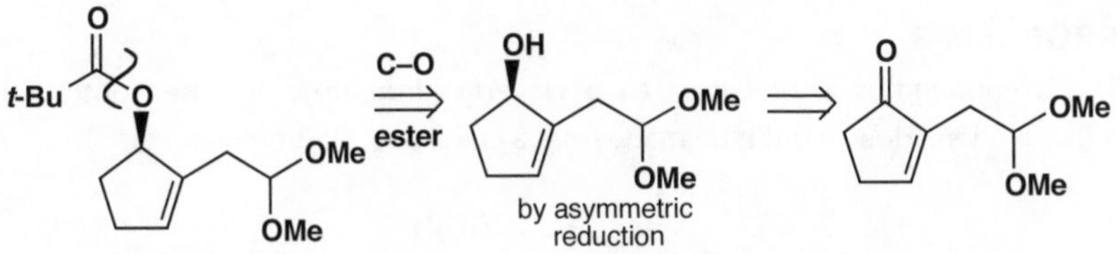

Reduction with Corey's CBS reducing agent gave the alcohol in 93% ee.

BH$_3$-THF

10% CBS
oxazaborolidine

(t-BuCO)$_2$O

TM

■ This compound was used to make a compound from the gingko tree by E. J. Corey and A. V. Gaval, *Tetrahedron Lett.*, 1988, **29**, 3201.

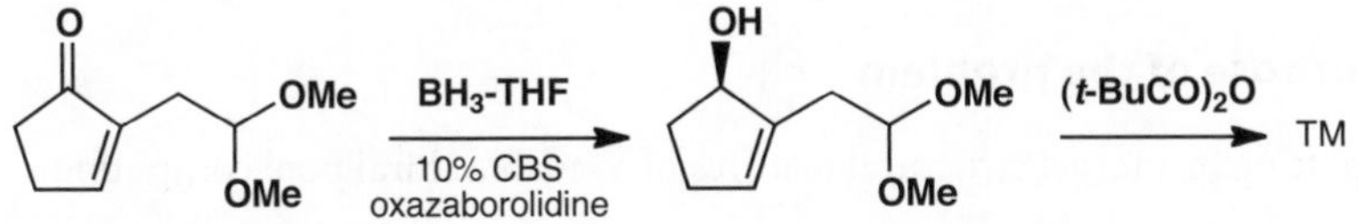

The second compound could be made by a Wittig reaction with a stabilized ylid and the required diol aldehyde derived from an epoxy-alcohol and hence an allylic alcohol by Sharpless epoxidation.

Wittig

A + Ph$_3$P B

A ⟹ Sharpless epoxidation

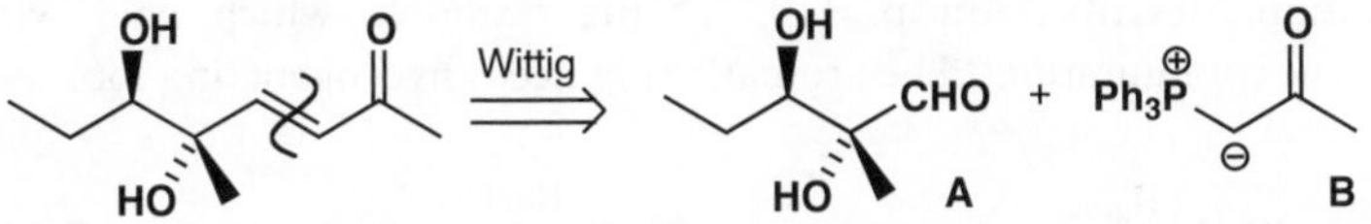

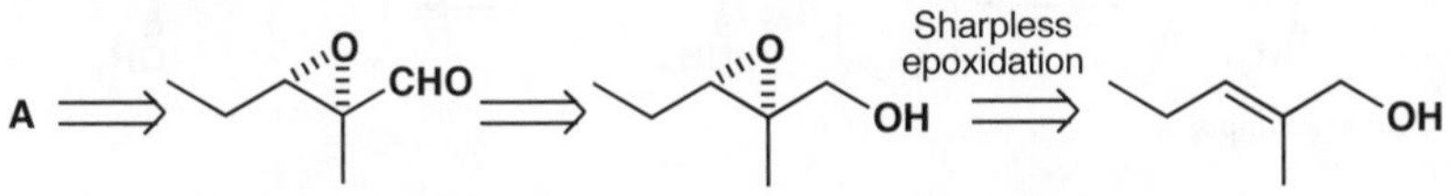

The first part of the synthesis gives an intermediate that had been used in the synthesis of the antibiotic methymycin. In practice the Wittig was carried out on the epoxy-aldehyde and treatment of the last intermediate with aqueous acid gave the target molecule.

■ S. Masamune *et al.*, *J. Am. Chem. Soc.* 1975, **97**, 3512

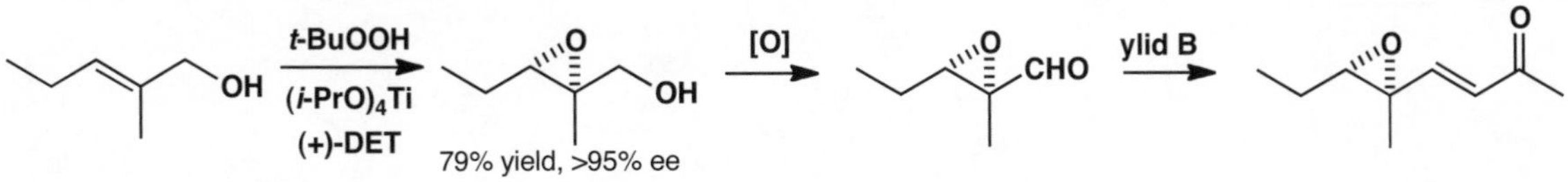

t-BuOOH

(i-PrO)$_4$Ti

(+)-DET

79% yield, >95% ee

[O]

ylid B

PROBLEM 8

This compound is a precursor to a Novartis drug used for the control of inflammation. How might it be made from a chiral pool starting material?

Purpose of the problem

Spotting in a target structural features of available chiral pool compounds.

Suggested solution

The hydrocarbon skeleton of the target is that of the amino acid phenylalanine. The configuration is (*S*), the same as the natural amino acid, so we can use the standard amino acid to hydroxy acid conversion via diazotization, described on p. 1105 of the textbook, which goes with retention of configuration. The aromatic ring needs hydrogenating too.

PROBLEM 9

Propose catalytic methods for the asymmetric synthesis of these four precursors to drug molecules.

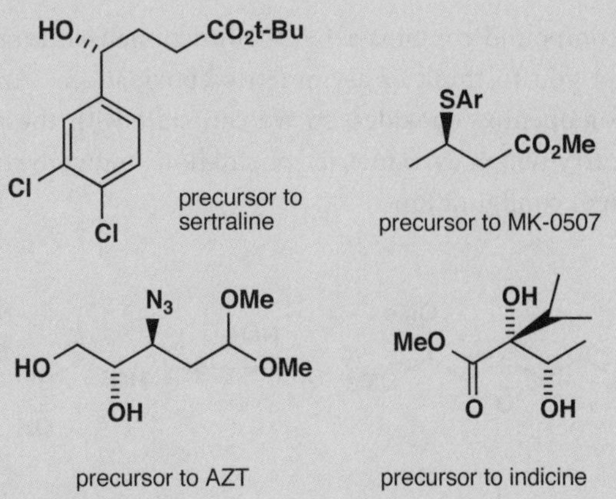

precursor to sertraline

precursor to MK-0507

precursor to AZT

precursor to indicine

Purpose of the problem

Identifying reliable catalytic reactions that give desired structural features.

Suggested solution

The sertraline precursor is a chiral alcohol with the stereogenic centre adjacent to an aromatic ring. An obvious approach is to make the hydroxyl group by asymmetric reduction of the corresponding ketone. CBS reduction is a possibility, as is a ruthenium-catalysed hydrogenation using the ligand TsDPEN (p. 1115 of the textbook).

■ You should not try to remember which enantiomer of the ligand you need for which enantiomer of the product: that can easily be looked up later. It is much more important to recognize the classes of molecules that can be reliably prepared by catalytic asymmetric reactions.

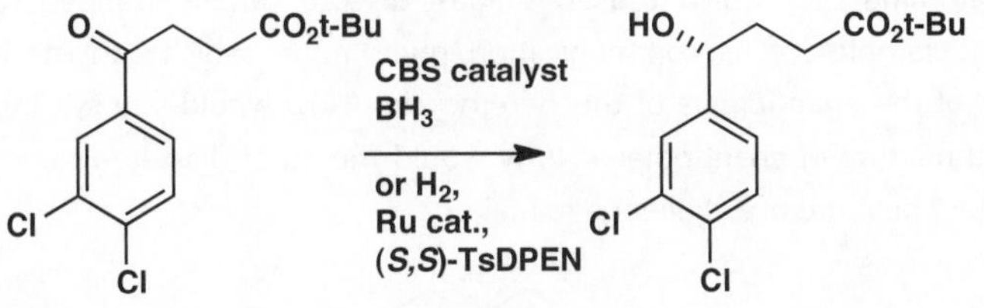

The second compound is a chiral sulfide. Although there are direct asymmetric ways of making chiral sulfur compounds, a reliable approach to sulfides is to use S_N2 substitution of a more readily made chiral precursor, because a thiolate is usually a good nucleophile. The S_N2 reaction goes with inversion, so we need the chiral alcohol shown below, converted to a derivative (such as a tosylate) capable of undergoing substitution. Care will be needed to avoid elimination, but thiolates are excellent nucleophiles and not too basic, so you would expect a successful outcome.

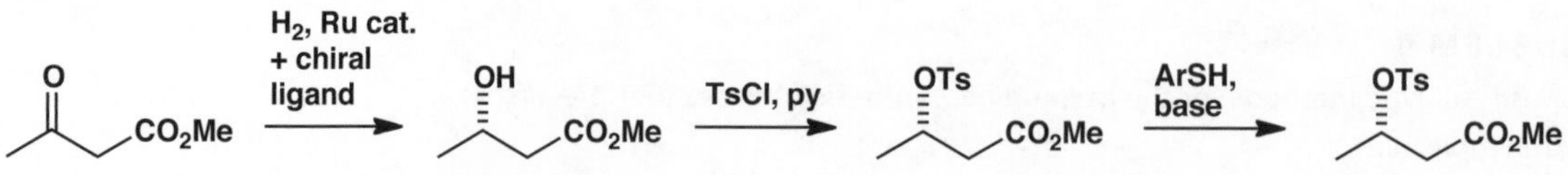

The third compound contains a 1,2,3-trifunctionalized arrangement that should prompt you to think of asymmetric epoxidation. Azide is a good nucleophile for opening epoxides, so we can start with the allylic alcohol shown here, carry out an asymmetric epoxidation, and convert to the target with inversion of configuration.

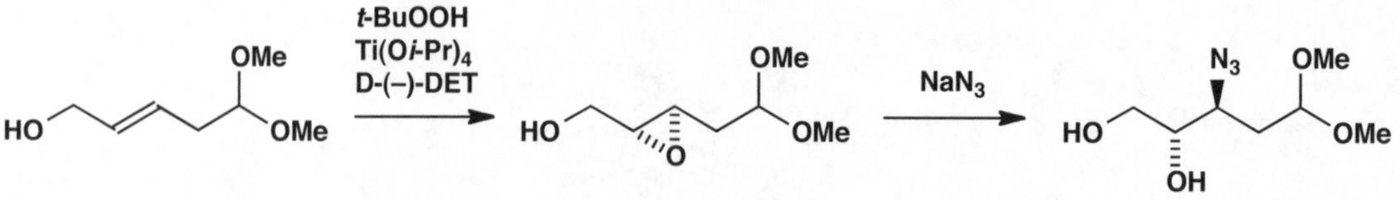

The final compound is a diol, so asymmetric dihydroxylation is a possible approach. The precursor is a rather unreactive alkene, but asymmetric dihydroxylation is a versatile reaction which can still perform well on challenging substrates.

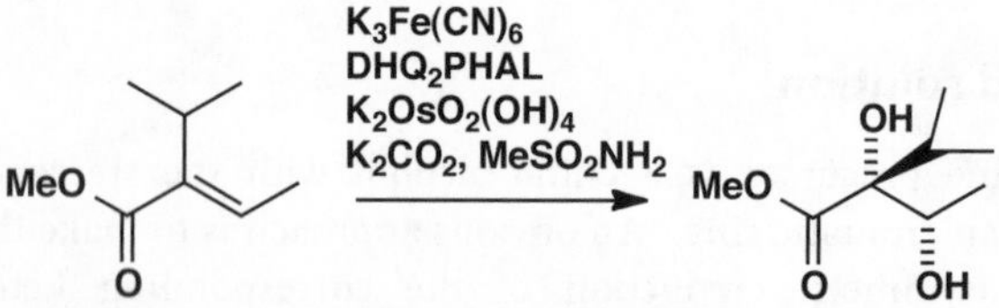

PROBLEM 10

The triatomine bug which causes Chagas' disease can be trapped by using synthetic samples of its communication pheromone, which consists of a 4:1 mixture of the enantiomers of this heterocycle. How would you synthesize the required mixture of enantiomers? Why would the other diastereoisomer of this compound be more of a challenge to make?

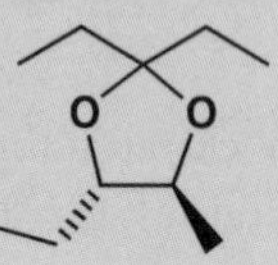

Purpose of the problem

Identifying structural features that can be made by asymmetric synthesis.

Suggested solution

To make a 4:1 mixture of enantiomers you need either to mix them in the right proportions, or to mix equal amounts of racemic mixture and a single enantiomer. In either case you need an asymmetric synthesis. The target compound is an acetal that can be made from a chiral diol, so you should immediately consider asymmetric dihydroxylation. The advantage of Sharpless' asymmetric dihydroxylation is that it can very easily give either enantiomer: in fact, it is one reaction where the enantioselective version is better than the racemic one, so you would be advised to make the two enantiomers using the two alternative chiral ligands, mix them in the correct proportions, then form the acetal. Note that the starting alkene is *trans*.

Making the other diastereoisomer would require the *cis* alkene. This is not a problem in itself, but more of a challenge for the catalyst, because now it has to distinguish between two similar groups (Et and Me) in order to oxidize one face of the alkene enantioselectively (for the *trans* alkene, the selection is between either Et and H or Me and H; switching Et for Me makes no difference to the outcome).

■ The discovery and synthesis of this pheromone is described by C. R. Unelius and co-workers in *Org. Lett.* 2010, **12**, 5601.

PROBLEM 11

This compound was developed by the Nutrasweet company as an artificial sweetener. Propose a strategy for its synthesis. Would your proposed approach still be suitable if the compound had turned out to be a successful product, required in multi-tonne quantities?

NC-00637

Purpose of the problem

Proposing an efficient synthetic route to a chiral target molecule: a common challenge in the pharmaceutical and related industries.

Suggested solution

The target can be best disconnected into three fragments at the amide bonds. The aminopyridine can be made by the standard methods of heterocycle synthesis (chapter 30), so we are more interested in the other two chiral fragments. The middle one is an amino acid, and you should recognize it as a member of the chiral pool, (S)-glutamic acid, so this poses no problem of synthesis. (Though it will need to be appropriately protected to form the correct amide).

The final fragment is a simple chiral carboxylic acid, so we need a method for its asymmetric synthesis. The most obvious choice is probably an asymmetric alkylation using Evans' oxazolidinone auxiliary: formation of the appropriate derivative of hexanoic acid is simple, and the enolate will be alkylated diastereoselectively by methyl iodide. You would probably take this approach if you need to make a few grams for initial studies.

If this compound were needed on the tonne scale then auxiliary chemistry is no good, however efficient recycling may be. A good alternative for the synthesis of compounds with unfunctionalized chiral centres adjacent to carboxylic acids or alcohols is the use of ruthenium-catalysed hydrogenation.

PROBLEM 12

The two aldehydes below are valuable products in the perfumery industry (Tropional® is a component of Issey Miyake's *L'Eau d'Issey* and Florhydral® is a component of *Allure* by Chanel). How would you make them as single enantiomers?

Purpose of the problem

Designing a synthesis where absolute stereochemistry must be controlled.

Suggested solution

Both targets have a single, simple chiral centre carrying a methyl group, so we need to devise a synthesis passing through an achiral precursor. For Tropional, you might imagine alkylating a derivative of Evans' auxiliary, followed by reduction to the aldehyde, but a more economical approach would be to use asymmetric reduction of an unsaturated carboxylic acid, since the compound required is readily made using an aldol-type condensation of the available aldehyde piperonal.

Florhydral has the methyl group β to the aldehyde. One possible approach is an asymmetric conjugate addition, but again asymmetric reduction of the acid (or allylic alcohol) is preferable, since the required alkene is easy to make by aldol chemistry. Here we show one example with the acid and one with the alcohol, but either are possibilities in both cases.

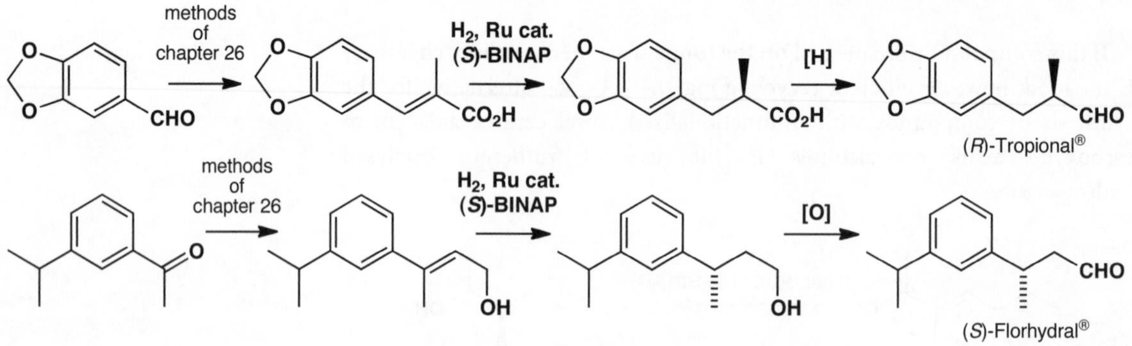

(R)-Tropional®

(S)-Florhydral®

Suggested solutions for Chapter 42

PROBLEM 1

Do you consider that thymine and caffeine are aromatic compounds? Explain.

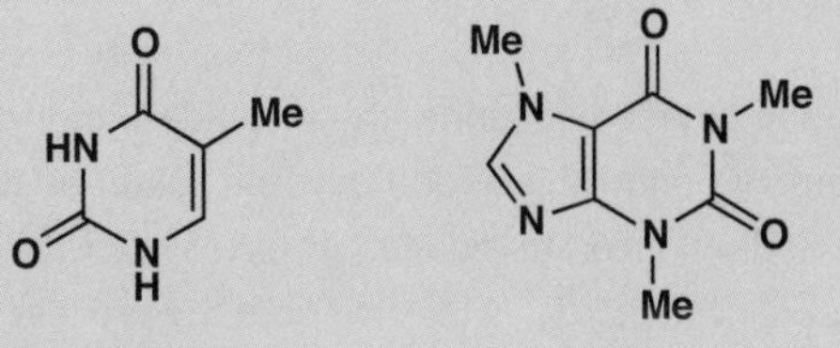

thymine   caffeine

Purpose of the problem

Revision of aromaticity and exploration of the structures of nucleic acid bases.

Suggested solution

Thymine, a pyrimidine, has an alkene and lone pair electrons on two nitrogens, making six in all for an aromatic structure. You may have shown this by drawing delocalized structures.

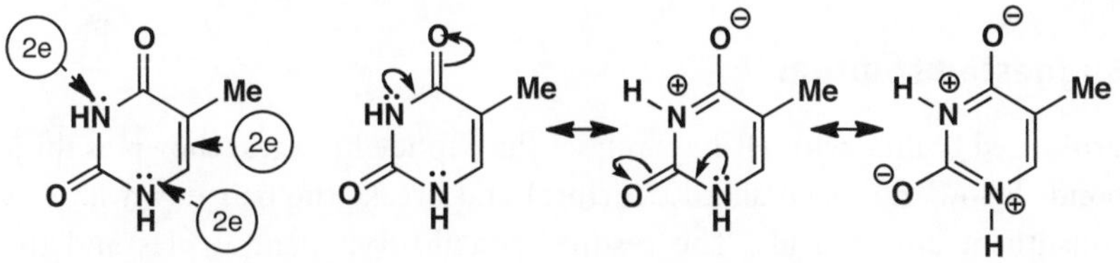

Caffeine, a purine, is slightly more complicated as it has two rings. You might have said that each ring is aromatic if you counted all the lone pairs on nitrogen *except* those on the 'pyridine-like' nitrogen (see p. 741 of the textbook for what we mean here) in the five-membered ring. Or you might have drawn a delocalized structure with ten electrons around its periphery.

six electrons in
six-membered ring

six electrons in
five-membered ring

ten electrons in
two rings together

PROBLEM 2

Human hair is a good source of cystine, the disulfide dimer of cysteine. Hair is boiled with aqueous HCl and HCO_2H for a day, the solution concentrated, and a large amount of sodium acetate added. About 5% of the hair by weight crystallizes out as pure cystine $[\alpha]_D$ –216. How does the process work? Why is such a high proportion of hair cystine? Why is no cysteine isolated by this process? Make a drawing of cystine to show why it is chiral. How would you convert the cystine to cysteine?

(*S*)-cysteine

Purpose of the problem

Some slightly more complicated amino acid chemistry including stereochemistry and the SH group.

Suggested solution

Prolonged boiling with HCl hydrolyses the peptide linkages (shown as thick bonds below in a generalized structure) and breaks the hair down into its constituent amino acids. The cystine crystallizes at neutral pHs and the mixture of HCl and NaOAc provides a buffer. Hair is much cross-linked by disulfide bridges and these are not broken down by hydrolysis.

one protein strand

another protein strand

disulfide
cross-link

No cysteine is isolated because (i) most of it is present as cystine in hair and (ii) any cysteine released in the hydrolysis will be oxidized in the air to cystine. The stereochemistry of cysteine is preserved in cystine which has C_2 symmetry and no plane or centre of symmetry so either of the diagrams below will suit. It is not important whether you draw the zwitterion or the uncharged structure. Reduction of the S–S bond by $NaBH_4$ converts cystine to cysteine.

■ The isolation of cystine is described in full detail in B. S. Furniss *et al.*, *Vogel's Textbook of Organic Chemistry* (5th edn), Longmans, Harlow, 1989 p.761.

PROBLEM 3

The amide of alanine can be resolved by pig kidney acylase. Which enantiomer of alanine is acylated faster with acetic anhydride? In the enzyme-catalysed hydrolysis, which enantiomer hydrolyses faster? In the separation, why is the mixture heated in acid solution, and what is filtered off? How does the separation of the free alanine by dissolution in ethanol work?

If the acylation is carried out carelessly, particularly if the heating is too long or too strong, a by-product is formed that is not hydrolysed by the enzyme. How does this happen?

Purpose of the problem

Rehearsal of some basic amino acid and enzyme chemistry plus revision of stereochemistry and asymmetric synthesis.

Suggested solution

The acylation takes place by the normal mechanism for the formation of amides from anhydrides, that is, by nucleophilic attack on the carbonyl group and loss of the most stable anion (acetate) from the tetrahedral intermediate. The two isomers of alanine are enantiomers and enantiomers *must* react at the same rate with achiral reagents.

■ You may feel that was a catch question. It was in a way but it is very important that you cling on to the fact that the chemistry of enantiomers is identical in every way except in reactions with enantiomerically pure chiral reagents.

In the enzyme-catalysed reaction, the acylase hydrolyses the amide of one enantiomer but not the other. This time the two enantiomers do *not* react at the same rate as the reagent (or catalyst if you prefer) is the single enantiomer of a large peptide. Not surprisingly, the enzyme cleaves the amide of natural alanine and leaves the other alone.

The purification and separation first requires removal of the enzyme. This is soluble in pH 8 buffer but acidification and heating denature the enzyme (this is rather like what happens to egg white on heating) and destroy its structure. The solid material filtered off is this denatured enzyme. The separation in ethanol works because the very polar amino acid is soluble only in water but the amide is soluble in ethanol.

■ The practical details of this process are in L. F. Fieser, *Organic Experiments* (second edn), D. C. Heath, Lexington Mass., 1968, 139.

Overheating the acid solution causes cyclization of the amide oxygen atom onto the carboxylic acid. This reaction happens only because the formation of a five-membered ring, an 'azlactone'. These compounds are dreaded by chemists making peptides because they racemize easily by enolization (the enol is achiral).

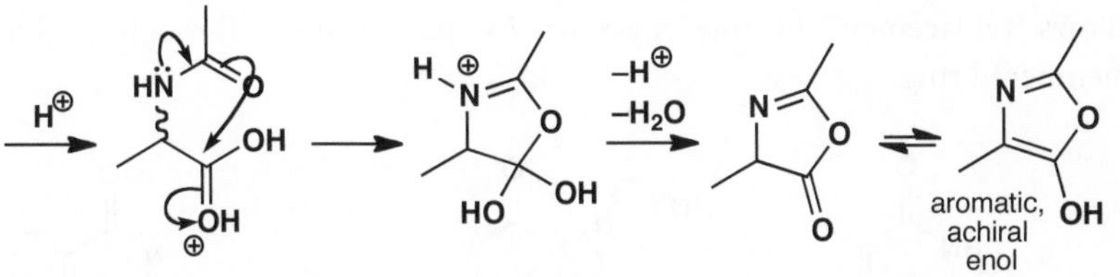

aromatic, achiral enol

PROBLEM 4

A patent discloses this method of making the anti-AIDS drug d4T. The first few stages involve differentiating the three hydroxyl groups of 5-methyluridine as we show below. Explain the reactions, especially the stereochemistry at the position of the bromine atom.

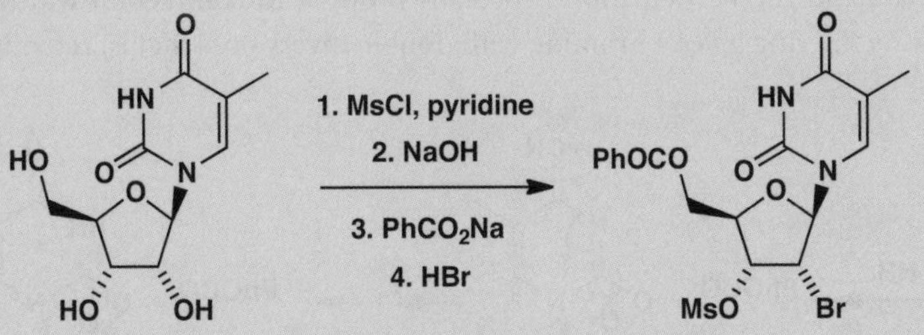

1. MsCl, pyridine
2. NaOH
3. PhCO$_2$Na
4. HBr

Suggest how the synthesis might be completed.

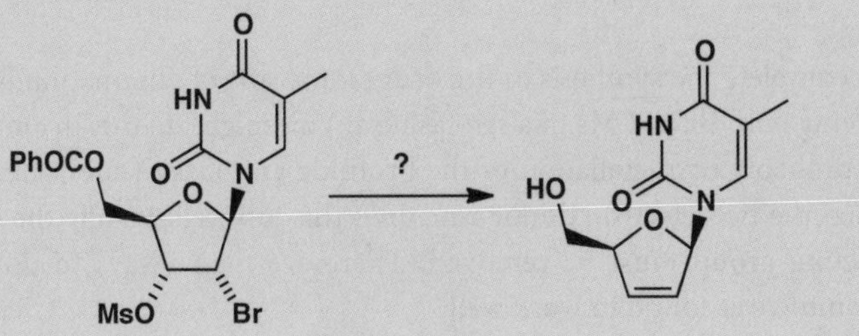

?

Purpose of the problem

A chance for you to explore nucleoside chemistry, particularly the remarkable control the heterocyclic base can exert over the stereochemistry of the sugar.

Suggested solution

There is a remarkable regio- and stereochemical control in this sequence. How are three OH groups converted into three different functional groups with retention of configuration? The first step must be the formation of the trimesylate. Then treatment with base brings the pyrimidine into play and allows replacement of one mesylate by participation through a five-membered ring.

Now the weakly nucleophilic benzoate can replace the only primary mesylate and the participation process is brought to completion with HBr. Opening the ring gives a bromide with double inversion—that is, retention.

■ The synthesis of d4T is reported by the Bristol-Myers Squibb company in 1997 by US patent 5,672,698. Some details are in B. Chen *et al.*, *Tetrahedron Lett.*, 1995, **36**, 7957.

To complete the synthesis of the drug, some sort of elimination is needed, removing both Br and Ms in a *syn* fashion. You might do this in a number of ways probably by metallation of the bromide and loss of mesylate. It turns out that the two-electron donor zinc does this job well. Finally the benzoate protecting group must be removed. There are many ways to do this but butylamine was found to work well.

How are phenyl glycosides formed from phenols (in nature or in the laboratory)?
Why is the configuration of the glycoside not related to that of the original sugar?

Purpose of the problem

Revision of the mechanism of acetal formation and the anomeric effect.

Suggested solution

The hemiacetal gives a locally planar oxonium ion that can add the phenol
from the top or bottom face. The bottom face is preferred because of the
anomeric effect and acetal formation is under thermodynamic control.

■ See pp. 801–803 of the textbook.

PROBLEM 6

'Caustic soda' (NaOH) was used to clean ovens and blocked drains. Many
commercial products for these jobs still contain NaOH. Even concentrated sodium
carbonate (Na_2CO_3) does quite a good job. How do these cleaners work? Why is
NaOH so dangerous to humans especially if it gets into the eye?

Purpose of the problem

Relating the structure of fats to everyday things as well as to everyday
chemical reactions.

Suggested solution

The grease in ovens and blockages in drains are generally caused by hard
fats that solidify there. Fats are triesters of glycerol (p. 1148 of the textbook)
and are hydrolysed by strong base giving liquid glycerol and the water-
soluble sodium salts of the acids.

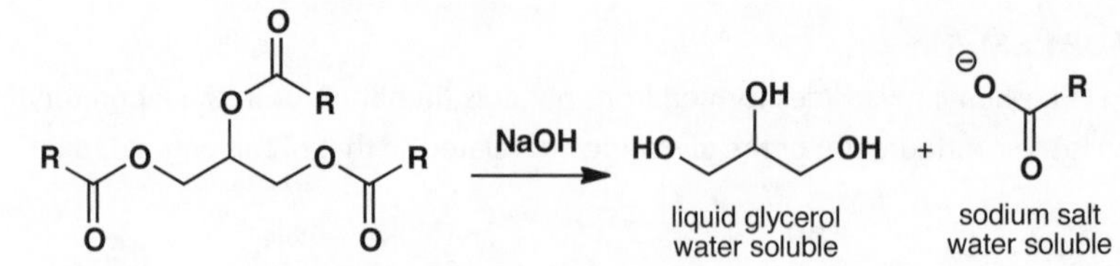

Sodium hydroxide is dangerous to humans because it not only hydrolyses esters but attacks proteins. It damages the skin and is particularly dangerous in the eyes as it quickly destroys the tissues there. Strong bases are more dangerous to us than are strong acids, though they are bad enough. The sodium salts from fats as well as glycerol are used in soaps.

PROBLEM 7

Draw all the keto and enol forms of ascorbic acid (the reduced form of vitamin C). Why is the one shown here the most stable?

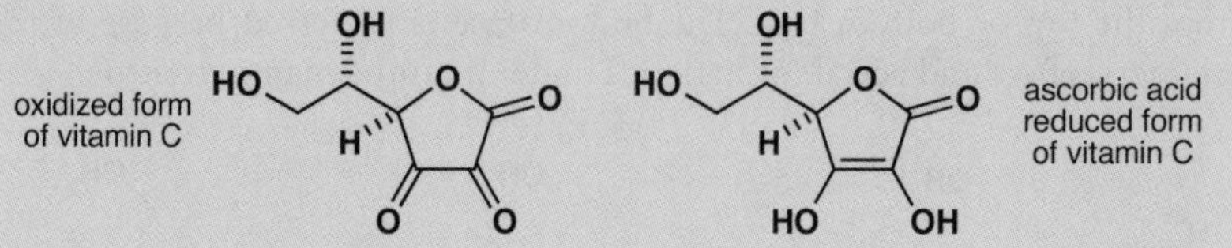

Purpose of the problem

Revision of enols and an assessment of stability by conjugation.

Suggested solution

There can be two keto forms with one carbonyl group and two keto (or ester) forms with two carbonyl groups.

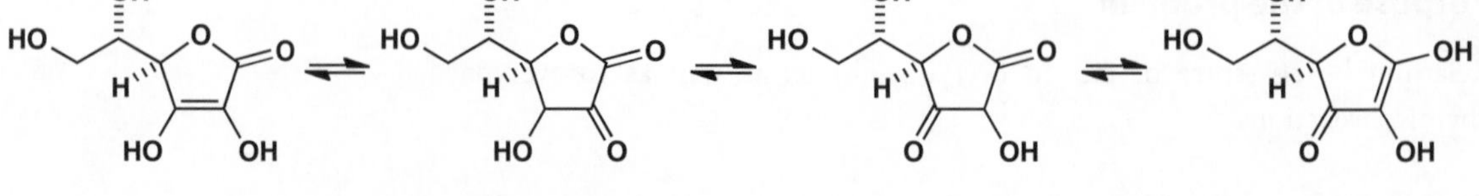

Two forms have greater conjugation than the other two and the favoured form preserves the ester rather than a ketone and so has extra conjugation.

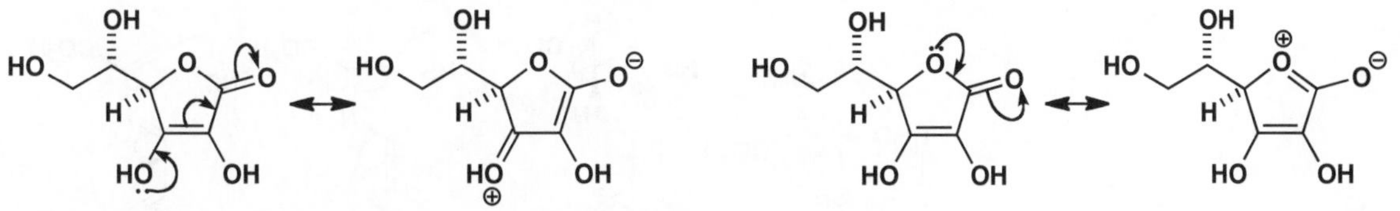

PROBLEM 8

The amino acid cyanoalanine is found in leguminous plants (*Lathyrus spp.*) but not in proteins. It is made in the plant from cysteine and cyanide by a two-step process catalysed by pyridoxal phosphate. Suggest a mechanism. We suggest you use the shorthand form of pyridoxal phosphate shown here.

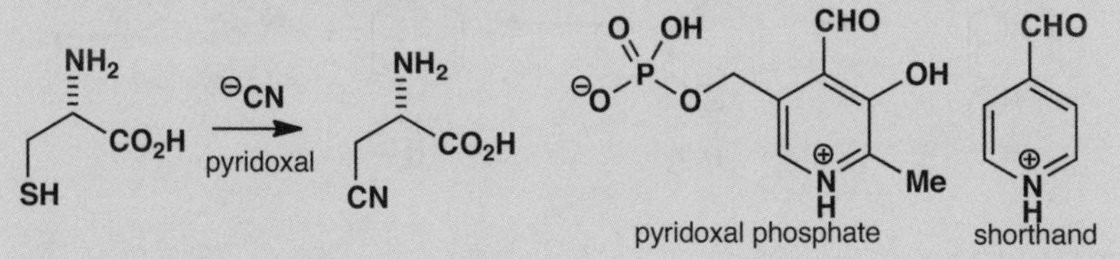

Purpose of the problem

Exploration of a new reaction in pyridoxal chemistry using pyridoxal itself rather than pyridoxamine.

Suggested solution

The reaction starts with the formation of the usual imine/enamine equilibrium but what looks like an S_N2 displacement of ^-SH by ^-CN turns out to be an elimination followed by a conjugate addition. Any attempt at an S_N2 displacement would simply remove the proton from the SH group. Notice that the pyridoxal is regenerated.

PROBLEM 9

Assign each of these natural products to a general class (such as amino acid metabolite, terpene, polyketide) explaining what makes you choose that class. Then assign them to a more specific part of the class (such as pyrrolidine alkaloid).

grandisol   polyzonimine   serotonin   scytalone   pelletierine

Purpose of the problem

Practice at the recognition needed to classify natural products.

Suggested solution

Grandisol and polyzonimine have ten carbon atoms each with branched chains having methyl groups at the branchpoints. They are terpenes and specifically monoterpenes. You might also have said that polyzonimine is an alkaloid as it has a basic nitrogen. Serotonin is an amino acid metabolite derived from tryptophan. Scytalone has the characteristic unbranched chain and alternate oxygen atoms of a polyketide, an aromatic pentaketide in fact. Pelletierine is an alkaloid, specifically a piperidine alkaloid.

■ They are also an insect pheromone (grandisol), a defence substance (polyzonimine), an important human metabolite (serotonin), a fungal metabolite (scytalone), and a toxic compound from hemlock (pelletierine).

PROBLEM 10

The piperidine alkaloid pelletierine, mentioned in problem 9, is made in nature from the amino acid lysine by pyridoxal chemistry. Fill in the details from this outline:

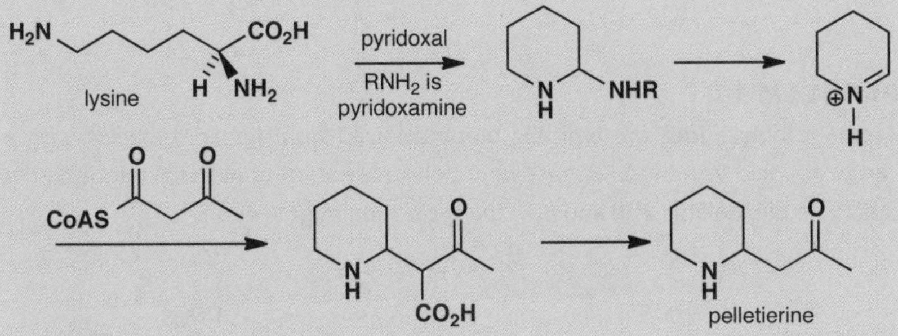

Purpose of the problem

A more thorough exploration of the biosynthesis of one group of alkaloids.

Suggested solution

The first stage produces the usual pyridoxal imine/enamine compound and decarboxylation gives a compound that can cyclize and give the cyclic iminium salt by loss of pyridoxamine.

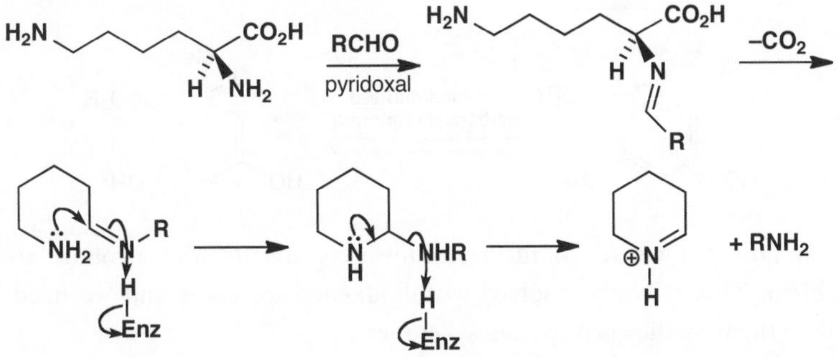

Now the enol of acetyl CoA adds to the iminium salt to complete the skeleton of the piperidine alkaloids. Hydrolysis and decarboxylation gives pelletierine.

PROBLEM 11

Aromatic polyketides are typically biosynthesized from linear ketoacids with a carboxylic acid terminus. Suggest what polyketide starting material might be the precursor of orsellinic acid and how the cyclization might occur.

Purpose of the problem

More detail on polyketide folding.

Suggested solution

Looking at this problem as if it were a chemical synthesis, we could disconnect orsellinic acid by aldol style chemistry.

But how are we to go further? Those *cis* alkenes and alcohols are a problem. This is easily resolved as the alkenes are enols and we need to replace them by the corresponding ketones.

■ See p. 1162 of the textbook.

We discover a linear polyketide derived from an acetate starter and three malonyl CoA units. The only C–C bond that needs to be made is the one that closes the six-membered ring. Enolization then gives aromatic orsellinic acid.

PROBLEM 12

Chemists like to make model compounds to see whether their ideas about mechanisms in nature can be reproduced in simple organic compounds. Nature's reducing agent is NADPH and, unlike $NaBH_4$, it reduces stereopecifically (p. 1150 of the textbook). A model for a proposed mechanism uses a much simpler molecule with a close resemblance to NADH. Acylation and treatment with Mg(II) causes stereospecific reduction of the remote ketone. Suggest a mechanism for this stereochemical control. How would you release the reduced product?

Purpose of the problem

An example of a model compound to support mechanistic suggestions.

Suggested solution

The ketone is too far away from the chiral centre for there to be any interaction across space. The idea was that the side chain would bend backwards so that the benzene ring would sit on top of the pyridine ring and that this could happen with NADH too.

This is a difficult problem but examination of the proposed mechanism should show you that binding to the magnesium holds the side chain over the pyridine ring. Enzymatic reactions often use binding to metals to hold substrates in position. Of course, in this example, the substrate is covalently bound to the reagent but simple ester exchange with MeO⁻ in MeOH releases it.

PROBLEM 13

Both humulene, a flavouring substance in beer, and caryophylene, a component of the flavour of cloves, are made in nature from farnesyl pyrophosphate. Suggest detailed pathways. How do the enzymes control which product will be formed?

farnesyl pyrophosphate   humulene   caryophyllene

Purpose of the problem

Some serious terpene biosynthesis for you to unravel.

Suggested solution

Judging from the number of carbon atoms (15) and the pattern of their methyl groups, these closely related compounds are clearly sequiterpenes. They can both be derived from the same intermediate by cyclization of farnesyl pyrophosphate without the need to isomerize an alkene. The eleven-membered ring in humulene can accommodate three E-alkenes.

Caryophyllene needs a second cyclization to give a four-membered ring—the stereochemistry is already there in the way that the molecule folds—and a proton must be lost. The enzymes control the processes so that the starting material is held in the right shape and, more subtly, to make the 'wrong' (more substituted) end of the alkene cyclize in the humulene synthesis. It might do this by removing the proton as the cyclization happens.

caryophyllene

PROBLEM 14

This experiment aims to imitate the biosynthesis of terpenes. A mixture of products results. Draw a mechanism for the reaction. To what extent is it biomimetic, and what can the natural system do better?

Purpose of the problem

Reminder of the weaknesses inherent in, and the reassurance possible from, biomimetic experiments.

Suggested solution

The relatively weak leaving group (acetate) is lost from the allylic acetate with Lewis acid catalysis to give a stable allyl cation. This couples with the other (isopentenyl) acetate in a way very similar to the natural process. However, what happens to the resulting cation is not well controlled. Loss of each of the three marked protons gives a different product. In the enzymatic reaction, loss of the proton would probably be concerted with C–C bond formation as a basic group, such as an imidazole of histidine or a carboxylate anion, would be in the right position to remove one of the protons selectively.

■ These experiments still give us confidence that the rather remarkable reactions proposed for the biosynthesis are feasible: M. Julia *et al.*, *J. Chem. Res.*, 1978, 268, 269